SUBSURFACE FLUID FLOW AND IMAGING

The practice of imaging has grown tremendously over the past few decades. At the same time, understanding fluid flow at depth has become increasingly important to activities such as hydrocarbon production, groundwater exploitation, environmental remediation, and sequestration of greenhouse gases.

This book introduces methodologies for subsurface imaging based upon asymptotic and trajectory-based methods for modeling fluid flow, transport, and deformation. It describes trajectory-based imaging and inversion from its mathematical formulation, through the construction and solution of the imaging equations, to the assessment of the accuracy and resolution associated with the image. Unique in its approach, it provides a unified framework for the complete spectrum of physical phenomena from wave-like hyperbolic problems to diffusive parabolic problems and non-linear problems of mixed character. The practical aspects of imaging, particularly efficient and robust methods for updating high resolution geologic models using fluid flow, transport, and geophysical data, are emphasized throughout the book.

Complete with online software applications and examples that enable readers to gain hands-on experience, this volume is an invaluable resource for graduate-level courses, as well as for academic researchers and industry practitioners in the fields of geoscience, hydrology, and petroleum and environmental engineering.

DONALD WYMAN VASCO is a Senior Scientist at the Lawrence Berkeley National Laboratory, recognized for his contributions to the development and application of asymptotic techniques to the modeling and inversion of fluid flow and geophysical data. Dr. Vasco is the recipient of the Society of Petroleum Engineers' Cedric K. Ferguson Certificate for the best peer-reviewed paper of 1999, and has authored or co-authored over 100 peer-reviewed articles on geophysical and hydrological inverse methods and modeling techniques.

AKHIL DATTA-GUPTA is Regents and University Distinguished Professor at Texas A&M University, and is known for his contributions to the theory and practice of streamline simulation in petroleum reservoir characterization, management and calibration of high-resolution geologic models. Professor Datta-Gupta is the recipient of the Society of Petroleum Engineers' John Franklin Carll Award (2009) and Lester C. Uren Award (2003) for his contributions related to reservoir characterization and 3-D streamline simulation. He served as a member of the Polar Research Board of the National Academy of Sciences, Technology Task Force of the National Petroleum Council and was elected to the US National Academy of Engineering in 2012.

SUBSURFACE FLUID FLOW AND IMAGING

With Applications for Hydrology, Reservoir Engineering, and Geophysics

DONALD WYMAN VASCO AND AKHIL DATTA-GUPTA

CAMBRIDGE
UNIVERSITY PRESS

CAMBRIDGE
UNIVERSITY PRESS

University Printing House, Cambridge CB2 8BS, United Kingdom

Cambridge University Press is part of the University of Cambridge.

It furthers the University's mission by disseminating knowledge in the pursuit of education, learning and research at the highest international levels of excellence.

www.cambridge.org
Information on this title: www.cambridge.org/9780521516334

First published 2016

Printed in the United Kingdom by TJ International Ltd. Padstow Cornwall

A catalogue record for this publication is available from the British Library

Library of Congress Cataloguing in Publication data
Names: Vasco, Donald Wyman, author. | Datta-Gupta, Akhil, 1960– author.
Title: Subsurface fluid flow and imaging : with applications for hydrology, reservoir engineering, and geophysics / Donald Wyman Vasco and Akhil Datta-Gupta.
Description: New York, NY : Cambridge University Press, 2016.
| © 2016 | Includes bibliographical references and index.
Identifiers: LCCN 2015037883| ISBN 9780521516334 (Hardback : alk. paper) |
ISBN 0521516331 (Hardback : alk. paper)
Subjects: LCSH: Fluid dynamics–Mathematical models. |
Groundwater flow–Mathematical models. | Hydrology. |
Petroleum engineering. | Geophysics.
Classification: LCC QC809.F5 V37 2016 | DDC 551.49–dc23
LC record available at http://lccn.loc.gov/2015037883

ISBN 978-0-521-51633-4 Hardback

Additional resources for this publication at www.cambridge.org/vasco

Contents

Preface

The practice of imaging has grown tremendously in the past few decades, both in sophistication and importance. There is a strong thread of commonality in the diverse quilt of applications of imaging in medicine, engineering, and the physical sciences. In particular, the same mathematical techniques, such as the use of trajectory-based and asymptotic methods, the central topic of this book, often serve as the underpinnings of each application. However, to the uninitiated, it might seem that each discipline has adopted a distinct formulation of the imaging problem. Thus, a sense of unity is lost in traversing the various applications. In addition, the development of imaging methods may be more extensive in one particular field compared to others. For example, in applied mathematics, trajectory-based imaging methods have been extended to a wide range of situations, such as diffusive and non-linear wave propagation. These advancements may not be appreciated or even known in other areas.

The goal of this book is to bring unity to the range of trajectory-based techniques for modeling fluid flow that may serve as the basis for efficient imaging algorithms. A secondary objective is to highlight the wide array of physical phenomena to which trajectory-based imaging methods lend themselves. It is widely known that a trajectory-based method, such as ray theory, is applicable to hyperbolic equations, typified by the wave equation. Less well known is the fact that trajectory-based methods may be used to study diffusive systems, governed by a parabolic equation. Similarly, ray methods for non-linear waves have been developed in applications such as gas dynamics and plasma physics, but are relatively unknown in such fields as hydrology. The fundamental techniques are then applied to important problems in the Earth sciences. Hopefully, after finishing this book the reader will glimpse the full range of trajectory-based imaging methods.

This book describes trajectory-based imaging from its mathematical formulation, through the formation and solution of the imaging equations, to the determination of the accuracy and resolution associated with the image. Our presentation is unique in that we cover a rather complete spectrum of physical phenomena. At the same time we have tried to focus on the practical side of imaging, emphasizing methods that are efficient and robust. Obtaining an image is not the end of the story, we need some measure of the reliability of our solution. We describe methods for assessing the solution, computing the resolution and uncertainty associated with an image. Finally, as illustrations, we include a wide range of applications and emphasize their similarity.

This book is intended for those involved in imaging research. It is hoped that the cross-fertilization between disciplines will spur innovation. The book is also appropriate for students involved in the physical sciences, engineering, medical imaging, and applied mathematics. Online resources including computer softwares and example data files have been provided for the reader to acquire hands on experience in the techniques and applications discussed throughout the book.

Acknowledgments

This book is a reality because of the direct and indirect contributions of many people. We are thankful to all of them, but a few warrant special mention. The Basic Energy Sciences geophysics group at Lawrence Berkeley National Laboratory and their groundbreaking work on poroelasticity was an inspiration and an education for us. Their impact on the book is clear. Particular thanks go to Dr. James Berryman, Professor Lane Johnson, and Dr. Steve Pride, who read through and commented on various chapters over the years that it took to complete this project. Many Petroleum Engineering graduate students at Texas A & M University have contributed to this effort through their thesis and dissertations. Many of the results presented here are due to their hard work and dedication. Special thanks to Dr. Lihua Zhao for reviewing the content and helping with word to LaTex conversion. We would like thank the Petroleum Engineering Department at Texas A & M, especially Dr. Dan Hill for allowing us to use the departmental resources. Thanks to the colleagues at Texas A & M, especially Dr. Michael J. King for countless discussions and for sharing his knowledge and insight on many topics in the book.

Special thanks to the U. S. Department of Energy, Basic Energy Sciences Program and Dr. Nicholas B. Woodward for longstanding financial support that led to the development of many of the concepts and techniques discussed here. Many of the applications have also been supported by the members of the industrial consortium, Model Calibration and Efficient Reservoir Imaging (MCERI) at Texas A & M.

This book would not have been possible without the help and sacrifice of our families. Thank you Dawn, Caroline, Savannah, John, Leo, and Will, for your tolerance of the distracted weekends and evenings, and for keeping me (D. W. Vasco) grounded and focused on what is really important. Thanks to Mausumi, Alina and Antara for your unwavering support and patience (A. Datta-Gupta).

1

The propagation of a disturbance in relation to imaging

1.1 Background and motivation

In the study of the Earth one encounters a wide array of physical processes. Examples which quickly come to mind are transient pressure variations associated with fluid flow, the propagation of elastic and electromagnetic waves, inelastic deformation, reactive chemical transport, and multiphase flow. This diversity rivals that of any other discipline. On the surface, these phenomena may seem unrelated, and it is difficult to envision techniques, other than numerical modeling, which could prove useful across such a wide range of processes.

Our understanding of the complex systems that comprise our planet is hampered by a difficulty characteristic of the Earth Sciences; our observations are, for the most part, remote and indirect. That is, measurements are typically gathered at the Earth's surface or from a relatively small number of boreholes penetrating the depths. Thus, the details of the physical system are shielded from view by the Earth itself. To be sure, great advances have been made in geophysical imaging and in a sense it is possible to 'see' within the Earth. However, as compared to a physics laboratory, a completely controlled field experiment is the exception rather than the rule in the geosciences. This necessitates tackling what is known as the inverse problem, the complement to the modeling of a natural system. In the inverse problem, remote observations are used to infer properties, usually model parameters, describing the system. For example, seismic amplitude and arrival time changes gathered in time-lapse monitoring are used to infer saturation and pressure changes in a reservoir due to fluid production. Increasingly, other fields such as medicine and non-destructive testing are adopting imaging and inversion for non-invasive evaluation.

In large measure this book is concerned with modeling techniques developed with the inverse problems in mind. To this end we shall be concerned with methods that place a premium on efficiency, flexibility, simplicity, and physical intuition. For example, finite-difference techniques, extremely versatile but computationally intensive methods that often do not scale well with problem size, will not play a large part in our discussion. Rather, we shall emphasize trajectory-based techniques perhaps best known for their use in medical and geophysical tomographic imaging. Though we shall be dealing with a methodology which is known and successful in solving particular classes of imaging problems, our setting will be much more general. That is, we shall use the approach not just to study high-frequency, wave-like behavior, but also in the study of diffusive and mixed propagation. Typically, diffusive behavior, such as a pressure change, is treated as having little in common with propagating high-frequency elastic or electromagnetic waves. This distinction, while valid in some sense, ignores the many traits that these transient phenomena share (Virieux et al., 1994). Their common characteristics are particularly useful in solving the inverse problem, for example, estimating properties such as permeability or hydraulic diffusivity within the Earth. We will also consider non-linear processes, such as multiphase flow. In that sense, parts of our book might be considered as a follow-up to texts concerned with the extension of ray methods to non-linear problems (Whitham, 1974; Anile et al., 1993; Maslov and Omel'yanov, 2001). Our emphasis will be on techniques that work in the face of the significant heterogeneity found within the Earth and most other natural systems.

Our intent is to provide methods which are applicable across a range of disciplines. One goal is to provide some degree of unity across the various specialties such as geochemistry, hydrogeology, geophysics, and reservoir engineering. A common framework and a shared methodology is particularly important with the emergence of fluid flow monitoring and time-lapse imaging. Due to the increasing expense of identifying and exploiting petroleum and geothermal reservoirs, and the increasing importance of water resources, there is less room for failed wells and inefficient extraction. It is necessary to take full advantage of large geophysical data sets in understanding fluid flow at depth. With the increasing role of unconventional hydrocarbon resources, it is important to understand the interaction between hydraulic fractures and natural fractures in the subsurface. The trajectory-based methods in this book are useful in both geophysical and flow-related modeling. The efficiency of this approach and its favorable scaling properties mean that it is useful in treating large data sets and large models. The visual and intuitive nature of the methods make them useful for interpreting observations. Also, the rapid turn-around time for an inversion means that the techniques are appropriate for time-lapse monitoring. The approaches described in this book transcend applications in the Earth sciences. In particular, the techniques and governing equations for dispersive, dissipative, and non-linear propagation occur in other areas, such as non-destructive testing and engineering, as well as in medical imaging.

The trajectory-based solutions that we shall describe provide insight into the physics of propagating disturbances and introduce flexibility into the modeling.

For example, the trajectories facilitate visualization of the movement of material and energy in complex systems. Furthermore, the formulation of a solution defined along a trajectory typically partitions into two distinct problems: a propagation time problem and another calculation involving the evolution of the amplitude. As we shall see, this partitioning provides flexibility, particularly in the treatment of the inverse problem. Thus, one may use the travel time of a disturbance as a basis for imaging an object. This leads to the idea of travel time tomography for a wide range of processes and the images that it can provide. As we shall see, the relationship between travel times and the properties of an object, such as the porosity and permeability, is often relatively direct and simple. This contrasts with the more complicated dependence of the amplitude of the disturbance on the internal structure of an object. Intuitively, the amplitudes depend upon the behavior of all trajectories in a given neighborhood, particularly how they diverge or converge. Thus, the amplitudes will not only depend upon the properties along a given trajectory, but also upon how the properties vary with distance away from the trajectory.

In the remainder of this chapter we shall discuss applications of a few of these ideas in a somewhat intuitive manner. We will examine a transient solution to the diffusion equation and how one might define a propagation time. The relationship between the propagation time and the properties of the medium will be described for a simple homogeneous medium. We will discuss the computation of a travel time for multiphase flow. Finally, a variety of applications will be noted to illustrate the power and utility of the trajectory based methods.

1.2 A propogating disturbance

At the most basic level we are concerned with the propagation of a disturbance or a change in an observable quantity, such as fluid pressure or elastic strain in a solid. Ultimately, we wish to relate the characteristics of the disturbance, such as its arrival time and amplitude to the properties of the medium through which it propagates. Some of the other processes of interest are the advective transport of a conservative or reactive tracer, multiphase flow, reactive chemical transport, electromagnetic processes, and heat flow. Typically, the disturbance is man made, as due to the injection or extraction of a volume of fluid, giving rise to transient phenomena, with propagation from a source to an observation point. One simple mathematical representation of a one-dimensional disturbance that maintains its shape as it propagates is

$$u(x,t) = f(x - ct) \tag{1.1}$$

where the profile of the disturbance at $t = 0$ is $f(x)$. The disturbance propagates to the right, in the positive x direction, with velocity c. The reader may consider the argument $\eta = x - ct$ to be a time dependent translation of the x-axis, where the rate of translation is specified by c.

We present a simple derivation of the governing equation here. A differential equation with (1.1) as a solution follows from differentiating $u(x, t)$ with respect to x and t. First, define the variable η representing the argument of f:

$$\eta = x - ct. \tag{1.2}$$

Then, differentiating $u(x, t)$ with respect to x gives, upon using the chain rule

$$\frac{\partial u}{\partial x} = \frac{\partial \eta}{\partial x}\frac{df}{d\eta} = \frac{df}{d\eta} \tag{1.3}$$

while the time derivative is

$$\frac{\partial u}{\partial t} = \frac{\partial \eta}{\partial t}\frac{df}{d\eta} = -c\frac{df}{d\eta} \tag{1.4}$$

and so one may infer from these two equations that $u(x, t)$ satisfies

$$\frac{\partial u}{\partial t} + c\frac{\partial u}{\partial x} = 0. \tag{1.5}$$

Equation (1.5) is the simplest possible linear wave equation (Whitham, 1974, p. 6), describing propagation through a uniform medium. Equation (1.5) describes one-way wave propagation, that is movement in only one direction. In many cases it is also possible to have movement in the reverse direction, leading to the classical wave equation which is a second-order partial differential equation (Whitham, 1974, p. 6).

Most physical processes cause the disturbance to change shape as it propagates and translational invariance is not maintained. For example, there is dispersion in which the propagation velocity depends upon the slope of the waveform, which is quite commonly observed. Then there is dissipation in which the attenuation of a propagating pulse depends upon its amplitude or perhaps its curvature. This behavior is characteristic of diffusive phenomena. Thus, features in a waveform are preferentially decreased in amplitude at a rate that is proportional to the height or possibly the curvature, leading to the 'smoothing' of a pulse as it propagates. Finally, the effects of dispersion and dissipation, present even when a medium behaves linearly, can be confounded by the effects of non-linear behavior. Non-linearity can lead to a sharpening or steepening of a disturbance, think of a breaking wave, counteracting the effects of dispersion and dissipation (Whitham, 1974). We shall touch on some of these effects in this chapter and in the chapters that follow.

1.3 An example involving dissipation

A large class of physical processes, from heat flow and fluid pressure diffusion, to electromagnetic wave propagation (Virieux et al., 1994), involve some form of dissipation. As an example, consider the change in fluid pressure $p(\mathbf{x}, t)$ as a function of the spatial coordinates $\mathbf{x}$ and time t, associated with the injection of a slightly compressible fluid, governed by the diffusion equation

$$\nabla \cdot \left(\frac{k}{\mu} \nabla p \right) = S \frac{\partial p}{\partial t} + f \tag{1.6}$$

where $k(\mathbf{x})$ is the intrinsic or absolute permeability, μ is the fluid viscosity, $S(\mathbf{x})$ is the specific storage coefficient, and $f(\mathbf{x}, t)$ is the fluid source or sink (de Marsily, 1986). Equation (1.6) follows directly from mass conservation and the relationship between fluid flow velocity $\mathbf{q}$ and the fluid pressure gradient

$$\mathbf{q} = \frac{k}{\mu} \nabla p \tag{1.7}$$

first derived by Henri Darcy in a study of the fountains of Dijon, France (de Marsily, 1986, p. 58). Equation (1.7) simply states that the flow velocity $\mathbf{q}$ is proportional to the pressure gradient with the proportionality constant k/μ. Darcy's law (1.7) is akin to Fouriers equation (Fourier, 1822) describing the conduction of heat in a solid, later used by Fick (1855) to develop a quantitative basis for diffusion (Crank, 1975, p. 2). As we shall see in Chapter 2, the governing equation for the evolution of pressure, Equation (1.6), follows from the equation for the conservation of the fluid mass. Substituting Darcy's law, Equation (1.7), into the mass conservation equation results in Equation (1.6) if the porous medium is assumed to behave in a linear elastic fashion.

Let us consider a numerical simulation as an example: the injection of fluid at a constant rate into a well intersecting a heterogeneous formation (Figure 1.1). As fluid is pumped into the well indicated by the bulls-eye in the figure, the pressure increases in the surrounding formation. The pressure increase migrates away from the well in a diffusive fashion, leading to a gradual change in pressure throughout the formation. In response to the sudden injection or withdrawal of fluid at the well, as with the initiation of pumping, the fluid pressure will evolve over time according to Equation (1.6). The pressure change will propagate out into the reservoir in a diffusive manner, as determined by the spatially varying coefficients $S(\mathbf{x})$ and $k(\mathbf{x})$ of Equation (1.6). If one examines snapshots of the pressure field at various times, there is no clear pressure 'front' per se, just a gradual change in pressure over time (Figure 1.2). The calculation of the evolving pressure, given the source term $f(\mathbf{x}, t)$ and the medium properties $S(\mathbf{x})$ and $k(\mathbf{x})$, constitutes the forward problem and requires the solution of Equation (1.6). Solving the forward problem, calculating

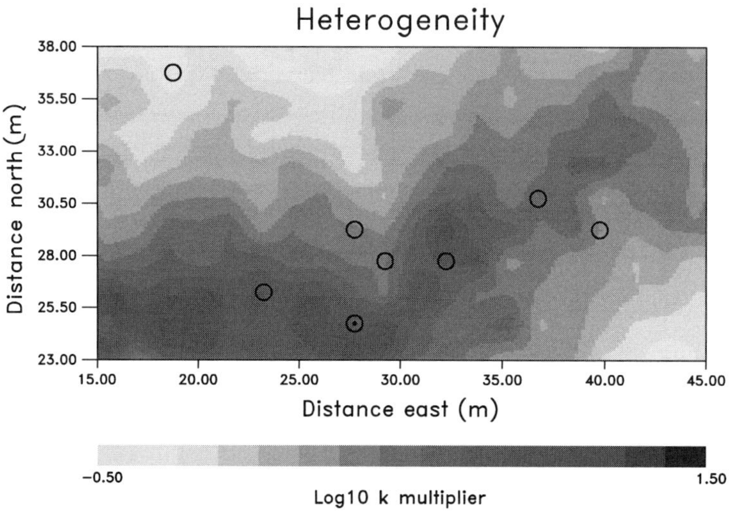

Figure 1.1 Permeability variation used in modeling a pumping experiment. The injection well is indicated by the bulls-eye.

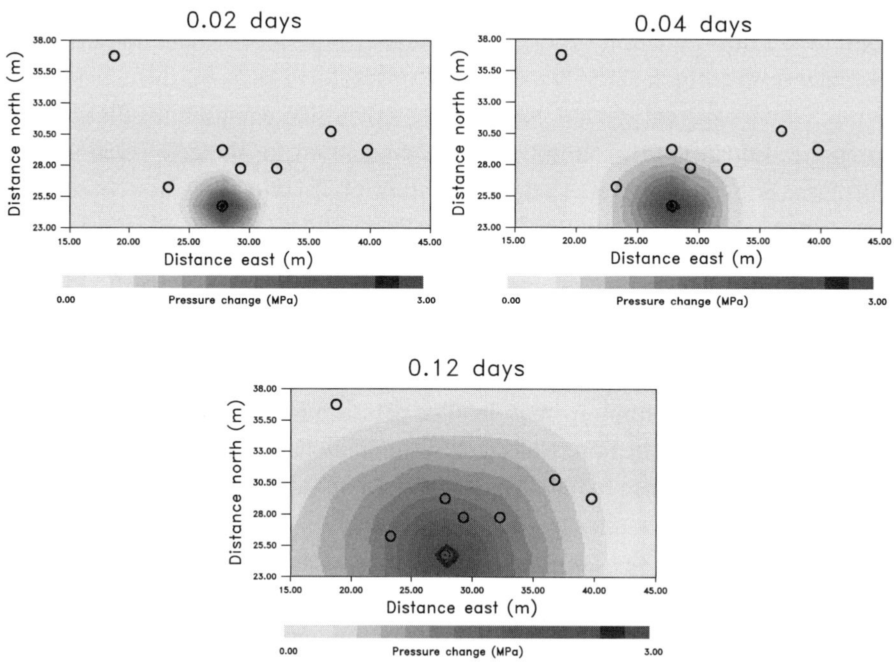

Figure 1.2 Normalized change in fluid pressure at various times due to injection.

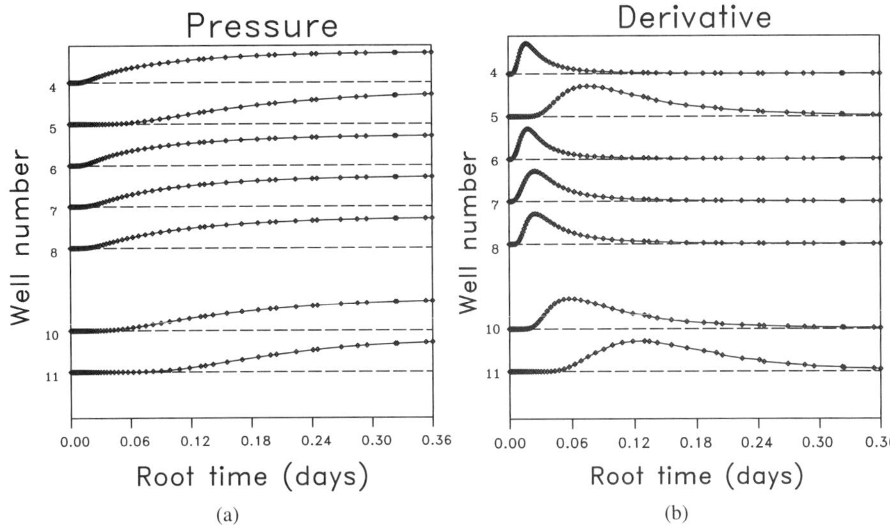

Figure 1.3 The change in fluid pressure (a) and its time derivative and (b) for the seven observing wells.

the pressure $p(\mathbf{x}, t)$ given the flow properties at all points on a simulation grid, is a stable and well-posed task, usually accomplished using a numerical technique such as a finite-difference or a finite-element routine (Peaceman, 1977; Pruess et al., 1999; Datta-Gupta and King, 2007). Figure 1.2 displays the pressure changes over time, as calculated by a integral finite-difference numerical simulator (Pruess et al., 1999), for a medium with the spatially varying permeability indicated in Figure 1.1. The pressure changes resulting at the seven observation points, denoted by open circles in Figures 1.1 and 1.2, are shown in Figure 1.3.

The pressure changes associated with a step function source do not display well-defined onsets. Rather, the pressure increases monotonically with time and it is not possible to define an 'arrival time' for the pressure disturbance at each station. It is also difficult to define an 'amplitude' because the pressure increases gradually over time and no clear maximum is attained. The situation improves if we consider the time-derivative of pressure rather than the pressure itself. Consider the pressure derivatives with respect to time, shown in Figure 1.3, for each of the seven observation wells. The pressure time-derivative has a greater resemblance to a propagating pulse, with a well-defined peak, denoting the maximum rate of pressure change. The peak can be used to define an arrival time and an amplitude for the propagating pulse at each observation point. The time-derivative of the pressure, normalized by its amplitude, more closely resembles a propagating wave (Figure 1.4). The fact that the time-derivative of the pressure variation is

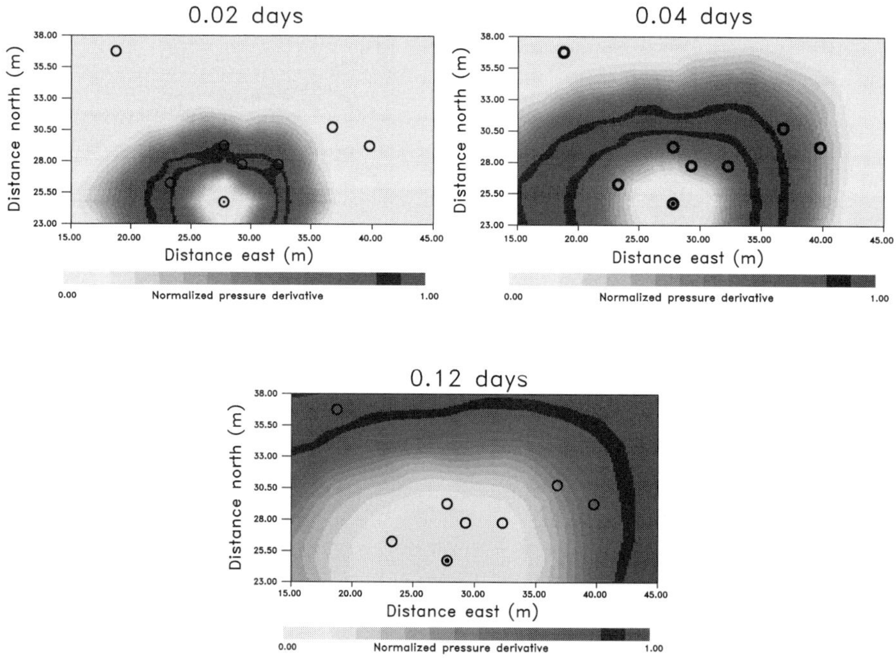

Figure 1.4 Normalized pressure derivative field at various times due to fluid injection.

pulse-like in nature makes physical sense. Because the source is a step-like function, as required to propagate pressure changes away from the source well, its time-derivative is an impulsive function. The time-derivative maps the pressure variation back into the equivalent of an impulse response. The resulting pulse-like feature is such that we can measure the amplitude of the change and we can define a time at which the rate of change is a maximum – the arrival time of the disturbance.

Motivated by the numerical simulation, let us return to the pressure Equation (1.6) and to the notion of a propagating pulse. We can produce one of the simplest solutions if we assume a homogeneous medium with an intrinsic permeability k_0, viscosity μ_0, and a specific storage coefficient S_0. In Chapter 4 we shall consider propagation in a fully heterogeneous medium. For a homogeneous medium without a source or sink, we may write Equation (1.6) as

$$D\nabla \cdot \nabla p = \frac{\partial p}{\partial t} \tag{1.8}$$

where $D = k_0/\mu_0 S_0$ is the diffusion coefficient (Crank, 1975, p. 11). Equation (1.8) implies that the rate-of-change of p at a point is proportional to the spatial curvature of the pressure field. Thus, sharp spatial variations will decay rapidly, leading to

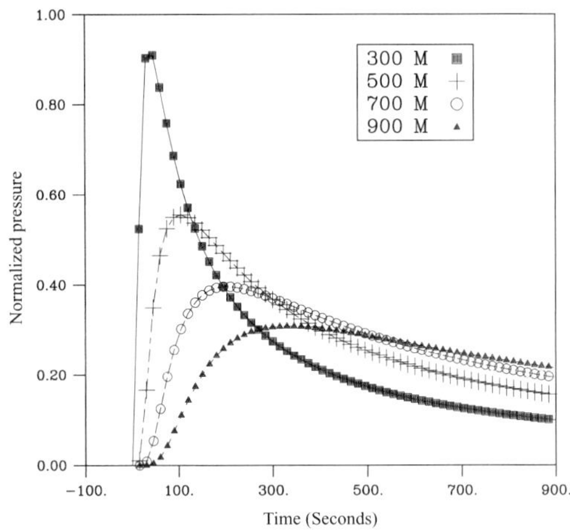

Figure 1.5 Pressure variation $p(r,t) = \frac{C}{\sqrt{t^3}} \exp\left(-\frac{1}{4D}\frac{r^2}{t}\right)$, associated with an impulsive source corresponding to four observation points located at successively greater distances from the source.

a smoothing of the pressure field with time. For an impulsive or delta-function source one may solve Equation (1.8) using either the Fourier transform (Bracewell, 2000) or separation of variables (Miller, 1977). The solution is (Crank, 1975, p. 11; de Marsily, 1986, p. 162)

$$p(r,t) = \frac{C}{\sqrt{t^3}} \exp\left(-\frac{1}{4D}\frac{r^2}{t}\right) \tag{1.9}$$

where r is the radial distance of the observation point from the injection location, C is an arbitrary constant which is determined from the initial amplitude of the pressure pulse source. In Figure 1.5, the pressure variation at four different distances are plotted as a function of time.

We can dig a little deeper in an attempt to better define what we mean by the arrival time of a propagating pressure pulse. Notice that we have switched from the step-like source, used in the numerical simulation, to an impulsive, or delta-function like source. For the pressure Equation (1.8) this is not a significant change because, as a consequence of its linearity, we can easily switch between these two source types by either differentiation or integration. The most identifiable features of the curves in Figure 1.5 are the locations of the peaks in the pressure variations. Thc condition for a pressure peak is given by the vanishing of the time derivative for a particular observation point r_o:

$$\frac{\partial p}{\partial t} = \left[-\frac{3}{2t} + \frac{r_o^2}{4Dt^2}\right] \frac{C}{\sqrt{t^3}} \exp\left(-\frac{1}{4D}\frac{r_o^2}{t}\right) = 0. \tag{1.10}$$

For finite values of t and a non-vanishing pressure variation, the exponential term is non-zero and condition (1.10) is satisfied when the expression in the square brackets vanishes, leading to

$$t_o = \frac{r_o^2}{6D}, \tag{1.11}$$

where t_o signifies the time at which the peak pressure is observed at the point r_o. Equation (1.11) reveals that the arrival time of the peak pressure at r_o is determined by the quantity

$$D = \frac{k_0}{\mu_0 S_0},$$

providing a relationship between an observable quantity and the properties of the medium and the fluid. We can turn Equation (1.11) around and use the observed arrival time, t_o, at the location r_o to infer the value of D for a particular medium, thus solving the simplest of inverse problems.

As indicated earlier, a transient solution of the diffusion equation can be viewed as a propagating disturbance. This brings us to one of the central points of this book – we can interpret both conventional wave-like processes, such as elastic and high-frequency electromagnetic propagation and diffusive phenomena using a common framework. This is useful because many real-world physical systems do not fall neatly into either category and can display aspects of both. Another point, demonstrated in Chapter 4, is that we can use trajectory-based methods to develop semi-analytic solutions and to model such processes. Such solutions complement purely numerical approaches by providing insight, both from visualization and from the semi-analytic expressions themselves. Furthermore, the trajectory-based solutions provide additional flexibility by subdividing the modeling into a travel time calculation and an amplitude calculation. We shall have much more to say about these ideas in the chapters that follow.

1.4 A non-linear example

Linear governing equations, as exemplified by the diffusion Equation (1.6), can be used to model important physical processes, but they are by no means the entire story. In fact, one would not get very far in modeling fluid flow under the restriction of linearity. Compressible flow, multiphase flow, and reactive transport are but a few examples of non-linear behavior. Trajectory-based modeling techniques are equally applicable to physical processes described by non-linear equations. In fact, some of the earliest applications of the method of characteristics were to problems involving compressible gas dynamics (Courant and Friedrichs, 1948). In this section we will simply touch upon some of these ideas. For the interested reader there are excellent expositions that can provide additional details (Courant and Friedrichs, 1948; Whitham, 1974; Debnath, 2005).

Perhaps the simplest way to introduce non-linearity into the equation governing the propagation of a disturbance is to allow the velocity c in Equation (1.5) to be a function of the amplitude. An important example is provided by the Buckley–Leverett equation. This equation provides a simple model of waterflooding, a technique whereby water is injected into a reservoir to displace the in situ oil and to maintain the fluid pressure. We can write the Buckley–Leverett equation as a kinematic wave equation governing the evolution of the water saturation $S(x,t)$

$$\frac{\partial S}{\partial t} + c(S)\frac{\partial S}{\partial x} = 0 \tag{1.12}$$

(Whitham, 1974; Peaceman, 1977; Lake, 1989; Debnath, 2005), where

$$c(S) = \frac{q}{\phi}\frac{df_w}{dS} \tag{1.13}$$

is the saturation velocity, q is the total fluid velocity obtained by solving the two-phase pressure equation (Peaceman, 1977, p. 19), and ϕ is the porosity. A key feature of two-phase flow is that one fluid can resist the flow of the other. Thus, the ease at which a fluid flows will depend upon its current saturation S at that location. The fractional flow function, $f_w(S)$ describes the ability of the phase to flow for a given saturation at the point x and at time t. The fractional flow function depends strongly upon the water saturation S. As required by our governing Equation (1.12), we will only consider one-dimensional flow, However, as we shall see in Chapter 6, everything works just as well in a three-dimensional setting.

We will seek a solution $S(x,t)$ along a curve $x(t)$, known as a characteristic. Along the trajectories $x(t)$, the time derivative of the saturation is given by

$$\frac{d}{dt}(S(x(t),t)) = \frac{\partial S}{\partial t} + \frac{dx}{dt}\frac{\partial S}{\partial x}. \tag{1.14}$$

From Equation (1.14) we can see that if one considers trajectories $x(t)$ defined by

$$\frac{dx}{dt} = c(S), \tag{1.15}$$

then it follows from Equation (1.12) that the saturation remains constant along these curves:

$$\frac{d}{dt}(S(x(t),t)) = \frac{\partial S}{\partial t} + c(S)\frac{\partial S}{\partial x} = 0. \tag{1.16}$$

Given sufficient initial or boundary conditions we can solve Equations (1.15) and (1.16) for a trajectory-based solution of Equation (1.12). The trajectory-based formulation provides an expression for the velocity of the saturation along the trajectory

$$\frac{dx}{dt} = c(S) = \frac{q}{\phi}\frac{df_w}{dS}. \tag{1.17}$$

Up to this point we have written the position x as a function of time t but we could easily write both of these variables as functions of another parameter, r, such as arc-length along the path, or the equivalent single-phase travel time based on the total fluid velocity (also called the time-of-flight) along the path. We will discuss this transformation in detail in Chapters 5 and 6. For now, we rearrange (1.17) and integrate along the trajectory from the injection point to the observation point x_o to derive an expression for the travel time of a saturation front:

$$T(x_o) = \int_{\mathbf{X}} \frac{\phi}{q}\left(\frac{df_w}{dS}\right)^{-1} dr. \tag{1.18}$$

Notice that the travel time is a function of the properties of the porous medium and the fluid through the total velocity q, the fractional flow function, and Darcy's law in Equation (1.7). Thus, the arrival time of the injected phase at various observation points provides us with a mechanism for inferring the spatial variation of flow properties via transport tomography. This typically involves computing the water front propagation using the Buckley–Leverett solution along trajectories, or streamlines (Datta-Gupta and King, 2007). This is illustrated in Figure 1.6 for an injector-producer pair with multiple perforation intervals and will be discussed in detail in Chapter 6.

1.5 Heterogeneity and imaging

If the Earth were laterally homogeneous it would simplify a great many activities such as energy production and waste disposal. However, heterogeneity is present at many scales within the Earth and we typically must expend significant effort in order to characterize the distribution of properties at depth. Properties, such as hydraulic and electrical conductivity, often vary by orders of magnitude, significantly altering flow patterns in the subsurface. Thus, we are forced to consider the influence of heterogeneity on the propagation of mass and energy fluxes. Tomographic imaging is one method for characterizing the heterogeneity in a porous medium.

1.5.1 Tomography: an intuitive introduction

The goal of computational tomography is simple: to image the spatial variation of the properties of a region. For example, the travel time associated with pressure

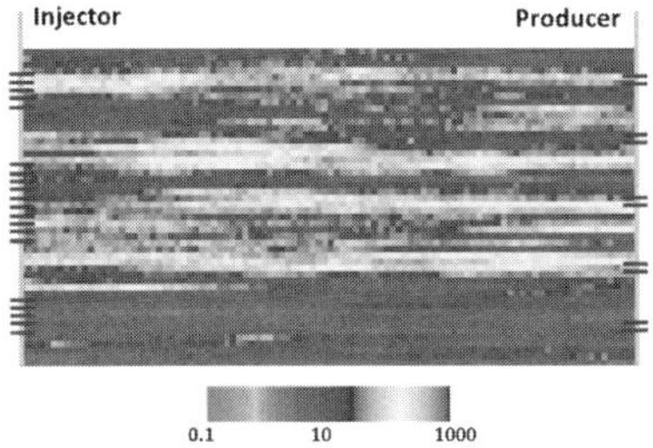

(a) Heterogenous permeability field.

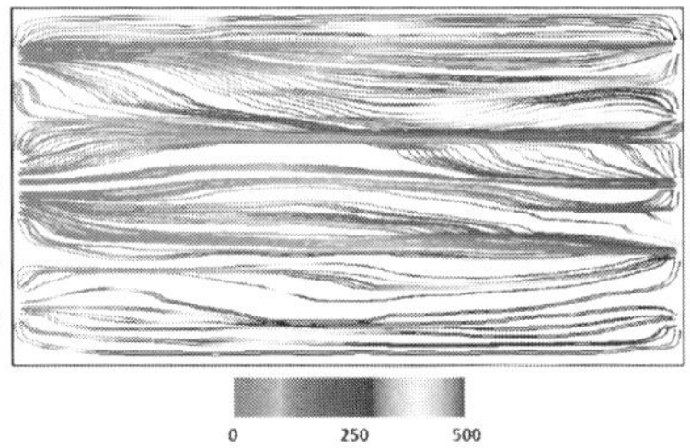

(b) Trajectories depicting the flow paths and time of flight.

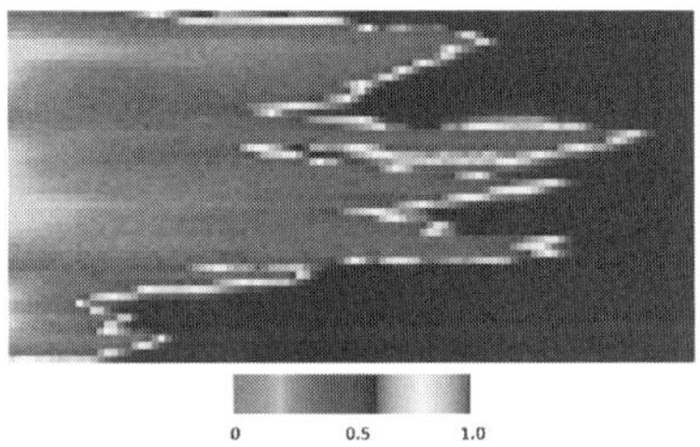

(c) The propagation of water saturation along the trajectories.

Figure 1.6 An illustration of multiphase transport calculations in heterogeneous media using the Buckeley–Leverett solution along one-dimensional characteristic curves [from Kam and Datta-Gupta (2015)]. Reprinted with permission. Copyright SPE. For the colour version, please refer to the plate section.

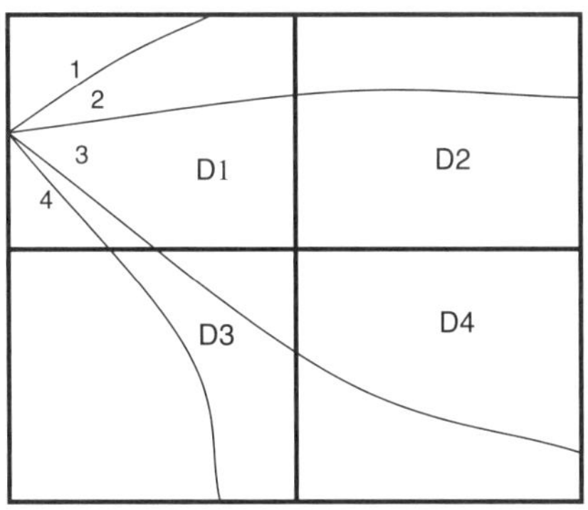

Figure 1.7 Illustration of tomographic imaging.

propagation along a trajectory $\mathbf{X}(s)$, may be used to estimate the spatial variation in diffusion coefficients. Tomographic imaging is accomplished by sub-dividing a region of interest and propagating pulses through a sufficient number of sub-divisions such that one can estimate the properties throughout the region. The basic idea is illustrated in Figure 1.7 for an area with four sub-divisions. The first pulse only propagates along a trajectory though grid block 1, allowing for

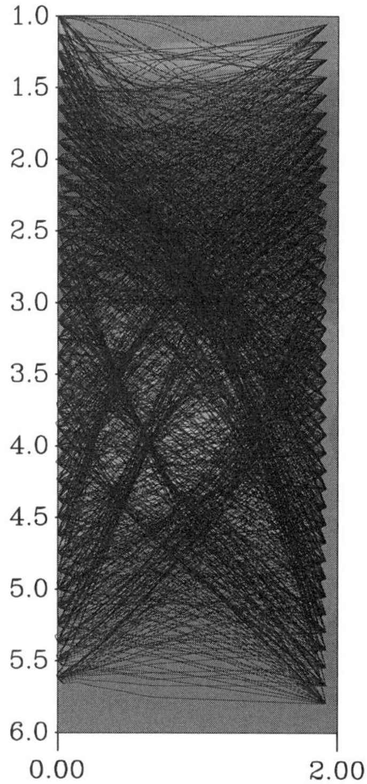

Figure 1.8 Trajectories associated with a seismic tomographic imaging experiment.

the determination of property $D1$. Pulse number 4 samples both grid block 1 and grid block 3, and, after correcting for the propagation through grid block 1, can be used to estimate property $D3$. A similar procedure enables one to determine property $D2$ and property $D4$ from the propagation along trajectories 2 and 3, However, there usually are regions that are not well resolved, such as the upper- and lower-most cross borehole regions in Figure 1.8. Note that the uniqueness of the estimates depends upon the number of grid blocks used to represent the variation within the region and the spatial coverage of the trajectories. With a fine enough discretization and limited coverage, essentially all grid blocks could be poorly determined. Typically, in a real experiment many more raypaths are used (Figure 1.8) and a fine discretization may still be resolvable. In addition, there will likely be grid blocks intersected by a large number of trajectories. Due to errors in the data, the propagation time associated with these trajectories may be inconsistent and no single distribution of property values may satisfy all of the observations exactly. That is why a least squares approach is usually adopted in which one minimizes the sum of the squares of the misfit to the data (Parker, 1994). Furthermore, in the face of non-uniqueness, functions that penalize model roughness and deviations from the geologic model (penalty functions) are often added to the misfit function to stablize the inversion (Menke, 1989). We will discuss these topics in greater detail in Chapters 5 and 6.

1.5.2 Examples

Let us view a few examples in which external flow-related data are used to image internal properties such as permeability. We hope to convey some idea of the diversity of current applications. Some of these examples will be revisited in the chapters to come.

Transient pressure tomography

We have already indicated that the arrival time of a pressure disturbance is sensitive to the flow properties of a porous medium. That discussion culminated in Equation (1.11) relating t_o, the travel time to a point at a distance of r_o from the injection site, to the diffusivity D of the medium. We can combine that expression with the tomographic approach in order to estimate the diffusivity D. In Chapter 4, we shall discuss this methodology in a rigorous fashion. There we will consider the crosswell pressure experiment portrayed in Figure 1.9 in which we have nine sources in the injection well on the left. An impulsive source is activated successively in each source location, giving rise to a pressure variation that is recorded by ten sensors, located in the adjacent observation well. The sensors are isolated by a packer system so that the pressure does not propagate up the well. The arrival time is associated with the observed peak pressure at each of the receivers. The paths taken by the transient pressure pulses from the sources to the receivers are indicated by the trajectories in Figure 1.9. Note the significant curvature of the paths due to the large variations in the diffusivity of the medium. The travel time variations may be used to infer the large-scale variation in propagation velocity, as shown in Figure 1.9. This requires iterative updating of a starting (prior) model to reconcile the model predicted and observed travel times.

Tracer response and transport tomography

In tracer experiments, a trace chemical is introduced into a fluid in motion. Typically, a flow field is prescribed using an injection well and one or more extraction wells. The technology has advanced significantly and now there are experiments utilizing multi-level samplers and other downhole devices. For example, in one such application each well segment can have a different chemical tracer. With such novel technologies analysis of the tracer effluent is indicative of both the location and the breakthrough time along the wellbore (Kam and Datta-Gupta, 2015). The appearance of the tracer and its breakthrough at an extraction well can be used to infer the flow properties and the distribution of a reactive pore fluid between

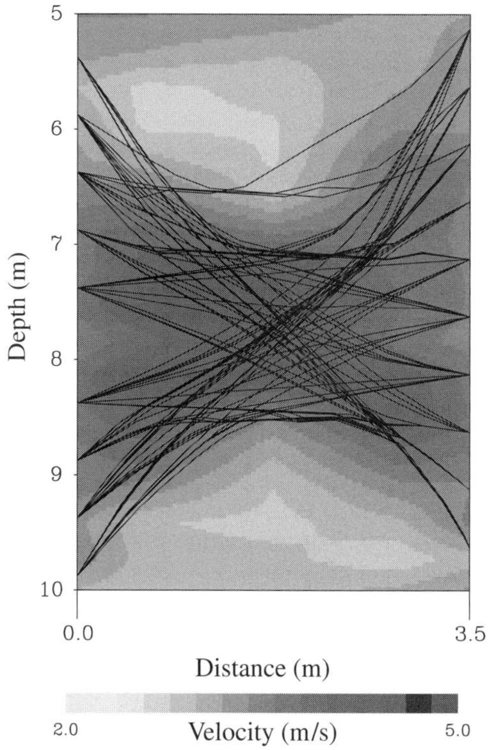

Figure 1.9 An example of transient pressure tomography. Trajectories from the nine sources to the ten receivers. The background color variation represents the imaged velocity variations associated with transient pressure propagation.

the wells (Jin et al., 1995). As in the case for transient pressure, one can derive a relationship between the breakthrough time and the porosity and permeability of the medium. A trajectory-based model of tracer transport provides an efficient approach for inverting tracer breakthrough data (Press et al., 1992; Vasco and Datta-Gupta, 1999; Datta-Gupta et al., 2002; Vasco et al., 2015). The techniques are particularly useful in the presence of a large data set, as might be produced by an array of multi-level samplers, or from the geophysical monitoring of tracer flow in a porous medium.

As an example, we refer to the multi-level tracer tests illustrated in Figure 1.10 (Annable et al., 1998). In this experiment, multiple tracers were injected into an isolated test cell using four injection wells. Tracer breakthroughs were measured at three extraction wells and twelve multi-level samplers located within the test cell. The tracer particle trajectories and travel time along the trajectories are also shown based on a prior conceptual model. Trajectory-based methods are particularly well-suited to rapidly identify the changes required to reconcile the prior model and the

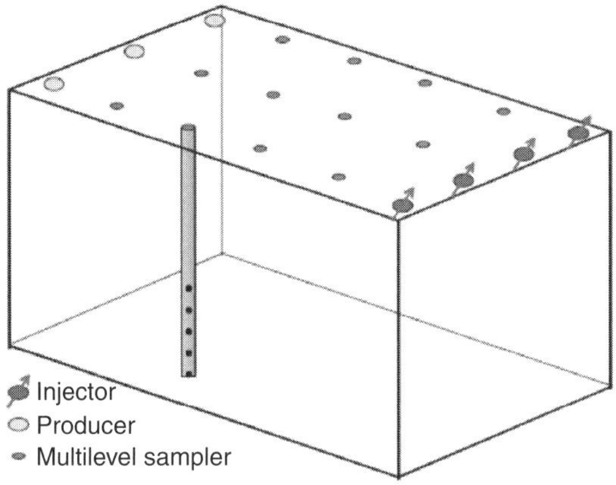

(a)

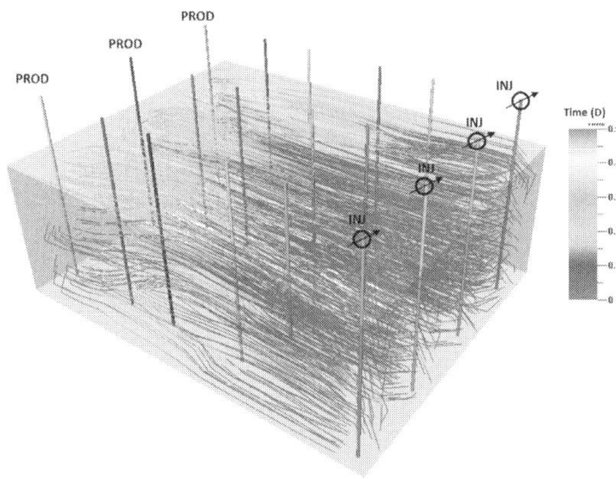

(b)

Figure 1.10 (a) Test cell diagram with multi-level samplers. (b) Tracer particle trajectories and travel times [from Yoon et al., (2001)]. Reprinted with permission. Copyright SPE. For the colour version, please refer to the plate section.

tracer response. These changes provide useful insight into the geologic heterogeneity in transport properties and flow mechanisms. We will have more to say about this in Chapter 5.

Multiphase fluid flow and production tomography

A major challenge in the characterization of a hydrocarbon reservoir is the reconciliation of high resolution geologic models and multiphase fluid production and pressure data. Here, we consider an example involving the arrival times of the injected water at the extraction wells in a hydrocarbon reservoir and the fraction of produced water, also called water-cut, as a function of time. The characterization of hydrocarbon reservoirs typically starts with the construction of one or more geologic models through the integration of a variety of static (independent of time) data sources: cores, wells logs, 3D seismic and prior geologic information. Very often the modeling of fluid flow through these models fails to reproduce the observed dynamic (time-varying) response of the reservoir such as fluid production rates and wellbore pressures.

Inverse modeling, also known as history matching, is used to update the prior geologic models to match the observed dynamic data. The process may require a significant computational effort because these models can consist of several million cells, resulting in a very large inverse problem. The trajectory-based methods discussed in this book are particularly well-suited for such large-scale inverse problems because we can compute the parameter sensitivities in an efficient, semi-analytic fashion. Model parameter sensitivities are partial derivatives, relating changes in the reservoir properties to changes in observed data. Sensitivities form the basis for many different inverse modeling schemes as we will see in subsequent chapters. Trajectory-based inversions utilizing semi-analytic sensitivities share the efficiencies of tomographic imaging, leading to a form of production tomography (Vasco and Datta-Gupta, 2001a,b).

As an illustration, we consider the application of inverse modeling to an offshore turbidite oil reservoir (Hohl et al., 2006). The prior model describing the spatial variations in permeability and the well locations in this reservoir is shown in Figure 1.11(a). The model consists of about one million grid blocks. A snapshot of the trajectories, depicting the flow paths in the reservoir at a given instant, is shown in Figure 1.11(b). These trajectories are used to construct semi-analytic sensitivities needed in the inversion of dynamic flow data. Inverse problems related to calibration of high-resolution geological models are typically ill-posed with many possible solutions. A common way to tackle this issue is to regularize the inverse problem, for example, by anchoring to our prior model. This is illustrated in Figure 1.12 for the off-shore turbiditic reservoir (Hohl et al., 2006). Trajectory-based tomography provides a rapid means to identify discrepancies between the prior geologic model and the actual field performance, often leading to new geologic insights and improved understanding of reservoir flow.

Geophysical time-lapse imaging

Geophysical field studies routinely provide information pertaining to the structure of the Earth, assisting in exploration for and exploitation of resources such as water, petroleum, geothermal fluids, and minerals. Recently, we have seen an increased interest in the geophysical monitoring of changes in the Earth associated with activities such as fluid extraction and waste disposal. The primary means of geophysical monitoring is time-lapse imaging in which observations from multiple surveys, gathered at distinct times, are differenced to estimate changes in a geophysical attribute. Attributes can consist of, among others, the amplitudes of transmitted and

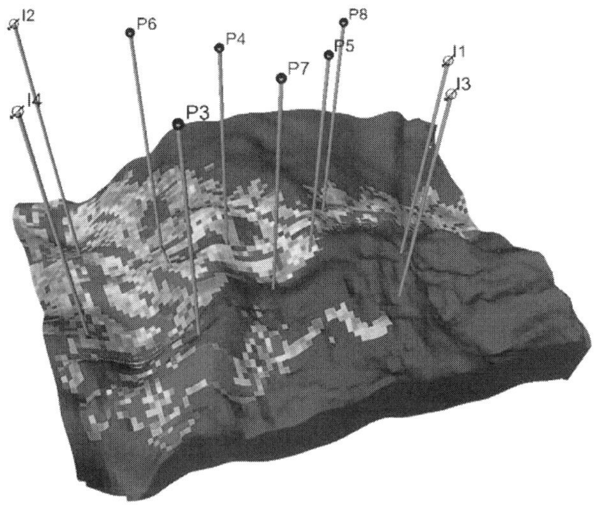

(a) Permeability distribution.

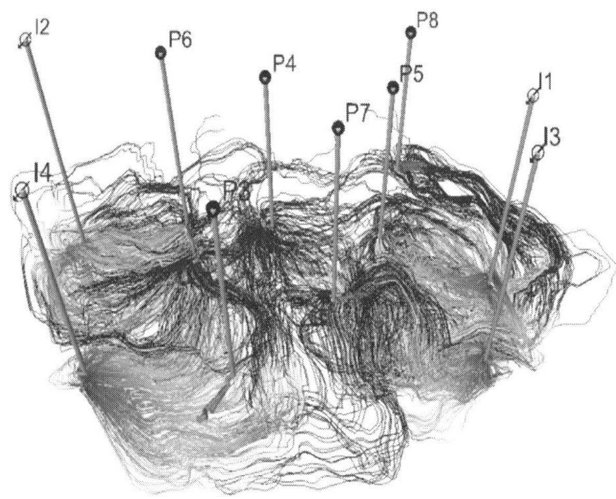

(b) Streamlines or flow paths.

Figure 1.11 Geologic model describing permeability distribution and a snapshot of streamlines for an offshore turbidite reservoir [from Hohl et al., (2006)]. Reprinted with permission. Copyright SPE. For the colour version, please refer to the plate section.

reflected elastic waves, electric field amplitude and phase information, and travel times from a tomographic survey. For example in Figure 1.13 we show reflection amplitude data from the Norne field (Watanabe et al., 2014). We will discuss the analysis of these data in more detail in Chapter 6. An advantage of time-lapse monitoring is that the differencing cancels out much of the static heterogeneity, enhancing signals due to changes in fluid pressure and saturation. Thus, time-lapse

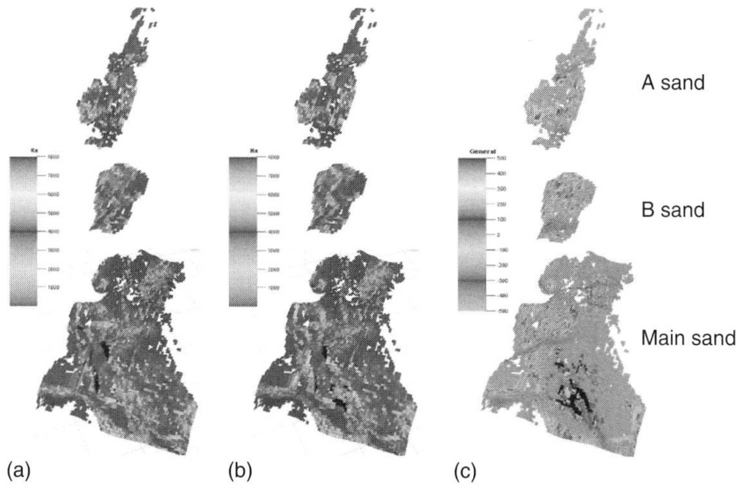

Figure 1.12 An illustration of geologic model calibration using multiphase flow data (a) Prior geologic model, (b) Updated model, and (c) Changes made during calibration [from Hohl et al., (2006)]. Reprinted with permission. Copyright SPE. For the colour version, please refer to the plate section.

differences are sensitive to changes in the state of a reservoir. As such, they are useful for estimating flow properties such as reservoir porosity and permeability. We will discuss how this is accomplished in Chapter 6.

Seismic reflection amplitude changes are a common example of time-lapse geophysical data used to monitor fluid flow. Seismic surveys provide high spatial resolution and are capable of estimating saturation and pressure changes within a reservoir (Tura and Lumley, 1998; Landro, 2001). When combined with fluid flow observations such as well bore flowing pressure and fluid production data, the changes in seismic attributes can provide valuable constraints on flow properties (Huang et al., 1998; Behrens et al., 2002; Calvert, 2005; MacBeth and Al-Maskeri, 2006). Trajectory-based methods are useful because of the efficient sensitivity computation of the seismic response to reservoir properties. In most cases, such sensitivities can be formulated semi-analytically and require little computational overhead beyond the flow simulation (Vasco et al., 2004; Rey et al., 2012; Watanabe et al., 2014).

Another potential strength of time-lapse monitoring is associated with high-resolution temporal sampling, providing information on the time variation of processes at depth. That is, given sufficient sampling in time, it should be possible to capture the propagation of transient phenomena. Contrast this with the far more common case of a few static 'snapshots', widely spaced in time, missing most of the temporal variations. Such sampling typically aliases all of the changes due to fluid flow. For example, signals due to the pressure variation, saturation changes, and thermal fluctuations will be indistinguishable without adequate temporal resolution.

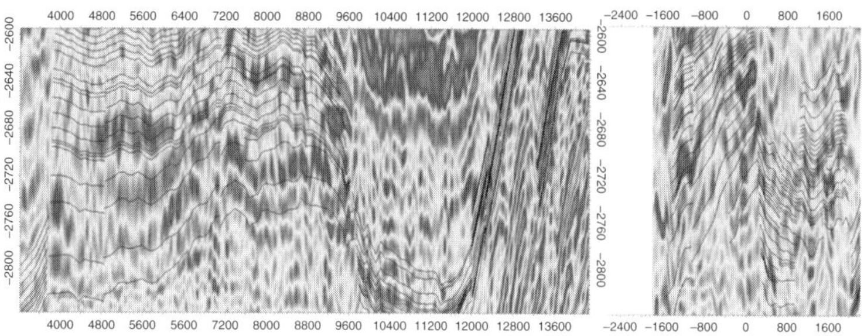

(a) Inline and crossline sections of seismic amplitudes from the Norne field.

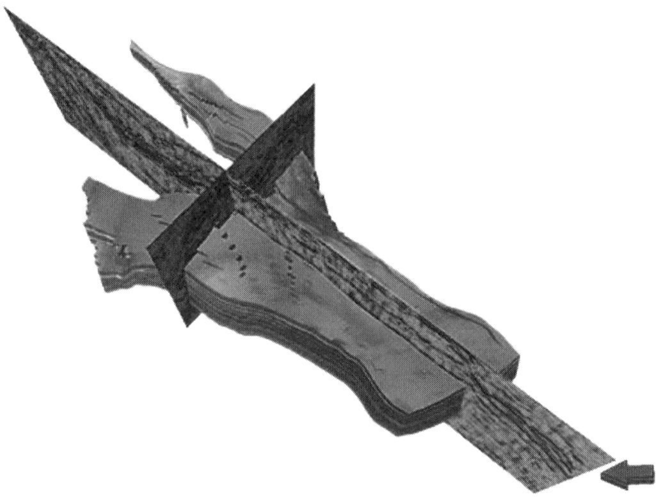

(b) A three-dimensional perspective indicating the locations of the inline and crossline sections within the geologic model.

Figure 1.13 Seismic reflection observations from the Norne field (Watanabe et al., 2014). Reprinted with permission. Copyright SPE. For the colour version, please refer to the plate section.

Fortunately, we are beginning to see improvements in the temporal sampling associated with the geophysical monitoring of flow and transport. Permanent arrays of seismic instruments are providing cost-effective, long-term monitoring for several major oil fields (van Gestel et al., 2008; Bertrand et al., 2014). Automated seismic source-receiver arrays, such as the continuous active seismic source monitoring (CASSM) system, provide temporal sampling on the order of minutes (Daley et al., 2011), making it possible to monitor processes such as injection into a fracture (Ajo-Franklin et al., 2012). In Chapters 5 and 7 we shall explore the advantages of using the onset of changes in geophysical attributes to better understand fluid flow in a porous medium. As an example, consider the crosswell configuration in Figure 1.14, with a single source in the well on the left and six receivers in the adjacent borehole (Daley et al., 2011; Vasco et al., 2014). The source is triggered at four pulses per second and these are stacked (averaged) into a single composite pulse

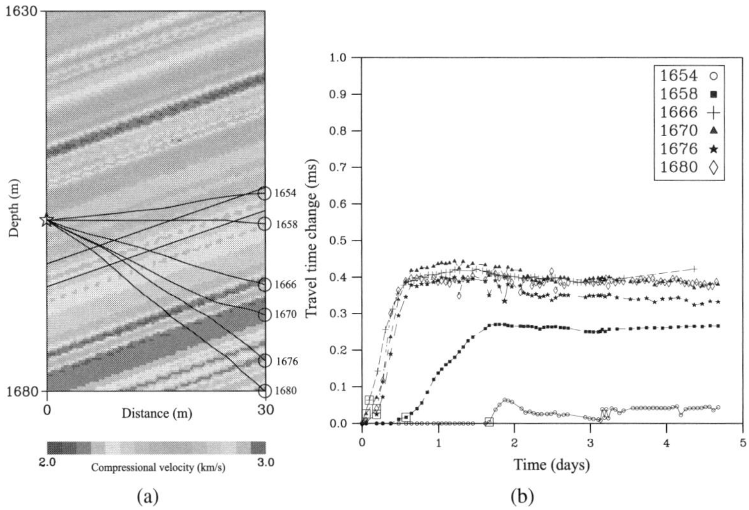

Figure 1.14 (a) Continuous active source seismic monitoring (CASSM) system with a single active source, indicated by the star on the left, and six receivers in the well on the right, denoted by open circles. The layer containing the injected carbon dioxide is denoted by two dipping parallel lines. The raypaths for the given velocity model are also plotted. (b) Seismic travel time changes observed at the receivers due to the injection of carbon dioxide. For the colour version, please refer to the plate section.

to increase the signal relative to the noise. As the system was operating, carbon dioxide was injected into a porous layer slightly below the source. The layer containing the carbon dioxide is denoted by the two parallel lines in Figure 1.14. Over time the carbon dioxide migrated away from the injection well, primarily traveling up dip toward the observation borehole containing the receivers. This migration, and the resulting substitution of carbon dioxide for the in situ water, gave rise to a decrease in the seismic velocity within the layer and an increase in the travel time to the six receivers (Figure 1.14). While we can use the magnitude of the travel time change to infer saturation changes within the layer, the effects of pressure changes are small, there are advantages in using the onset time of the changes. We define the onset time as the calendar time at which the observed quantity begins to deviate from its background value. The onset times in Figure 1.14, denoted by the unfilled squares, are those points in time when the travel time deviation reaches 5 percent of its peak change. In general, onset times are more sensitive to flow-related changes, such as saturation changes, and less influenced by variations in the rock physics model relating seismic velocities and fluid saturations (Vasco et al., 2014, 2015).

Imaging using deformation and strain

Geodetic data, involving the displacement of points on and within the Earth, form yet another set of time-lapse measurements. This class of observations can be quite diverse, ranging from time strains derived from time-lapse seismic reflection data, through tilt, leveling, and displacement data from the Global Positioning System, to interferometric synthetic aperture radar (InSAR) data derived from passing satellites. One advantage of geodetic data is that it is usually gathered more often than other geophysical measurements, providing improved temporal sampling. Another advantage is that deformation and strain observations are often sensitive to volume and aperture changes frequently associated with fluid pressure variations. We will consider two common types of geodetic data, tilt measurements, and range change observations from interferometric synthetic aperture radar (InSAR).

Similar in principle to a bubble level, a tiltmeter measures angular changes over time. Each tiltmeter provides two horizontal components, the gradient of the vertical displacement in two directions. The technology for measuring tilt has advanced to a significant degree and it is now possible to obtain measurements in shallow boreholes or even wells several kilometers deep. A distinct advantage of tilt data is the temporal sampling that it can provide, of the order of a sample per minute. Under favorable circumstances, such as a sufficient signal and a target that is resolvable, one can use tilt data to infer volume or aperture changes due to fluid injection (Vasco et al., 2001, 2002). The temporal sampling is sufficient to estimate the onset of volume change induced by flow within a fracture (Vasco 2004b). The estimated onset time can be used to calculate flow paths within the fracture zone and from these quantities one may estimate effective permeability variations within the fracture zone.

At best, individual instruments, such as tiltmeters, provide a dispersed network of point measurements. A different acquisition strategy is provided by instruments that scan an area, reflecting electromagnetic waves off the Earth's surface and measuring the returning energy. Both laser ranging (LIDAR) and interferometric synthetic aperture radar (InSAR) are examples of this approach. While scanning system surveys are typically less frequent, current satellite-based systems have return times between one week and a few months but this is not a fundamental limitation. For example, there are laser-scanning engineering systems that could gather local data every few minutes if necessary.

Even the monthly, or almost weekly, sampling by current satellite systems is useful for monitoring large-scale and long-term fluid injection or production. (Figure 1.15). For example, InSAR has proven valuable in monitoring the geological storage of carbon dioxide at the In Salah project in Algeria (Vasco et al., 2008a, 2010) and nearby gas production (Rucci et al., 2010). The temporal sampling varied from one to several months, depending upon the variations in the satellite orbit. However, given the large volumes of injected carbon dioxide and the planned fifteen year life of the project, the sampling proved sufficient to image associated reservoir volume changes (Vasco et al., 2008a). We present an analysis of injection data from the In Salah project in Chapter 7. It also proved feasible to detect range

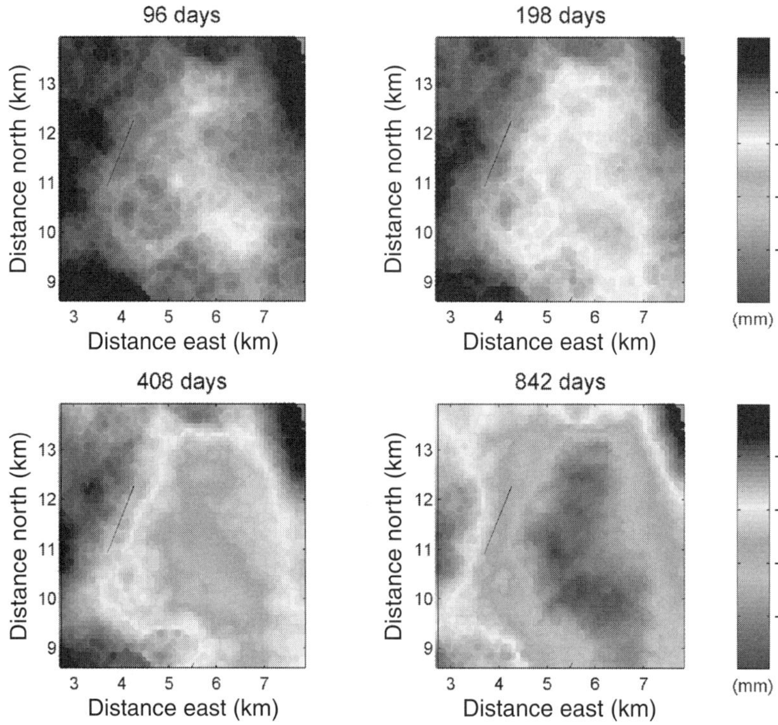

Figure 1.15 Range change in millimeters for four time intervals since the start of production at well KB-11. The segment of the production well within the reservoir is indicated by the black line. For the colour version, please refer to the plate section.

change associated with production from an adjacent gas field. The peak range changes are observed at rather large distances, from 1 to 2 km, from the production well. One would expect the largest production-related fluid pressure change to occur in the vicinity of the well. This discrepancy may be due to variations in the porosity and compressibility of the reservoir leading to larger volume changes, and corresponding range changes, far from the well. This points to one of the difficulties of trying to relate the magnitudes of reservoir volume and pressure changes within the reservoir. In order to do this successfully, we typically must know coupling coefficients that are likely to vary spatially. Because the reservoir was thin, essentially two-dimensional, one could solve for the history of reservoir volume change using the InSAR range change data. Assuming a model of reservoir compressibility it is possible to estimate reservoir pressure associated with the volume change. Rucci et al. (2010) used two models to estimate the pressure changes, a uniform reservoir compressibility model and a variable compressibility model based upon the magnitudes of the long-term volume changes. Using the

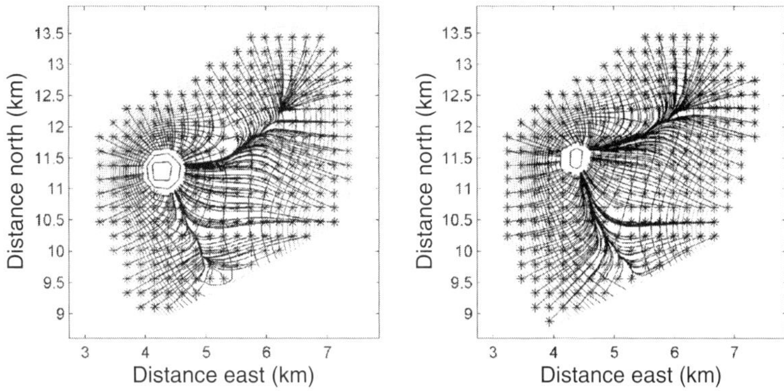

Figure 1.16 Contour plot of the square root of the onset time of reservoir pressure change due to gas production at well KB-11. The trajectories denote the propagation path of the transient change in reservoir pressure. Values in the left panel were computed using a uniform model or reservoir compressibility. Values in the right panel are based upon a spatially variable model of reservoir compressibility derived from the long term deformation data.

entire history of pressure estimates, one can map the onset time of the pressure changes in the reservoir (Figure 1.16). The onset times were calculated in two ways. First using a uniform compressibility model to map the volume change into pressure, and second using a spatially varying compressibility model. The pair of onset time estimates are plotted in Figure 1.16. The two estimates agree rather well, an indication that the onset time is not very sensitive to spatial variations in the reservoir compressibility (Rucci et al., 2010) used in the calculation. Furthermore, the earliest onset times are in the region around the production well, in agreement with our intuition that the pressure changes start at the well and propagate outward. The robustness of the onset times with respect to changes in the reservoir compressibility was also demonstrated analytically in Rucci et al. (2010).

The trajectories plotted in Figure 1.16 form the basis for an inversion of the onset times. Specifically, we can break up the reservoir model into grid blocks and perform what is akin to diffusive traveltime tomography. The intuitive idea is rooted in the discussion in Section 1.3. The basis for our approach to diffusion tomography is given in Chapter 4 for fluid pressure. In Chapter 7, we extend the discussion to coupled poroelastic propagation and consider a problem that is very similar to what we face here. There we note that having the field of onset times allows us to calculate the trajectories from the observations, effectively linearizing the inverse problem for diffusivity. Furthermore, the temporal sampling allows us to specify the incremental travel time change along the trajectory, providing resolution along the ray path. So, in spite of the absence of crossing rays, we can still resolve the travel time increments along the path and hence calculate the diffusivity necessary to produce the incremental changes.

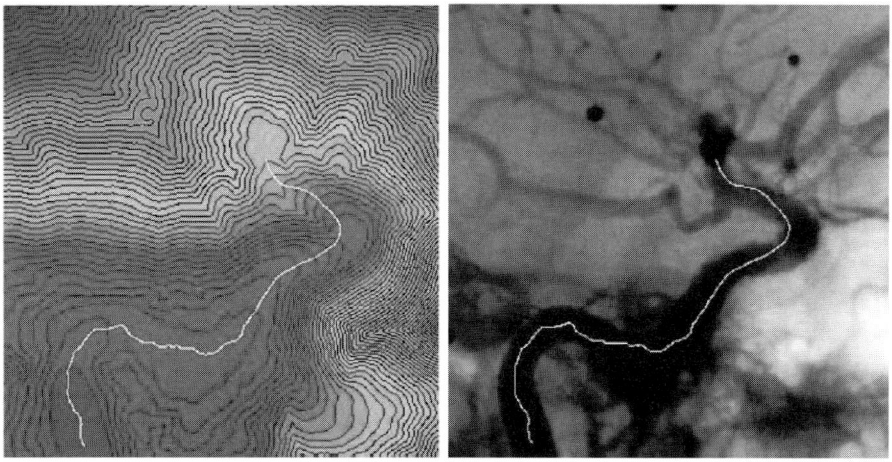

(a) (b)

Figure 1.17 An application of a trajectory-based method to medical image analysis (a) minimal path between two points using the Fast Marching Method and trajectory construction (Deschamps and Chen, 2001) and (b) an angiographic image of brain vessels. Reprinted with Permission. Copyright Elsevier Limited.

1.6 Summary

In this chapter we have emphasized that processes associated with fluid flow, transport, and associated deformation can be viewed as propagating disturbances. Furthermore, one can define paths forming the basis for the trajectory-based modeling of such propagation. The paths may be used to our advantage in tackling the inverse problem and in performing tomography using observations related to fluid flow and transport. It is worth pointing out that the trajectory-based methods discussed in the book have found applications in numerous other fields, such as medical imaging and image analysis. For example, in Chapter 4 Section 4.3.1 we discuss the fast computation of front propagation in diffusive flow using Fast Marching Methods (Sethian, 1999) and the construction of trajectories from the fronts. In Figure 1.17, an identical approach is used to compute the minimal path on an angiographic image of brain vessels (Deschamps and Chen, 2001). Such minimal paths can be used for virtual endoscopy whereby the clinicians can generate perspective views of the image along user defined paths without invasive intervention. The power and utility of the trajectory-based methods lie in their broad range of applicability, computational efficiency and the visual and intuitive appeal. The reader is encouraged to attempt the software examples in the subsequent chapters to acquire hands-on experience and appreciation of these methods.

2

Principles and equations governing fluid flow and deformation

2.1 Introduction

In this chapter we review the basic governing equations for fluid flow in permeable media. Our treatment is general and includes coupled deformation and fluid flow. Starting with the basic conservation laws we derive the pressure and transport equations for single phase and multiphase flow in porous media. We cover physical processes involving both miscible and immiscible displacements as well as mass transfer from compositional effects and present the equations describing these processes. We have included a fairly extensive discussion on the underlying principles behind elastic deformation and how the concepts carry over for modeling fluid flow in deformable porous media. We conclude the chapter with a discussion on poroelasticity and the associated equations for fluid flow in deformable porous matrix in the presence of one or more fluid phases. In summary, this chapter provides the foundation and notation for subsequent developments. However, in each of the subsequent chapters we will start with the relevant governing equations so the reader may comfortably skip this chapter and refer back to it as necessary.

2.2 Underlying principles

2.2.1 Motion, force, momentum, work, and energy

At the most basic level there are the concepts of motion and force. We can all agree when an object is stationary with respect to a 'stable' reference point. First, define a reference point in a chosen coordinate system by $\mathbf{x}_0$ and the position of our object in that coordinate system by the vector $\mathbf{x}$. An object is considered stationary when its position relative to the reference point is unchanging, that is,

$$\mathbf{x} - \mathbf{x}_0 = \mathbf{c},$$

where $\mathbf{c}$ is a constant. Such a motionless object is termed **static**. An object in motion changes its position over time and one may think of the position vector as a function of time $\mathbf{x}(t)$. In order to describe the motion of an object with some precision we define the **velocity** of an object $\mathbf{v}(t)$, the rate of change of an object's position with time, formally defined as the limit

$$\mathbf{v}(t) = \lim_{\Delta t \to 0} \frac{\Delta \mathbf{x}(t)}{\Delta t},$$

where $\Delta \mathbf{x} = \mathbf{x}(t + \Delta t) - \mathbf{x}(t)$. In addition, we also need a measure of the change of an object's velocity over time because the deviation from motion at a constant velocity is an important indicator that a force is pushing or pulling on the object, acting to change its motion. Thus, we define the **acceleration**, $\mathbf{a}(t)$, as the rate of change of the velocity of an object

$$\mathbf{a}(t) = \lim_{\Delta t \to 0} \frac{\Delta \mathbf{v}(t)}{\Delta t}.$$

The notions of velocity and acceleration provide a quantitative description of the motion of an object and their formal definition required the invention of the calculus, a significant advancement in mathematics.

Quantitative descriptions of position and motion appeared soon after the formulation of a coordinate system by Descartes and others. They were particularly influential in the development of astronomy. But, determining how and why an object moves, for example in the Earth's gravitational field, took some time to figure out. To say that a **force** is something that induces the motion of an object, a push or a pull, is simply a qualitative definition and does not advance our knowledge significantly. It took the genius of Galileo and Newton (Newton, 1729), among others, to determine that motion at a constant velocity was the 'natural' state of an object, and that a force was needed to change the situation in a very specific way, giving rise to the acceleration of an object, a change in its current velocity.

Let us begin with an easy situation, an object at rest with respect to a reference point. Thus, the resultant force on the object is zero. 'Resultant' is the key word here because we might have several non-zero forces acting on the object, balancing out to produce zero net force. Consider the case of M such forces

$$\bar{\mathbf{F}} = \sum_{i=1}^{M} \mathbf{F}_i = \mathbf{0}, \tag{2.1}$$

where $\mathbf{F}_i$ is the i-th force, such as gravity, frictional resistance, or an electromagnetic attraction, and $\bar{\mathbf{F}}$ is the resultant or net force. When the imposed forces sum to zero, the system is said to be in static equilibrium. If the forces do not balance, the object undergoes an acceleration and it begins to move. Specifically, Equation (2.1) is replaced by

$$\bar{\mathbf{F}} = \sum_{i=1}^{M} \mathbf{F}_i = \frac{d(m\mathbf{v})}{dt}, \tag{2.2}$$

where $\mathbf{v}(t)$ is the velocity of the particle and m is a proportionality factor which is the inertial mass of the object. This is **Newton's equation of motion**, the bedrock of classical dynamics. In words, Equation (2.2) states that a net force gives rise to a non-zero acceleration. Equation (2.2) connects the forces acting on a point mass to its motion in a very precise yet simple manner, a profound achievement indeed! The quantity in parenthesis in Equation (2.2) is known as the **momentum** of the object and is often denoted by the vector $\mathbf{p} = m\mathbf{v}$.

The French mathematician d'Alembert moved the dynamics of Equation (2.2) back to the realm of statics by defining the force of inertia $\mathbf{A}$

$$\mathbf{A} = -\frac{d(m\mathbf{v})}{dt}$$

and rewriting Equation (2.2) as

$$\bar{\mathbf{F}} + \mathbf{A} = \mathbf{0}. \tag{2.3}$$

Equation (2.3), known as d'Alembert's principle, states that the sum of the imposed forces and the force of inertia results in a system at equilibrium. One way to think of d'Alembert's approach is to imagine allowing the reference point to vary as a function of time $\mathbf{x}_0(t)$ and to accelerate along with the object. A useful treatment of this viewpoint is given in Lanczos (1962, p. 88). Because the velocity is the time derivative of position, Equation (2.3) leads to a second-order differential equation for the position of the particle as a function of time. This is one motivation for the study of second-order differential equations. Indeed, the theory of differential equations tells us that, if the forces depend upon time, the spatial coordinates, and the first time derivative of the spatial coordinates, solutions to Equation (2.3) exist and depend uniquely upon the initial conditions (Courant and Hilbert, 1962).

Along with the concepts of motion and force we have the associated idea of the energy of an object or system. As with the other ideas, the exact definition of energy and related concepts has a long history (Lanczos, 1962). Perhaps because of the subtle nature of energy there were several competing attempts to relate it to the motion of an object. It may be easier to conceive of the **work** done by a force upon a particle. The precise definition of work is

$$W = \mathbf{F}_i \cdot \Delta\mathbf{x} = |\mathbf{F}_i|\,|\Delta\mathbf{x}| \cos\theta_i, \tag{2.4}$$

where $\mathbf{F}_i$ denotes the particular force under consideration, $\Delta\mathbf{x}$ signifies a permissible displacement, θ_i represents the angle between the force vector and the displacement vector and provides a quantitative measure of the interaction between a force and the motion of an object. The expression is most useful in the study of the many forces that are functions of the spatial coordinates $\mathbf{x}$, such as the pull of gravity or the tug of a magnetic force. To enhance our understanding of this definition, consider the example of a spherical object, such as a ball bearing, sitting on a polished floor. Assuming that the floor is level, as the ball bearing rolls the force of gravity does no work upon it because the motion is perpendicular to the gravitational force vector. Thus, for permissible displacements within the plane defined by the floor, the ball bearing is in equilibrium with respect to the force of gravity. Were the floor to suddenly give way and the ball bearing to fall, it would accelerate under the force of gravity and the force of gravity would do measurable work upon the bearing. Given measurements of the force and the displacement, one can compute the exact value of the work done for the particular motion of an object subject to a force.

We can use the idea of work to formulate an exact definition of equilibrium for an object under the influence of a set of forces. To do this, we consider the virtual work due to a permissible virtual displacement $\Delta\mathbf{x}$. For our ball bearing example the permissible virtual displacements are those within the plane defined by the floor. The **principle of virtual work** states that a given object will be in equilibrium if, and only if, the total virtual work of all impressed forces vanishes:

$$\delta W_T = \bar{\mathbf{F}} \cdot \delta\mathbf{x} = 0. \tag{2.5}$$

Note that the virtual work is zero when the contributing forces cancel and $\bar{\mathbf{F}}$ vanishes, or when the resultant force is perpendicular to all permissible displacements. Equation (2.5) applies to the static situation in which there is no motion in the direction of the forces. Using d'Alembert's principle, we can extend the idea of equilibrium to the dynamic situation in which an object moves steadily in response to a force. We can do this by defining an effective force

$$\mathbf{E} = \bar{\mathbf{F}} + \mathbf{A}, \tag{2.6}$$

incorporating both the impressed forces and the force of inertia. Let us further generalize the situation by considering a system containing N particles in which case the virtual work associated with the permissible displacements is given by

$$\delta W = \sum_{k=1}^{N} \mathbf{E}_k \cdot \delta\mathbf{x}_k = \sum_{k=1}^{N} (\bar{\mathbf{F}}_k + \mathbf{A}_k) \cdot \delta\mathbf{x}_k = 0. \tag{2.7}$$

The conceptual point of view of d'Alembert underlies all of mechanics and is the genesis of many advanced techniques.

The definition of work can be used to motivate an expression for the energy associated with the motion of a particle, that is, the **kinetic energy**. Kinetic energy is perhaps the easiest form of energy to quantify because it is defined directly in terms of an object's velocity and mass. Consider the work done by a force on a particle during a finite motion. Because the force may vary for each increment of distance traveled, the total work is an integral over all portions of the path:

$$W = \int \bar{\mathbf{F}} \cdot d\mathbf{x}. \tag{2.8}$$

For the sake of illustration consider one-dimensional motion, for which Equation (2.8) becomes the integral

$$W = \int_{x_0}^{x} \bar{F} dx \tag{2.9}$$

that can be explicitly evaluated by incorporating Newton's law, Equation (2.2), and writing the acceleration as

$$a = \frac{dv}{dt} = \frac{dv}{dx}\frac{dx}{dt} = v\frac{dv}{dx}. \tag{2.10}$$

Substituting Equation (2.2) and Equation (2.10) into the integral expression for work, Equation (2.9), we find that

$$W = \int_{x_0}^{x} \bar{F} dx = \int_{x_0}^{x} mv\frac{dv}{dx}dx = \int_{x_0}^{x} mvdv = \frac{1}{2}mv^2 - \frac{1}{2}mv_0^2. \tag{2.11}$$

Thus, the total work associated with the path from x_0 to x is given by the change in kinetic energy, where the kinetic energy K is defined by

$$K = \frac{1}{2}mv^2, \tag{2.12}$$

a result known as the **work-energy theorem**.

In addition to the kinetic energy associated with the motion of an object, there are numerous types of stored or internal energy to consider. There are various important ways to categorize these different types of energy. One type of energy is the **potential energy** of objects situated in a force field or subject to a force. For example, a rock taken to the top of a building has potential energy due to its placement within the Earth's gravitational field, or a pinball situated at the end of a coiled spring. There is **internal energy** due to micro-processes such as molecular, atomic, and sub-atomic reactions, including thermal and acoustic processes. Alternatively, we may categorize energy as due to conservative forces and non-conservative forces. For a **conservative force**, such as the force of gravity in interstellar space, the amount of work done on a particle acted upon by the force and traveling from a point $\mathbf{x}_0$ to a point $\mathbf{x}$, does not depend on the path taken by the particle. Thus, for a conservative force the work

$$W(\mathbf{x}, \mathbf{x}_0) = \int_{\mathbf{x}_0}^{\mathbf{x}} \bar{\mathbf{F}} \cdot d\mathbf{x} \tag{2.13}$$

only depends on the initial and final points of the path. Analogous to the kinetic energy, we can define the potential energy as the work in going from the initial point to the final point of the path, if we imagine moving so slowly that the square of the velocity is negligible. The potential energy is given by

$$U(\mathbf{x}) - U(\mathbf{x}_0) = \int_{\mathbf{x}_0}^{\mathbf{x}} \bar{\mathbf{F}} \cdot d\mathbf{x}. \tag{2.14}$$

One typically defines a reference point $\mathbf{x}_R$ at which the potential energy is zero, then we can imagine traveling from the reference point to the point $\mathbf{x}_0$ and then onto the point $\mathbf{x}$ resulting in the expression

$$U(\mathbf{x}) - U(\mathbf{x}_0) = \int_{\mathbf{x}_R}^{\mathbf{x}} \bar{\mathbf{F}} \cdot d\mathbf{x} - \int_{\mathbf{x}_R}^{\mathbf{x}_0} \bar{\mathbf{F}} \cdot d\mathbf{x}. \tag{2.15}$$

The overlapping portion of the path from $\mathbf{x}_R$ to $\mathbf{x}_0$ cancels in Equation (2.15) resulting in the expression (2.14) for the change in potential energy. It is common practice to take the reference position to be a location at which the force is zero. For example, the reference point for the gravitational force of the Earth is usually an infinite distance from the center of the Earth. For a **non-conservative** force the work is path-dependent. A well-known non-conservative force is friction. Because a conservative force only depends on the initial and final points of its journey, for a fixed starting point we can write it as a function of the end point $\mathbf{x}$: $U(\mathbf{x})$. Note that, because the position vector is often considered to be a function of time, $\mathbf{x}(t)$, the potential energy of a system is also an implicit function of time. However, conservative forces cannot depend explicitly upon either time or the particle velocity. If the force did depend upon time, a longer path would produce a different result because the forces would have changed during the additional time taken to travel along the path. Similarly, traveling the same path at a different velocity does not change the work done by a conservative force.

Continuous objects and fields: continuum mechanics

We do not typically encounter point masses in our daily lives. Rather, one is faced with continuous fluids or extended solid bodies, that are in reality vast collections of tightly coupled atomic masses. Trying to decompose such objects into constituent point masses would be a computational nightmare. A successful approach for modeling extended objects involves smoothing over the individual point masses and treating the body as a continuous distribution of material, leading to the notions of continuum mechanics. Thus, for a given volume of material we assume an average density, mass per unit volume, as well as other average properties such as electrical and thermal conductivity and viscosity. The issue of averaging or smoothing is particularly acute for the study of flow through a porous media, where the process under consideration occurs in an intrinsically heterogeneous medium.

There are mathematical techniques for effectively scaling up from atomic, molecular, or grain-sized particles to a macroscopic sized body, methods known as effective medium theories. Three approaches are prominent in the study of poroelasticity: mixture theory (Truesdell, 1962; Green and Naghdi, 1965; Garg, 1971; Morland, 1972; Drumheller, 1978; Berryman, 1986; Berryman et al., 1988), volume averaging (Slattery, 1968, 1981; Whitaker, 1969; Pride et al., 1992) and the related averaging over a representative elementary volume (Bear, 1972; de Marsily, 1986), and homogenization techniques such as the method of multiple scales and the two-space method (Keller, 1977; Auriault, 1980; Burridge and Keller, 1981). While we shall have a bit more to say about such techniques later, in the section on poroelasticity, we shall not invoke such formal techniques. Rather we take it for granted that at some intermediate length scale very large collections of particles have definable average properties (Figure 2.1).

For example, we assume that one can, at least in theory, estimate the density in the vicinity of a point by taking a very small cube of material surrounding the point and weighting it in order to estimate the mass ΔM. If the volume of the cube is ΔV then the density is given by the ratio

$$\rho(\mathbf{x}) = \lim_{\Delta V \to 0} \frac{\Delta M(\mathbf{x})}{\Delta V(\mathbf{x})}, \tag{2.16}$$

where it is assumed that this ratio is relatively constant for a range of intermediate length scales. We assign the estimated density to the reference point. Furthermore, it is assumed that we can, again in theory, repeat this process for every point $\mathbf{x}$ of the body, thereby constructing a continuous distribution of density $\rho(\mathbf{x})$. The length scale is termed intermediate because it is taken to be large enough to contain a sufficient number of particles to form a stable average and small enough that it is not influenced by the large scale heterogeneity of the body. Note that the construction of averages, as described here, can require some thought and ingenuity, particularly

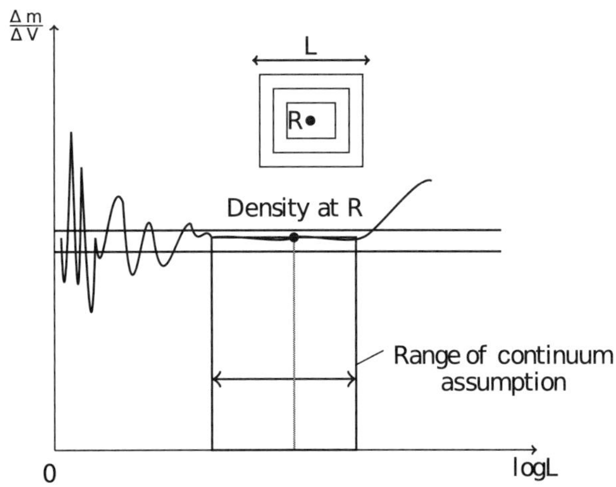

Figure 2.1 Averaging of properties in continuum assumption. We assume that materials have definable average properties over a range of length scales.

in the presence of pervasive structure such as small-scale layering. For example, averaging a layered structure leads to an apparently anisotropic material.

Just as properties such as mass are averaged, leading to a distribution, the forces acting upon a body can be averaged to produce resultant 'body forces' acting on all points of an object. For example, the force of gravity will pull all points of a body towards the Earth's center, similar to its action on a point mass. However, because an object is a vast collection of constituent points there will be an additional complication due to the interaction of the points. Such interactions cannot be modeled as single forces acting upon each small piece of the body. For example, the pressure in a fluid due to the weight of any overlying mass does not act in a single direction. Rather, it is best visualized as three mutually perpendicular force couples impressed upon a small sample. Each pair of force couples is squeezing a small portion of the sample along one axis. If we imagine a cut, oriented perpendicular to that direction, the force couple would act to squeeze the two sides together. However, if we rotate the cut so that the force couple lies parallel to it, then its sides will not be forced together. Thus, the effect of a force $\mathbf{f}$ on the surface of the cut depends upon its orientation with respect to the surface, denoted by S. The forces on each side of the cut must be equal and opposite because we cannot have a non-zero resultant force acting upon the cut, as it is in static equilibrium. We can define the traction as the force per unit area, our first encounter with the concept of **stress.** As with density, we can define the **surface traction vector t** as the ratio

$$\mathbf{t}(\mathbf{x}) = \lim_{\Delta S \to 0} \frac{\Delta \mathbf{f}(\mathbf{x})}{\Delta S(\mathbf{x})}, \tag{2.17}$$

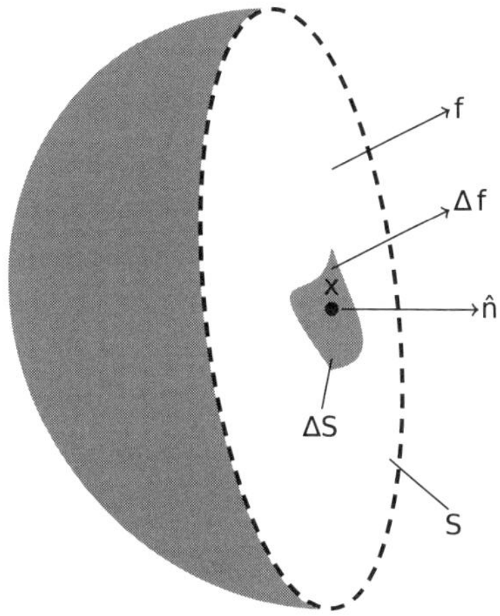

Figure 2.2 Surface traction vector at point **x** is given by $\mathbf{t}(\mathbf{x}) = \lim_{\Delta S \to 0} \Delta f(\mathbf{x}) / \Delta S(\mathbf{x})$ where $\Delta \mathbf{f}$ is the force acting on the surface ΔS. The orientation of the surface is defined by the unit vector $\hat{\mathbf{n}}$ perpendicular to S.

where **f** is the force associated with the cut that has a surface area of ΔS (Figure 2.2).

As mentioned above, the traction vector is not just a function of the area of the surface, it also depends upon the orientation of the surface with respect to the force vector $\Delta \mathbf{f}$. The orientation of the surface is usually characterized by, $\bar{\mathbf{n}}$, the unit vector perpendicular to the surface S. Because the orientation of the surface is specified by the normal vector, $\bar{\mathbf{n}}$ and a vector in three dimensions may be completely specified by a sum of multiples of the unit vectors along each axis, one might postulate that the traction vector **t** in any direction is given by a weighted sum of traction vectors associated with planes perpendicular to each axis. We denote the traction vector associated with the surface perpendicular to the i-th axis by

$$\mathbf{t}_i = \begin{pmatrix} \sigma_{1i} \\ \sigma_{2i} \\ \sigma_{3i} \end{pmatrix}. \tag{2.18}$$

We might suppose that we can write **t** in terms of the three vectors $\mathbf{t}_1$, $\mathbf{t}_2$, and $\mathbf{t}_3$. One can show that this is indeed the case by considering a force balance on a tetrahedral element, as shown in Fung (1969, p. 52), resulting in

$$t_j \Delta S - \sigma_{ji} \hat{n}_i \Delta S = 0 \tag{2.19}$$

or, using vector-matrix notation and dividing by ΔS,

$$\mathbf{t} = \sigma \cdot \hat{\mathbf{n}}, \tag{2.20}$$

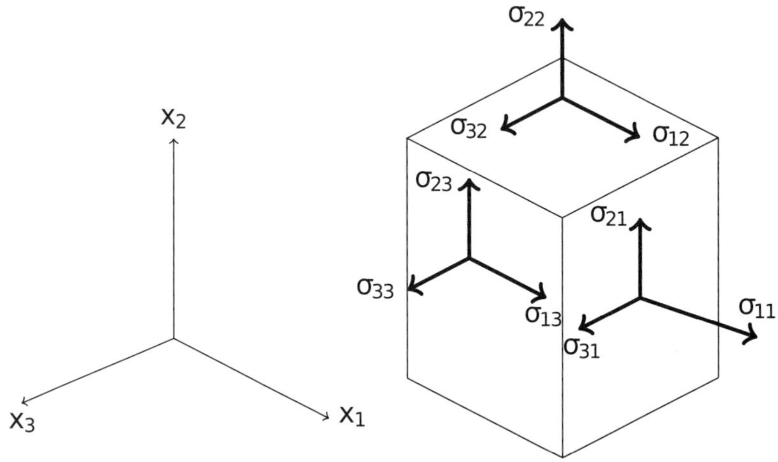

Figure 2.3 The nine components of the stress tensor acting on a cubic element. Note that the first subscript denotes the direction and the second subscript denotes the plane the component is acting on. The components acting perpendicular to the plane ($\sigma_{11}, \sigma_{22}, \sigma_{33}$) are called the normal stresses, while the components tangential to the plane ($\sigma_{12}, \sigma_{13}, \sigma_{21}, \sigma_{23}, \sigma_{31}, \sigma_{32}$) are called the shear stresses.

where σ is the **stress tensor**, a matrix whose columns are given by (2.18)

$$\sigma = \begin{pmatrix} \sigma_{11} & \sigma_{12} & \sigma_{13} \\ \sigma_{21} & \sigma_{22} & \sigma_{23} \\ \sigma_{31} & \sigma_{32} & \sigma_{33} \end{pmatrix}.$$

This matrix provides a general representation of the internal tractions in a body (Figure 2.3). Note that, in general, the stress cannot be represented by a single vector. A vector is a physical quantity that is specified by a magnitude and a direction. The stress tensor, on the other hand, requires specification of a magnitude and two directions: one for the force and one for the plane of interest. For example, consider the pressure p within a fluid at a particular depth. As we shall see [refer to Equation (2.67)] the corresponding stress tensor is given by

$$\sigma = \begin{pmatrix} p & 0 & 0 \\ 0 & p & 0 \\ 0 & 0 & p \end{pmatrix}.$$

Multiplication by the unit vectors along each axis will return different traction vectors. Because stress is defined in terms of a force (a vector) and a plane (characterized by a vector, the normal to the surface), it seems sensible that the stress cannot be related to any single vector. There are special cases, such as the compression of a long, stiff rod, where one component of the stress tensor, such as σ_{11}, dominates and a vector may provide a sufficient representation.

All of the quantities described above, scalars such as density and temperature, vectors such as a body force, and tensors such as the stress are functions of position **x** and are refered to as **fields**. Such objects are the basic quantities providing the observations gathered in field experiments and in the laboratory.

Conservation principles

We have already encountered a number of ideas from mechanics, such as Newton's relationship between a force and a change in $m\mathbf{v}$, as well as d'Alembert's principle. These relationships should not depend on the units used to specify the physical quantities, such as distance, time, or mass, involved in the physical laws. Furthermore, the principles should hold independent of the coordinate system used in the specification. Thus, for example if we use a translated or rotated set of coordinates, the principles should still be valid. Equivalently, the laws of physics must not depend upon our location or orientation in space. Such invariance can be shown to lead to conservation equations related to the particular transformation. For example, one can show that a translational invariance leads to the conservation of linear momentum. We should note that the phrase 'linear momentum' refers to the fact that it corresponds to straight line motion, as opposed to angular (rotational) momentum. A physically motivated route to the **conservation of linear momentum** begins with Newton's law of motion, Equation (2.2). The basic idea is that for a system which is not subject to external forces the right-hand-side of Equation (2.2) vanishes and so

$$\frac{d(m\mathbf{v})}{dt} = 0 \tag{2.21}$$

or

$$m\mathbf{v} = a\ constant. \tag{2.22}$$

Thus, the momentum of the object isolated from external forces is a constant.

Let us generalize the conservation of linear momentum to an extended body, as discussed in the previous sub-section, subject to a distribution of body (volumetric) forces and surface forces (surface tractions). Due to the presence of external forces, the right-hand-side of Equation (2.21) is no longer zero, rather the rate of change of the momentum balances the sum of the external forces. Because the body and the forces are distributed, we have to sum, or integrate to get the total force on an extended object. Consider an object of volume V, bounded by a surface S (Figure 2.4). The mass distribution of the object is characterized by the density $\rho(\mathbf{x})$, defined by Equation (2.16). The total external volumetric forces **f** and surface tractions **t** balance the rate of change of the total linear momentum, thus

$$\int_V \mathbf{f}dV + \int_S \mathbf{t}dS = \frac{d}{dt}\int_V \rho\mathbf{v}dV. \tag{2.23}$$

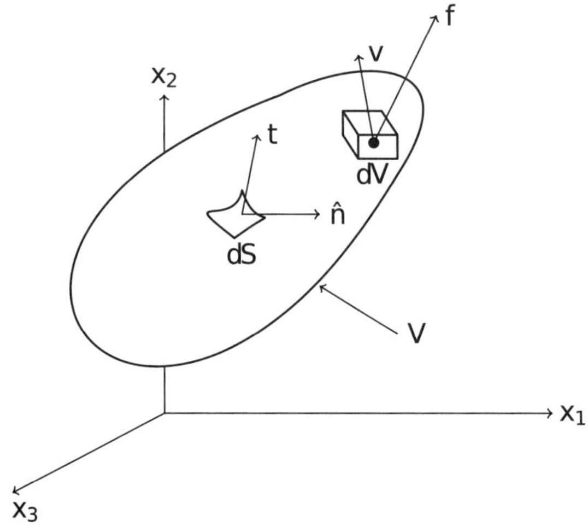

Figure 2.4 The conservation of linear momentum requires that the total external volumetric forces **f** and the surface traction force **t** balance the rate of change of linear momentum over the volume V.

Box 2.1 **The divergence theorem**

The divergence theorem, also known as Gauss' theorem in recognition of his early proof, relates the divergence of a vector field **F** contained within a volume V to the flux of the field through the surface S bounding the volume. The exact mathematical statement is

$$\iiint_V \nabla \cdot \mathbf{F} dV = \iint_S \mathbf{F} \cdot \hat{\mathbf{n}} dS$$

where $\hat{\mathbf{n}}$ is the normal vector to the surface S. A general proof of Gauss' theorem is better left to a book on vector calculus such as Marsden and Tromba (1976). However, following Schey (1973) we can provide a brief motivation by considering a simple volume that may be subdivided into N subvolumes (think small cubes).

Now for any two adjacent subvolumes (cubes) sharing a common internal surface S_i, the contribution $\mathbf{F} \cdot \mathbf{n}_i$ across that surface from one cube will be canceled by the contribution back across the surface from the other cube. This cancellation occurs because the normal vectors are equal and opposite for an internal surface while **F** remains the same on the surface. Thus, we can write the surface integral in the above expression as a sum over the surfaces of all the subvolumes

$$\iint_S \mathbf{F} \cdot \hat{\mathbf{n}} dS = \sum_{l=1}^{N} \iint_{S_l} \mathbf{F} \cdot \hat{\mathbf{n}} dS,$$

with the understanding that the contributions from all the internal surfaces will cancel in the sum. Dividing and multiplying each term in the sum by the volume of the respective cube give

$$\iint_S \mathbf{F} \cdot \hat{\mathbf{n}} dS = \sum_{l=1}^{N} \left[\frac{1}{\Delta V_l} \iint_{S_l} \mathbf{F} \cdot \hat{\mathbf{n}} dS \right] \Delta V_l$$

In the limit as the subvolumes shrink in size the quantity in square brackets approaches the divergence of the vector field **F** at the center point of the cube, thus we have

$$\lim_{N\to\infty, \Delta V_l \to 0} \sum_{l=1}^{N} \left[\frac{1}{\Delta V_l} \iint_{S_l} \mathbf{F} \cdot \hat{\mathbf{n}} dS \right] \Delta V_l = \iiint_V \nabla \cdot \mathbf{F} dV,$$

ending our argument.

The integrals are over the internal volume of the object and over the external surface of the object. We can write the surface traction vector **t** in terms of the stress tensor, using Equation (2.20):

$$\int_V \mathbf{f} dV + \int_S \sigma \cdot \hat{\mathbf{n}} dS = \frac{d}{dt} \int_V \rho \mathbf{v} dV. \tag{2.24}$$

Applying the divergence theorem [see Box 2.1], the surface integral in (2.24) can be written as a volume integral

$$\int_V \mathbf{f} dV + \int_V \nabla \cdot \sigma dV = \frac{d}{dt} \int_V \rho \mathbf{v} dV. \tag{2.25}$$

A similar conservation law holds for angular momentum. At this point let us simply note that the conservation of angular (rotational) momentum leads to the conclusion that the stress tensor σ is symmetric

$$\sigma = \sigma^T, \tag{2.26}$$

where the superscript T signifies the transpose of the matrix (Fung, 1969). In terms of components, Equation (2.26) means that

$$\sigma_{ij} = \sigma_{ji}. \tag{2.27}$$

Similar principles apply to other physical quantities such as the **conservation of the mass** of the system. At its simplest, the conservation of mass states that the mass contained in an isolated volume V is constant

$$\frac{dm}{dt} = \frac{d}{dt} \int_V \rho dV = 0. \tag{2.28}$$

As we shall see, there are alternative forms for this conservation equation, known as the equation of continuity, depending on the nature of the coordinates used to state the principle. There is also a conservation principle for energy that is important in certain situations (Goldstein, 1950) but will not be part of our discussion.

Constitutive equations

The various conservation principles lead to equations governing fluid flow, deformation, and other processes of interest. However, the resulting equations typically contain more variables than equations and cannot be solved uniquely. Therefore, one must rely on the physics of the process under study to invoke a constitutive relation between the various unknowns. For example, in Equation (2.25) we can write the velocity in terms of the time derivative of the displacements, a straightforward definition. However, we must invoke a relationship between the stresses in the body and the displacements in order to produce a set of equations solely in terms of the displacements. We will illustrate this in the material that follows.

2.3 Deformation

2.3.1 The basics

In response to forces and surface tractions, objects can move and change shape, that is deform. In this section we shall concentrate on deformation and strain, a change in the distance between the parts of a body, and especially on a quantitative description. In discussing deformation we idealize a hypothetical volume of material, denoted by V, bounded by a surface S. We need to account for the initial and final positions of the material of the body, the positions before and after the deformation. So as not to confuse the various states and coordinate systems, some rather precise language is required. For example, the word **point** will mean a location in space, the point of a coordinate system. Conversely, the word **particle** denotes a small volume element of the body. The deformation evolves as a function of time, with the time t_0 denoting the start of the deformation. At a later time, t, the deformation has progressed and the particles of the body have moved along various paths. A continuous one-to-one function relates the current position (at time t), $\mathbf{x}$, of a particle that was initially (at time t_0) at point $\mathbf{X}$

$$\mathbf{x} = \mathbf{x}(\mathbf{X}, t), \tag{2.29}$$

constituting the **Lagrangian** description of the deformation. In the Lagrangian approach the spatial coordinates $\mathbf{x}$ are dependent variables, functions of the material point $\mathbf{X}$ and the time. An alternative description of the deformation keeps track of the particle $\mathbf{X}$ currently occupying the location $\mathbf{x}$ via functions of the form

$$\mathbf{X} = \mathbf{X}(\mathbf{x}, t), \tag{2.30}$$

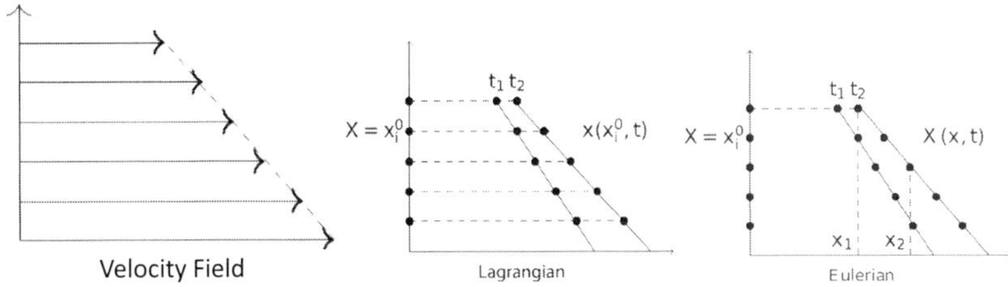

Figure 2.5 Illustration of the Lagrangian and Eulerian description for a group of passive particles flowing in a velocity field, $\mathbf{v}$. The Lagrangian description tracks the spatial position of a particle, identified by $\mathbf{X} = x_i^0$ with time: $\mathbf{x}(\mathbf{X}, t)$. The Eulerian description keeps track of the particle $\mathbf{X} = x_i^0$ that occupies a given position $\mathbf{x}$ at a given time t: $\mathbf{X}(\mathbf{x}, t)$.

being the **Eulerian** description of the deformation. In the Eulerian description the spatial coordinates are independent variables, as is the time t. This is illustrated in Figure 2.5.

The function in (2.29) is continuous if the partial derivatives of each component of $\mathbf{x}$ with respect to the components of the independent variables $\mathbf{X}$ exist and are finite. Furthermore, at a fixed time t, the mappings are one-to-one if the determinant of the matrix of partial derivatives

$$\mathbf{F} = \begin{pmatrix} \dfrac{\partial x_1}{\partial X_1} & \dfrac{\partial x_1}{\partial X_2} & \dfrac{\partial x_1}{\partial X_3} \\ \dfrac{\partial x_2}{\partial X_1} & \dfrac{\partial x_2}{\partial X_2} & \dfrac{\partial x_2}{\partial X_3} \\ \dfrac{\partial x_3}{\partial X_1} & \dfrac{\partial x_3}{\partial X_2} & \dfrac{\partial x_3}{\partial X_3} \end{pmatrix} \tag{2.31}$$

does not vanish. The matrix can be represented more compactly in terms of its components

$$F_{ij} = \left(\frac{\partial x_i}{\partial X_j} \right). \tag{2.32}$$

In what follows, we shall invoke the Einstein summation convention [see Box 2.2] and sum over repeated indices. The matrix $\mathbf{F}$ is referred to as the **material deformation gradient**. The companion matrix $\mathbf{H}$, the **spatial deformation gradient** related to the transformation (2.31) is defined as

$$H_{ij} = \left(\frac{\partial X_i}{\partial x_j} \right). \tag{2.33}$$

Using the chain rule, one can show that H_{ij} satisfies

$$F_{ij}H_{jk} = \frac{\partial x_i}{\partial X_j}\frac{\partial X_j}{\partial x_k} = \delta_{ik}, \tag{2.34}$$

where δ_{ik} is the Kronecker delta function, equal to 1 if i equals k and is zero otherwise. Thus, the matrix $\mathbf{F}$ is the inverse of the matrix $\mathbf{H}$.

We can define the **displacement vector**, a measure of the change in position that occurs between the initial time 0 and the final time t. Because at time 0 the body has not yet deformed, the particles are in their original positions given by $\mathbf{X}$. Thus, the displacement vector is given by

$$\mathbf{u} = \mathbf{x} - \mathbf{X}, \tag{2.35}$$

a measure of the movement that occurred between time 0 and time t. Note that

$$\mathbf{x} = \mathbf{X} + \mathbf{u},$$

and one may think of $\mathbf{x}$ and $\mathbf{X}$ as two coordinate systems and of $\mathbf{u}$ as the map from one to the other (Figure 2.6). The **strain** is a measure of the change in distance between two nearby particles due to the deformation. Thus, it involves two displacements, one for each particle, over the specified time interval. Let us denote the

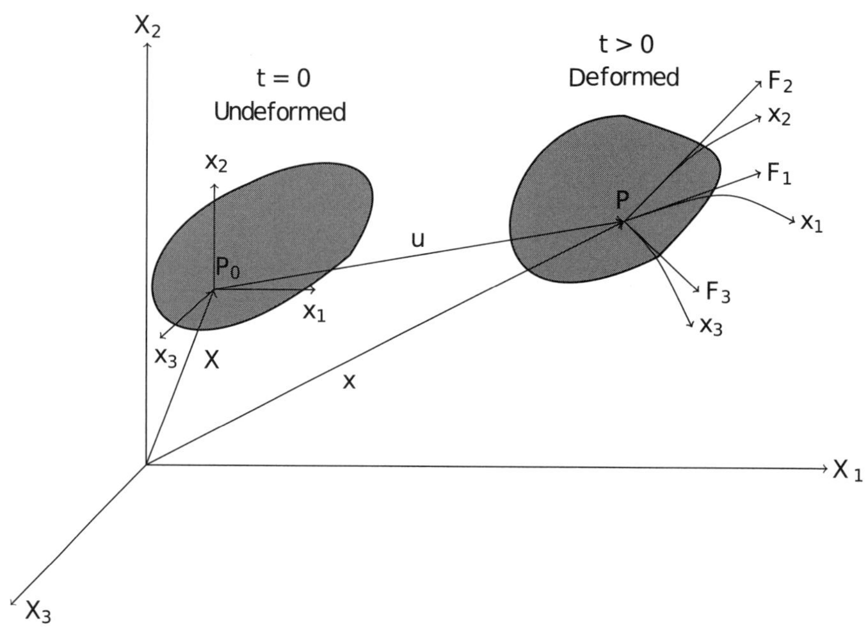

Figure 2.6 The undeformed configuration of a material body is shown together with its deformed configuration at a later time after displacement $\mathbf{u}$. The deformation of a material element at its initial position P_0 is related to its final position P by the deformation gradient, F.

original distance between the particles by dX and the distance after the deformation by dx. The change in squared distance $(dx)^2 - (dX)^2$ is used as a measure of the deformation. Using the functional relationship (2.29) and the chain rule, we can write

$$dx_i = \frac{\partial x_i}{\partial X_j} dX_j \tag{2.36}$$

and hence the change in squared distance may be written as

$$(dx)^2 - (dX)^2 = \left(\frac{\partial x_k}{\partial X_i} \frac{\partial x_k}{\partial X_j} - \delta_{ij} \right) dX_i dX_j. \tag{2.37}$$

The quantity in the parenthesis on the right-hand-side provides a measure of the strain.

Box 2.2 **Einstein's summation convention**

The Einstein summation convention implies that when an index occurs more than once in the same expression, the expression is summed over all possible values of that index. For example, from Equation (2.29), we obtain a relationship between dx_i and dX_j:

$$dx_i = \sum_{j=1}^{3} \frac{\partial x_i}{\partial X_j} dX_j.$$

Using Einstein's convention, this sum can be written succinctly as (2.36), where we omit the summation sign over the repeated indices j. Similarly, (2.37) implies

$$(dx)^2 - (dX)^2 = \sum_{k=1}^{3} \sum_{i=1}^{3} \sum_{j=1}^{3} \left(\frac{\partial x_k}{\partial X_i} \frac{\partial x_k}{\partial X_j} - \delta_{ij} \right) dX_i dX_j.$$

A useful measure of strain is given in terms of the components of the displacement vector **u**. We can derive such a measure by taking the partial derivative of the displacement vector, Equation (2.35) with respect to the components of **X** and solving for

$$\frac{\partial x_k}{\partial X_i} = \frac{\partial u_k}{\partial X_i} + \delta_{ki} \tag{2.38}$$

and

$$\frac{\partial x_k}{\partial X_j} = \frac{\partial u_k}{\partial X_j} + \delta_{kj}. \tag{2.39}$$

Substituting these expressions into Equation (2.37) results in an expression for the strain in terms of the spatial derivatives of the components of the displacement vector:

$$(dx)^2 - (dX)^2 = \left(\frac{\partial u_i}{\partial X_j} + \frac{\partial u_j}{\partial X_i} + \frac{\partial u_k}{\partial X_i}\frac{\partial u_k}{\partial X_j} \right) dX_i dX_j. \tag{2.40}$$

Thus, we can define the **Lagrangian or Green's finite strain tensor**

$$L_{ij} = \frac{1}{2}\left(\frac{\partial u_i}{\partial X_j} + \frac{\partial u_j}{\partial X_i} + \frac{\partial u_k}{\partial X_i}\frac{\partial u_k}{\partial X_j} \right). \tag{2.41}$$

In a similar fashion, we can derive the **Eulerian finite strain tensor**

$$E_{ij} = \frac{1}{2}\left(\frac{\partial u_i}{\partial x_j} + \frac{\partial u_j}{\partial x_i} + \frac{\partial u_k}{\partial x_i}\frac{\partial u_k}{\partial x_j} \right). \tag{2.42}$$

The reader should show this as an exercise. If the displacement gradients are small the product terms can be neglected, and L_{ij} and E_{ij} reduce to the Lagrangian

$$l_{ij} = \frac{1}{2}\left(\frac{\partial u_i}{\partial X_j} + \frac{\partial u_j}{\partial X_i} \right), \tag{2.43}$$

and the Eulerian

$$\varepsilon_{ij} = \frac{1}{2}\left(\frac{\partial u_i}{\partial x_j} + \frac{\partial u_j}{\partial x_i} \right), \tag{2.44}$$

infinitesimal strain tensors, respectively. These tensors provide useful measures of the strain for small deformations (Fung, 1969, p. 98). If the displacements are small, there will be very little difference between the material and the spatial coordinate systems and hence $l_{ij} \approx \varepsilon_{ij}$ and one need not worry about which coordinate description to use. The elements of the infinitesimal strain tensor have geometric interpretations (Fung, 1969, p. 99). For example, the diagonal elements of the tensor represent extensions along each of the axes, respectively. The off-diagonal elements represent the change in angle between the originally orthogonal pairs of axes. No such simple interpretation is possible if the components of strain are not small. On many occasions we shall be interested in the time derivative of the strain tensors, for example when dealing with fluids. A useful quantity in fluid mechanics is the **rate-of-deformation tensor**, which is simply the time derivative of the strain tensor. In the case of the Eulerian coordinates, the rate-of-deformation tensor is defined as

$$\dot{\varepsilon}_{ij} = \frac{d\varepsilon_{ij}}{dt} = \frac{1}{2}\frac{d}{dt}\left(\frac{\partial u_i}{\partial x_j} + \frac{\partial u_j}{\partial x_i} \right) = \frac{1}{2}\left(\frac{\partial v_i}{\partial x_j} + \frac{\partial v_j}{\partial x_i} \right), \tag{2.45}$$

where v_i is the time derivative of u_i.

While the assumption of infinitesimal strain is often applicable, there are situations in which the strain is significant and we must be mindful of the Lagrangian and Eulerian coordinate descriptions. For example, in viscoelastic processes there can be significant flow during the deformation. Therefore, let us revisit the two coordinate systems and examine how physical properties and their changes are described. Remember the Lagrangian or material description follows the movement (position) of a particle, via the function $\mathbf{x} = \mathbf{x}(\mathbf{X}, t)$. When the interest is in the change of a particular quantity, some care is required in order to produce a useful measure. For example, one could keep a record of the density of a specific particle as it winds its way through space. If these values are used to measure the change in density, then we are considering the **material or convective derivative** of the density. There are two different ways to compute the rate of change of a property of a specific particle. Let us consider this carefully, fixing the particle and denoting it by $\mathbf{X}_0 = \mathbf{X}(\mathbf{x}, 0)$. If we consider Lagrangian specification and use the particle identity and the time as independent variables, then the total derivative with respect to time is the same as the partial derivative because time only appears once

$$\frac{d\rho(\mathbf{X}_0, t)}{dt} = \frac{d\rho(\mathbf{X}(\mathbf{x}, 0), t)}{dt} = \frac{\partial\rho(\mathbf{X}_0, t)}{\partial t}. \tag{2.46}$$

However, if we consider Eulerian specification in which $\mathbf{x}$ and t are the independent variables, then the density of a specific particle is both an implicit and an explicit function of time, $\rho(\mathbf{x}(\mathbf{X}_0, t), t)$ and the derivative must account for the change in the position of the particle as well as for the change in the density of the particle so that

$$\frac{d\rho(\mathbf{x}(\mathbf{X}_0, t), t)}{dt} = \frac{\partial\rho}{\partial t} + \frac{\partial\rho}{\partial x_i}\frac{\partial x_i}{\partial t}. \tag{2.47}$$

Because the identity of the particle is fixed, the spatial position only depends upon t, $\mathbf{x} = \mathbf{x}(\mathbf{X}_0, t)$, and we can write the partial derivative of x_i as a total derivative, and Equation (2.47) becomes

$$\frac{d\rho(\mathbf{x}(\mathbf{X}_0, t), t)}{dt} = \frac{\partial\rho}{\partial t} + \frac{\partial\rho}{\partial x_i}\frac{dx_i}{dt}, \tag{2.48}$$

or, in terms of the components of the velocity vector, v_i,

$$\frac{d\rho(\mathbf{x}(\mathbf{X}_0, t), t)}{dt} = \frac{\partial\rho}{\partial t} + v_i\frac{\partial\rho}{\partial x_i} = \frac{\partial\rho}{\partial t} + \mathbf{v}\cdot\nabla\rho. \tag{2.49}$$

The derivative given above is known as the **material derivative operator** of the density and the second term on the right is called the **convective rate of change**, expressing the contribution of the motion of the particle to the rate of change of the density.

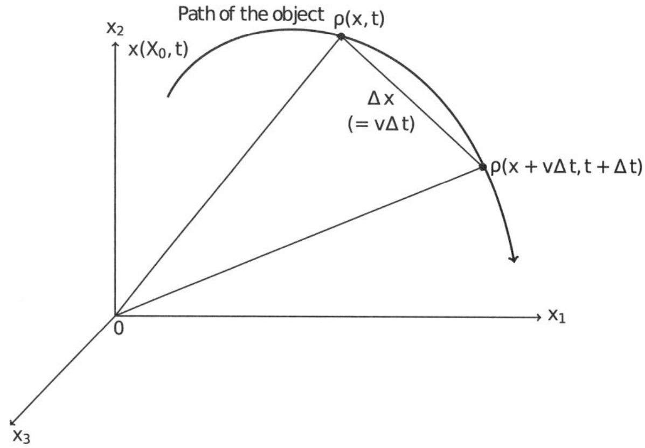

Figure 2.7 Material derivative in Eulerian coordinates is $d\rho/dt = \lim_{\Delta t \to 0} \rho(\mathbf{x} + \Delta\mathbf{x}, t + \Delta t) - \rho(\mathbf{x}, t)/\Delta t$, while the material derivative in Lagrangian coordinates is $d\rho/dt = [\lim_{\Delta t \to 0} \rho(\mathbf{X}_0, t + \Delta t) - \rho(\mathbf{X}_0, t)]/\Delta t$.

Box 2.3 **Physical motivation of the material time derivative**

A more physically motivated derivation of the material time derivative (2.49) is possible if we consider a particle and its motion over a time increment Δt (Figure 2.7). During this time interval the particle will move a distance $\Delta\mathbf{x} = \mathbf{v}\Delta t$ where $\mathbf{v} = d\mathbf{x}(\mathbf{X}_0, t)/dt$ is the particles velocity vector. The change in density during the motion is given by

$$\Delta\rho = \rho(\mathbf{x} + \mathbf{v}\Delta t, t + \Delta t) - \rho(\mathbf{x}, t).$$

We can expand the first term on the right-hand-side as a Taylor series and, assuming that Δt is small, neglect terms of order $(\Delta t)^2$ and higher:

$$\Delta\rho = \rho(\mathbf{x}, t) + \frac{\partial \rho}{\partial x_i} v_i \Delta t + \frac{\partial \rho}{\partial t}\Delta t - \rho(\mathbf{x}, t)$$

$$= \left[\frac{\partial \rho}{\partial x_i} v_i + \frac{\partial \rho}{\partial t}\right]\Delta t.$$

Hence, in the limit as Δt approaches zero we have

$$\lim_{\Delta t \to 0} \frac{\Delta\rho}{\Delta t} = \frac{d\rho}{dt} = \frac{\partial \rho}{\partial t} + v_i \frac{\partial \rho}{\partial x_i},$$

producing the result (2.49).

Now consider the time derivative of the position vector for a particle, the velocity of the particle. We can make use of the definition of the displacement vector,

Equation (2.35), and the fact that the identity of the particle is fixed and does not depend on time to write the velocity as

$$\mathbf{v} = \frac{d(\mathbf{u} + \mathbf{X}_0)}{dt} = \frac{d\mathbf{u}}{dt}. \tag{2.50}$$

As noted above, one can express the velocity vector in terms of Lagrangian or Eulerian coordinates, depending on our needs. In Lagrangian coordinates, for a fixed particle denoted by $\mathbf{X}_0$, the total derivative is equivalent to the partial derivative with respect to time

$$\mathbf{v} = \frac{d\mathbf{u}(\mathbf{X}_0, t)}{dt} = \frac{\partial \mathbf{u}(\mathbf{X}_0, t)}{\partial t}. \tag{2.51}$$

In Eulerian form, the displacement is a function of the spatial location vector $\mathbf{x}$ and the time and hence the velocity vector is given by

$$v_j = \frac{du_j(\mathbf{x}(\mathbf{X}_0, t), t)}{dt} = \frac{\partial u_j}{\partial t} + v_i \frac{\partial u_j}{\partial x_i} \tag{2.52}$$

an implicit expression for the velocity. As the reader may want to show, similar considerations apply when computing the acceleration vector components, leading to the Lagrangian expression

$$\mathbf{a} = \frac{\partial^2 \mathbf{u}(\mathbf{X}_0, t)}{\partial t^2} \tag{2.53}$$

and, alternatively, the Eulerian expression

$$a_j = \frac{\partial v_j}{\partial t} + v_i \frac{\partial v_j}{\partial x_i},$$

or

$$\mathbf{a} = \frac{d\mathbf{v}}{dt} = \frac{\partial \mathbf{v}}{\partial t} + \mathbf{v} \cdot \nabla \mathbf{v}, \tag{2.54}$$

where $\nabla \mathbf{v}$ is the matrix of derivatives

$$\nabla \mathbf{v} = \begin{pmatrix} \frac{\partial v_1}{\partial x_1} & \frac{\partial v_2}{\partial x_1} & \frac{\partial v_3}{\partial x_1} \\ \frac{\partial v_1}{\partial x_2} & \frac{\partial v_2}{\partial x_2} & \frac{\partial v_3}{\partial x_2} \\ \frac{\partial v_1}{\partial x_3} & \frac{\partial v_2}{\partial x_3} & \frac{\partial v_3}{\partial x_3} \end{pmatrix}$$

known as a **dyadic**. Note that in Eulerian coordinates the acceleration is non-linearly related to the velocity due to the deformation. This fact renders the governing equations of finite deformation, such as the Navier–Stokes equations for fluid flow, non-linear. Such non-linearity leads to mathematical complexity but also provides a certain richness to the models, leading to such phenomena as vortices and finite-amplitude breaking waves (Chorin and Marsden, 1993). In addition to derivatives there are integrals of variables, and we shall need to represent such integrals in the appropriate coordinate systems. For example, we will encounter the material derivative of volume integrals at several points in this chapter. Instead of a detailed derivation of the expression for the **material derivative of an integral**, we simply present the result, motivated by the plausibility argument in Box 2.4 The reader can either accept the expression or seek a more analytical treatment in a book on continuum mechanics such as Truesdall (1985). The formal expression for a material derivative of the integral of a scalar quantity $\rho(\mathbf{x}, t)$, such as the density, is

$$\begin{aligned}\frac{d}{dt}\int_V \rho dV &= \int_V \left[\frac{\partial\rho}{\partial t} + \nabla\cdot(\rho\mathbf{v})\right] dV \\ &= \int_V \left[\frac{\partial\rho}{\partial t} + \mathbf{v}\cdot\nabla\rho + \rho\nabla\cdot\mathbf{v}\right] dV \\ &= \int_V \left[\frac{d\rho}{dt} + \rho\nabla\cdot\mathbf{v}\right] dV, \end{aligned} \tag{2.55}$$

where we have made use of the definition of the material derivative operator (2.49) in vector form.

Box 2.4 **Plausibility argument for the form of a material derivative of a volume integral**

As motivation, consider the underlying definition of the material derivative of the integral for the total mass in volume V:

$$\frac{d}{dt}\int_V \rho dV = \lim_{\Delta t\to 0}\frac{1}{\Delta t}\left[\int_V \rho(\mathbf{x}, t+\Delta t)dV - \int_{V_0} \rho(\mathbf{x}, t)dV\right], \tag{2.56}$$

where V_0 is the volume of integration at time t and V is the volume of integration at a slightly later time $t+\Delta t$. We denote the change in volume by $\Delta V = V - V_0$ and partition the first integral on the right-hand-side in Equation (2.56) into one over the original volume V_0 and another over the change in volume ΔV

$$\begin{aligned}\frac{d}{dt}\int_V \rho dV &= \lim_{\Delta t\to 0}\frac{1}{\Delta t}\left[\int_{V_0} \rho(\mathbf{x}, t+\Delta t)dV + \int_{\Delta V} \rho(x, t+\Delta t)dV - \int_{V_0} \rho(\mathbf{x}, t)dV\right] \\ &= \lim_{\Delta t\to 0}\left[\frac{1}{\Delta t}\int_{V_0} \{\rho(\mathbf{x}, t+\Delta t) - \rho(\mathbf{x}, t)\}dV + \frac{1}{\Delta t}\int_{\Delta V} \rho(x, t+\Delta t)dV\right]\end{aligned}$$

$$= \int_{V_0} \frac{\partial \rho}{\partial t} dV + \lim_{\Delta t \to 0} \frac{1}{\Delta t} \int_{\Delta V} \rho(x, t + \Delta t) dV.$$

The last integral on the right-hand-side requires some analysis because as Δt approaches zero the new volume V approaches the initial volume V_0 and hence the volume change ΔV approaches zero. Because the boundary of the volume is being deformed by the flow field $\mathbf{v}$, the volume change during the time interval Δt for a surface patch ΔS is given by

$$\Delta V = \mathbf{v} \cdot \mathbf{n} \Delta S \Delta t, \tag{2.57}$$

where $\mathbf{n}$ is the vector normal to the surface element ΔS. In the limit of Δt approaching zero, the expression for the material derivative of the integral is

$$\frac{d}{dt} \int_V \rho dV = \int_{V_0} \frac{\partial \rho}{\partial t} dV + \lim_{\Delta t \to 0} \frac{1}{\Delta t} \int_S \rho(x, t + \Delta t) \mathbf{v} \cdot \mathbf{n} \Delta t dS$$
$$= \int_{V_0} \frac{\partial \rho}{\partial t} dV + \int_{S_0} \rho \mathbf{v} \cdot \mathbf{n} dS, \tag{2.58}$$

which, after invoking Gauss's theorem, may be written as the volume integral

$$\frac{d}{dt} \int_V \rho dV = \int_{V_0} \frac{\partial \rho}{\partial t} dV + \int_{V_0} \nabla \cdot (\rho \mathbf{v}) dV. \tag{2.59}$$

One interpretation of the formula, particularly Equation (2.59), is that we must account for the variation in the range of integration which is a function of time, a generalization of the rule for differentiation under the integral sign.

The material derivative of an integral allows us to express the conservation laws for an extended body. For example, one can express the conservation of mass, which states that the mass of an object may be constant, unchanging with time:

$$\frac{d}{dt} \int_V \rho dV = \int_V \left[\frac{d\rho}{dt} + \rho \nabla \cdot \mathbf{v} \right] dV = 0.$$

If we assume that mass conservation holds for each sub-volume of the body, then the region of integration may be varied arbitrarily. In particular, the integration volume can be shrunk to lie around any point of the body. Thus, the integrand [the quantity in brackets in Equation (2.55)] must vanish and the conservation of mass leads to the **equation of continuity** for the body:

$$\frac{d\rho}{dt} + \rho \nabla \cdot \mathbf{v} = 0. \tag{2.60}$$

Using the definition of the material derivative operator, Equation (2.49) we can write the equation of continuity as

$$\frac{\partial \rho}{\partial t} + \mathbf{v} \cdot \nabla \rho + \rho \nabla \cdot \mathbf{v} = 0$$

or

$$\frac{\partial \rho}{\partial t} + \nabla \cdot (\rho \mathbf{v}) = 0. \tag{2.61}$$

We can also use the material derivative of an integral to convert the integral statement of the conservation of linear momentum, Equation (2.25)

$$\frac{d}{dt} \int_V \rho \mathbf{v} dV = \int_V [\nabla \cdot \sigma + \mathbf{f}] dV$$

into a partial differential equation. Noting again the expression for the material derivative of an integral, Equation (2.56), we can write this equation as

$$\int_V \left[\frac{\partial}{\partial t} (\rho \mathbf{v}) + \mathbf{v} \cdot \nabla (\rho \mathbf{v}) + (\rho \mathbf{v}) \nabla \cdot \mathbf{v} \right] dV = \int_V [\nabla \cdot \sigma + \mathbf{f}] dV. \tag{2.62}$$

Because the volume of integration is arbitrary we can assume that this relationship holds for each point of the volume V. In addition, we expand the time derivative and the gradient of $\rho \mathbf{v}$ in (2.62)

$$\mathbf{v} \frac{\partial \rho}{\partial t} + \rho \frac{\partial \mathbf{v}}{\partial t} + (\mathbf{v} \cdot \nabla \rho) \mathbf{v} + \rho \mathbf{v} \cdot \nabla \mathbf{v} + (\rho \mathbf{v}) \nabla \cdot \mathbf{v} = \nabla \cdot \sigma + \mathbf{f}.$$

Regrouping terms, we can write this equation as

$$\mathbf{v} \left\{ \frac{\partial \rho}{\partial t} + \rho \nabla \cdot \mathbf{v} + \mathbf{v} \cdot \nabla \rho \right\} + \rho \left\{ \frac{\partial \mathbf{v}}{\partial t} + \mathbf{v} \cdot \nabla \mathbf{v} \right\} = \nabla \cdot \sigma + \mathbf{f}, \tag{2.63}$$

and note that the terms in the first set of braces vanish by the equation of continuity, Equation (2.60), and the terms in the second set of braces are just the material derivative of the velocity vector $\mathbf{v}$. Hence, Equation (2.63) reduces to

$$\rho \frac{d\mathbf{v}}{dt} = \nabla \cdot \sigma + \mathbf{f}, \tag{2.64}$$

a system of partial differential equations known as **Cauchy's equations of motion**.

Constitutive equations revisited: a preview of linear elasticity and Newtonian viscosity

Before venturing on to the equations governing motion in an elastic medium, the topic of the next section, let us consider the role of constitutive equations in characterizing a material. A constitutive equation or relation is an equation describing a property of a material. As one might imagine, there are now thousands of constitutive equations for the many types of materials employed today. Here we will note two of the classic constitutive equations frequently invoked in the study of solids and fluids. Earlier, we alluded to the importance of constitutive equations in reducing the number of unknowns in a set of governing equations. For example, the conservation of linear momentum leads to Equation (2.64), a set of three coupled equations in the three unknown components of $\mathbf{u}$ and six independent components of the symmetric stress tensor σ. A constitutive equation relating the stress tensor to components of the displacement vector would enable us to write Equation (2.64) entirely in terms of $\mathbf{u}$. In this treatment, we shall most often deal with such stress-strain constitutive relationships. However, we note that constitutive equations exist for processes as diverse as heat transfer, electromagnetics, and mass transport.

One early constitutive equation was **Hooke's law**, originally derived to describe the mechanics of an elastic spring in which the stress was linearly proportional to the strain. It has subsequently been generalized to describe a linear relationship between the six independent components of the stress tensor σ [remember from the conservation of angular momentum we saw that the stress tensor is symmetric, see Equation (2.27)] and the six independent components of the strain tensor ε, such as the infinitesimal Eulerian strain tensor (2.44),

$$\sigma_{ij} = C_{ijkl}\varepsilon_{kl} = C_{ijkl}\frac{1}{2}\left(\frac{\partial u_k}{\partial x_l} + \frac{\partial u_l}{\partial x_k}\right), \tag{2.65}$$

where the matrix C_{ijkl} is a tensor of **linear elastic moduli** that are independent of stress or strain. After accounting for the symmetry of both the stress and strain tensors, one finds that there are thirty-six independent linear elastic moduli (Fung, 1969). Because the finite and infinitesimal strain tensors are defined in terms of the spatial derivatives of the components of the displacement vector [see Equations (2.42) and (2.44)], substituting Hooke's law into Equation (2.64). produces three equations for the three components of $\mathbf{u}$. The linear elastic body is one of the most useful physical models for the study of materials and one that we shall employ extensively, starting in the next section. However, there are many other such mathematical models of materials, and new ones are developed with some regularity.

Perhaps the most important constitutive relationship for the purposes of this book is that associated with a fluid. In a **Newtonian fluid** the shear stress is linearly proportional to the rate-of-deformation, the time derivative of the strain, a model for a viscous liquid. It is based on the idea that the faster one deforms a fluid, the greater is the required force. The exact constitutive equation, providing an expression for the stress tensor is

$$\sigma_{ij} = -p\delta_{ij} + D_{ijkl}\dot{\varepsilon}_{kl}, \tag{2.66}$$

where D_{ijkl} is a tensor of viscosity coefficients characterizing the fluid and p is the **fluid pressure** (Fung, 1969). Here the dot signifies the time derivative and ε_{ij} is given by Equation (2.44). For a non-viscous or inviscid fluid all components of D_{ijkl} vanish and Equation (2.66) reduces to

$$\sigma_{ij} = -p\delta_{ij}, \tag{2.67}$$

an expression noted previously, following Equation (2.20). The negative sign in expression (2.67) results from our adoption of the convention used in geophysics in which the traction vector, acting across a surface, has the direction of a force due to the material on the side to which the normal vector **n** points (Aki and Richards, 1980a, p. 13). It should be noted that other constitutive relationships may be required in reducing a governing equation, such as Equation (2.65), to a form with a unique solution. For example, the density can depend upon the pressure as well as the temperature. For an ideal gas, we have

$$p = RT\rho,$$

where R is a constant and T is the temperature, while for liquid water one often uses

$$\rho = \rho_0 e^{\beta(p-p_0)}$$

where ρ_0, β, and p_0 are constants. For a solid, the density is often assumed to be constant and thus treated as a parameter rather than a variable.

These models are perhaps the simplest possible mathematical abstractions of the behavior of real solids and liquids. Any real material is likely to deviate somewhat from these relationships. Fortunately, many solids do behave elastically over a limited range of stresses and strains. Furthermore, it is often possible to apply the elastic model incrementally over a deformation history, much as one would sub-divide a non-linear curve into a sequence of approximately linear segments. However, at some point a material will usually undergo failure and then an elastic model no longer applies. Even in this case one can try and isolate the non-linear source region and invoke a source model in that region, treating the rest of the body as elastic (Backus and Mulcahy, 1976). This step is often taken when modeling an explosion or earthquake source, for example. As might be imagined, the possible deviations from elastic behavior are too numerous to describe in our short treatment. Classic examples are brittle failure, which involves fracture, and plastic failure in which the body yields and experiences unrecoverable deformation (Jaeger et al., 2007).

2.4 Elastic deformation

We now consider a few specific constitutive equations that are useful in modeling a wide range of materials. In this section, we will be concerned with elastic deformation which is a good model for many solid materials undergoing short-term deformation. By short term we mean a time interval over which the elastic behavior dominates inelastic behavior, such as flow or fracture, although this time interval quite possibly may extend to several years. We choose to start with elastic deformation, because even when studying fluid flow we shall have to consider elastic behavior. That is because we will be interested in fluids flowing within a porous solid matrix. It is frequently assumed that the solid matrix behaves elastically while the fluids flow within it. Most elastic deformation can be modeled assuming infinitesimal strain and neglecting the distinction between the Lagrangian and Eulerian formulations.

Our goal is to derive an equation, or set of equations, governing the deformation of a heterogeneous elastic material. We begin with the appropriate conservation laws, the conservation of mass and the conservation of momentum. Augmenting these conservation laws with the constitutive equation for an elastic medium produces the desired governing equations. So let us start with the conservation of mass, where the total mass contained in our elastic body is given by the integral

$$m = \int_V \rho dV.$$

According to the conservation of mass, the total mass of the system is conserved during the deformation and so

$$\frac{dm}{dt} = \frac{d}{dt}\int_V \rho dV = \int_V \left[\frac{d\rho}{dt} + \rho\nabla\cdot\mathbf{v}\right]dV = 0, \tag{2.68}$$

where we have used the formula for the material derivative of an integral, Equation (2.55). Now Equation (2.68) holds for any portion of the body because we are not adding or subtracting mass anywhere within our volume V. Thus, the volume of integration is arbitrary and hence the equation holds point-wise, leading to the partial differential equation of continuity [see Equation (2.60)]

$$\frac{d\rho}{dt} + \rho\nabla\cdot\mathbf{v} = 0, \tag{2.69}$$

a reformulation of the conservation of mass as a differential equation. We can re-arrange Equation (2.69) to arrive at the expression

$$\frac{1}{\rho}\frac{d\rho}{dt} = -\nabla \cdot \mathbf{v} \tag{2.70}$$

which relates the fractional change in density to the divergence of the velocity field associated with the deformation. Noting the definition of the material derivative, Equation (2.49), we can write the equation of continuity (2.69) entirely in terms of partial derivatives

$$\frac{\partial \rho}{\partial t} + \nabla \cdot (\rho \mathbf{v}) = 0. \tag{2.71}$$

Next, we invoke the conservation of linear momentum for the volume V, leading to Cauchy's equations of motion, derived earlier [Equations (2.62) to (2.64)]:

$$\rho \frac{d\mathbf{v}}{dt} = \nabla \cdot \sigma + \mathbf{f}. \tag{2.72}$$

Equation (2.72) holds regardless of the type of material we are modeling. That is, the equations are independent of the constitutive equation.

The final step in our derivation incorporates a constitutive equation into Equation (2.72), reducing it to a solvable system of partial differential equations. For an elastic model the appropriate constitutive relationship is the generalized Hooke's law given previously, in Equation (2.65),

$$\sigma = \mathbf{C}\varepsilon \tag{2.73}$$

where $\mathbf{C}$ is the tensor of elastic moduli. The strain tensor ε could be the Eulerian finite strain tensor, given by E_{ij} in Equation (2.42), but is more likely to be the infinitesimal strain tensor

$$\varepsilon_{ij} = \frac{1}{2}\left(\frac{\partial u_i}{\partial x_j} + \frac{\partial u_j}{\partial x_i}\right) \tag{2.74}$$

because most work utilizing elasticity involves small strains. When we are interested in small strains, the material derivative in Equation (2.72), given in Equation (2.54) is well approximated by the partial derivative with respect to time:

$$\frac{d\mathbf{v}}{dt} = \frac{\partial \mathbf{v}}{\partial t} + \mathbf{v} \cdot \nabla \mathbf{v} \approx \frac{\partial \mathbf{v}}{\partial t} = \frac{\partial^2 \mathbf{u}}{\partial t^2} \tag{2.75}$$

because the displacements and the gradients of the displacements are small, meaning that the product terms are even smaller and may be neglected.

Substituting the expression (2.74) into the constitutive relationship (2.73) produces a relationship between the spatial derivatives of the displacement vector components and the stress tensor. In index notation we can write an expression for the components of the infinitesimal elastic stress tensor

$$\sigma_{ij} = \frac{1}{2} C_{ijkl} \left(\frac{\partial u_k}{\partial x_l} + \frac{\partial u_l}{\partial x_k} \right). \tag{2.76}$$

Equation (2.76) is appropriate for a fully anisotropic body, characterized by the entire set of thirty-six independent elastic moduli. In many cases the elastic materials of interest have a particular symmetry, such as transverse isotropy. For each specific symmetry one can reduce the number of elastic moduli required to characterize a given material. For an isotropic material, in which the elastic properties do not depend upon the propagation direction, only two moduli are required to characterize an elastic body (Fung, 1969; Aki and Richards, 1980a). When the body is isotropic Equation (2.76) reduces to

$$\sigma_{ij} = \lambda \nabla \cdot \mathbf{u} \delta_{ij} + G \left(\frac{\partial u_i}{\partial x_j} + \frac{\partial u_j}{\partial x_i} \right), \tag{2.77}$$

where λ is referred to as Lame's constant and G is known as the shear modulus (Figure 2.8). The shear modulus is a measure of the rigidity of the material and, as the name indicates, is the parameter that most strongly influences the shear velocity of a material. Be forewarned, there are a large number of material constants available for describing an elastic body, and each one is useful in a particular context or laboratory setting. This profusion of moduli becomes even more severe when we consider poroelasticity. Thus, there are several equivalent pairs of elastic constants that may be used to state the isotropic constitutive relationship (2.77). An important parameter for us is the bulk modulus (Turcotte and Schubert, 1982, p. 112)

$$K_b = \lambda + \frac{2}{3} G,$$

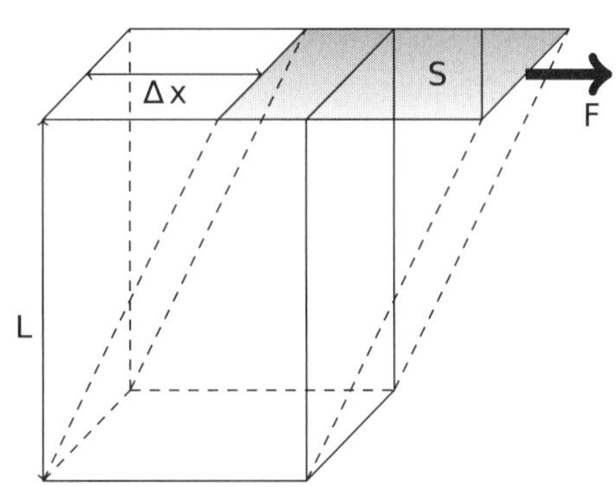

Figure 2.8 The shear modulus is defined as the ratio of shear stress (F/S) to the shear strain ($\Delta x/L$), or, written explicitly $G = FL/S\Delta x$.

relating the pressure p to the fractional volume change

$$p = K_b \frac{\Delta V}{V} \tag{2.78}$$

that arises when studying the interaction of fluid pressure and solid deformation. Writing the constitutive equation for an isotropic body (2.77) in terms of the shear and bulk moduli results in the stress relationship

$$\sigma_{ij} = K_b \nabla \cdot \mathbf{u} \delta_{ij} + G \left[\left(\frac{\partial u_i}{\partial x_j} + \frac{\partial u_j}{\partial x_i} \right) - \frac{2}{3} \nabla \cdot \mathbf{u} \delta_{ij} \right], \tag{2.79}$$

a form that appears prominently in studies of poroelastic material. There are a couple of points to note regarding Equation (2.79). First, we can write it in tensor or dyadic notation, similar to matrix notation:

$$\sigma = K_b (\nabla \cdot \mathbf{u}) \mathbf{I} + G \left[\nabla \mathbf{u} + \nabla \mathbf{u}^T - \frac{2}{3} (\nabla \cdot \mathbf{u}) \mathbf{I} \right],$$

where

$$\nabla \mathbf{u} = \begin{pmatrix} \frac{\partial u_1}{\partial x_1} & \frac{\partial u_2}{\partial x_1} & \frac{\partial u_3}{\partial x_1} \\ \frac{\partial u_1}{\partial x_2} & \frac{\partial u_2}{\partial x_2} & \frac{\partial u_3}{\partial x_2} \\ \frac{\partial u_1}{\partial x_3} & \frac{\partial u_2}{\partial x_3} & \frac{\partial u_3}{\partial x_3} \end{pmatrix},$$

and $\nabla \mathbf{u}^T$ is the transpose of this matrix. Second, if we take the trace of the σ, the sum of the diagonal elements of the stress matrix [denoted by $tr(\sigma)$], the second term on the right-hand-side vanishes. That is, the quantity in square brackets on the right-hand-side vanishes because $tr(\nabla \mathbf{u}) = tr\left(\nabla \mathbf{u}^T\right) = \nabla \cdot \mathbf{u}$ and $tr(\mathbf{I}) = 3$.

Substituting expression (2.79) into Cauchy's equation of motion (2.72) results in a system of partial differential equations for the components of the displacement vector. After regrouping terms, and assuming small strains so that the approximation (2.75) may be used, we can write this equation as

$$\rho \frac{\partial^2 \mathbf{u}}{\partial t^t} = G \nabla^2 \mathbf{u} + \left(K_b - \frac{1}{3} G \right) \nabla \nabla \cdot \mathbf{u} + \nabla \left(K_b - \frac{2}{3} G \right) \nabla \cdot \mathbf{u} + 2 \nabla G \cdot \varepsilon + \mathbf{f}, \tag{2.80}$$

where ε is the infinitesimal strain tensor given in Equation (2.74). The reader may show this as an exercise.

2.5 Fluid flow

A derivation of the equations of motion for a fluid would follow the steps taken in our presentation on elastic deformation. For example, the fluid would be subject to the equation of continuity and the conservation of linear and angular momentum, leading to Cauchy's equations of motion (2.72). For a Newtonian fluid, we would incorporate the constitutive law (2.66) to derive the equations of motion, called the Navier–Stokes equations. While the derivation may be straight-forward, the results are very important. For fluid flow in the natural environment, one cannot typically assume small strains or even small strain gradients and must deal with a system of non-linear differential equations. The richness of the phenomena contained in these equations is truly impressive: solitons, vortices, and finite-amplitude waves are just three examples (Vallis, 2006). Anyone who has faced the ocean or encountered severe weather can testify to the power of fluid dynamics.

Our task is complicated by the fact that we are considering flow in a porous medium. Thus, our fluid will be interacting with an enveloping solid and we are actually modeling a two-component system. The most accurate approach involves coupled or simultaneous modeling of all components of the system. In the next section, we describe this approach in the case of a Newtonian fluid and an elastic solid, leading to the equations of poroelasticity. In that section we treat the case in which the porous matrix only plays a minor role in the dynamics of the system. This approximation allows us to focus primarily upon the fluid, limiting the complexity of the governing equations. Furthermore, we shall assume that the flow regime is such that the velocity gradients are small and the non-linear terms in the material derivative may be ignored. Even with these approximations, the equations can become quite complicated, particularly when multiple fluids are allowed to flow simultaneously, leading to multiphase flow. However, we begin with single phase flow in order to find our footing.

2.5.1 *The flow of a single fluid*

The literature on fluid flow in a porous medium is vast. Numerous books and papers describe flow in an idealized medium under a range of conditions. To keep from getting bogged down in the myriad of possibilities we shall limit ourselves to the flow of a single homogeneous fluid in a porous medium that behaves rather passively. One useful mental image is water flowing through a consolidated or semi-consolidated sand body. Water migrates through the connected pore space, between the sand grains, in a possibly tortuous path. In keeping with our general approach, we treat the porous medium as a continuum characterized by spatially varying properties. For us, the most important attribute of the solid, at least with respect to the fluid flow, is the amount of pore space accessible to the fluid. Accessible space means a greater fluid mass in a given volume. In addition, as we discuss later, greater pore space typically enhances the ability of the fluid to flow because there is less solid material blocking its path. The **porosity** is the percentage of empty space in a specified volume of the porous medium that may be occupied by the fluid. We denote porosity by φ and define it as the void space volume divided by the sample volume. Perhaps a more relevant measure is the **kinematic porosity**, defined as the ratio of the volume of water able to circulate to the volume of the sample. Note that the porosity is a spatially varying quantity and may even vary with time if we allow the porous medium to deform or fill with material.

We can incorporate porosity into our definition of the fluid mass m_f contained in the volume V. In particular, the fluid mass contained in V is given by the density of the fluid multiplied by the fluid volume. Since the fluid can only occupy the pore space, the fluid volume is given by the total volume multiplied by the porosity assuming a fully saturated medium. Thus, an appropriate measure of the fluid mass is the integral

$$m_f = \int_V \varphi \rho_f dV, \tag{2.81}$$

where ρ_f is the density of the fluid which may, along with the porosity, vary as a function of position and time. Thus, we may re-state the conservation of mass for a fluid contained in a porous medium

$$\frac{dm_f}{dt} = \frac{d}{dt} \int_V \varphi \rho_f dV = \int_V \left[\frac{d\left(\varphi \rho_f\right)}{dt} + \varphi \rho_f \nabla \cdot \mathbf{v} \right] dV = 0, \tag{2.82}$$

leading to the equation of continuity for the fluid

$$\frac{\partial \left(\varphi \rho_f\right)}{\partial t} + \nabla \cdot \left(\varphi \rho_f \mathbf{v}\right) = 0. \tag{2.83}$$

The equation of continuity is usually written in terms of the **filtration velocity vector**

$$\mathbf{q} = \varphi \mathbf{v}, \tag{2.84}$$

as explained below, resulting in an expression for the conservation of the mass of the fluid

$$\frac{\partial \left(\varphi \rho_f\right)}{\partial t} + \nabla \cdot \left(\rho_f \mathbf{q}\right) = 0. \tag{2.85}$$

It turns out that we can make considerable headway by simply working with Equation (2.85) and equations relating the fluid pressure to the fluid density, the filtration velocity vector, and the porosity. First, we expand the differentiations of the product terms in Equation (2.85)

$$\varphi \frac{\partial \rho_f}{\partial t} + \rho_f \frac{\partial \varphi}{\partial t} + \rho_f \nabla \cdot \mathbf{q} + \nabla \rho_f \cdot \mathbf{q} = 0. \tag{2.86}$$

Let us then recall the equation-of-state for a slightly compressible fluid, a constitutive law, relating the fluid density to the fluid pressure p

$$\rho_f = \rho_0 e^{\beta(p-p_0)},$$

where ρ_0, β, and p_0 are constants. From this relationship, we can derive expressions for the temporal and spatial derivatives of the fluid density:

$$\frac{\partial \rho_f}{\partial t} = \rho_0 e^{\beta(p-p_0)} \beta \frac{\partial p}{\partial t} = \rho_f \beta \frac{\partial p}{\partial t} \tag{2.87}$$

and

$$\nabla \rho_f = \rho_0 e^{\beta(p-p_0)} \beta \nabla p = \rho_f \beta \nabla p. \tag{2.88}$$

Substituting these expressions into Equation (2.86), we find, after dividing through by ρ_f, that

$$\varphi \beta \frac{\partial p}{\partial t} + \frac{\partial \varphi}{\partial t} + \nabla \cdot \mathbf{q} + \beta \nabla p \cdot \mathbf{q} = 0. \tag{2.89}$$

Box 2.5 **A derivation of Darcy's law**

A classic paper by Hubbert (1940) was the first to demonstrate that Darcy's law is equivalent to the Navier–Stokes equations for the conservation of the linear momentum of a fluid. A fairly simple derivation of Darcy's law from the conservation of momentum for a fluid was given by Hart and St. John (1986). The conservation of linear momentum, Equation (2.64), re-stated here

$$\rho \frac{d}{dt} \mathbf{v} = \nabla \cdot \sigma + \mathbf{f},$$

describes the balance of forces and the acceleration of a fluid. For a non-viscous, or otherwise inviscid fluid, the stress tensor is given by [Equation (2.67)],

$$\sigma_{ij} = -p \delta_{ij},$$

where p is the pressure. The body forces $\mathbf{f}$ acting on a fluid flowing in a porous medium are the force of gravity $\mathbf{g}$ and a force due to the interaction of the fluid with the walls of the pores, which we denoted by $\mathcal{I}$. Thus, the conservation equation may be written as

$$\rho \frac{d}{dt} \mathbf{v} = -\nabla p + \mathcal{I} - \rho \mathbf{g},$$

where **g** is a vector pointing in the direction of the gravitational force. The interaction force $\mathcal{I}$ is commonly written in the form

$$\mathcal{I} = -\frac{\mu}{k}\mathbf{q},$$

where **q** is the filtration velocity, μ is the fluid viscosity, and k is the permeability. Due to the low velocity of fluid flowing in a porous medium, one can often neglect the inertial term on the left-hand-side of the momentum equation and we have

$$-\nabla p - \frac{\mu}{k}\mathbf{q} - \rho\mathbf{g} = 0.$$

Solving for the velocity **q**, gives the generalized form of Darcy's law (containing a gravitational force):

$$\mathbf{q} = -\frac{k}{\mu}\left(\nabla p + \rho\mathbf{g}\right).$$

Next, we incorporate a relationship between the filtration velocity vector **q** and the fluid pressure gradient proposed by Henri Darcy around 1856, the result of his study of the fountains in the city of Dijon, France. As noted by Whitaker (1969) and outlined in Box 2.5, Darcy's law is actually an expression of the conservation of momentum for the porous medium. Specifically, the general equations for the conservation of momentum for a Newtonian fluid, the Navier–Stokes equations, reduce to Darcy's law under appropriate assumptions.

Neglecting the influence of gravity, Darcy's law states that the filtration velocity is proportional to the fluid pressure gradient:

$$\mathbf{q} = -\frac{k}{\mu}\nabla p, \tag{2.90}$$

where μ is the **fluid viscosity** and k is the **absolute permeability** of the medium, a measure of the ease with which fluid flows in the porous medium. The absolute permeability k depends upon both the amount and the spatial distribution of pore space within the medium. It is particularly sensitive to the connectivity of the pore space and very much influenced by the pore 'throats', those narrow constrictions acting to impede flow between pores. Darcy's law (2.90) is analogous to Fourier's law relating heat flow to the temperature gradient in a thermally conducting solid. Substituting Darcy's law into Equation (2.89) results in

$$\varphi\beta\frac{\partial p}{\partial t} + \frac{\partial \varphi}{\partial t} - \nabla\cdot\left(\frac{k}{\mu}\nabla p\right) - \frac{\beta k}{\mu}\nabla p\cdot\nabla p = 0. \tag{2.91}$$

Our final task is to relate changes in porosity (φ) to changes in fluid pressure. This is where the mechanical properties of the porous material comprising the matrix will be accounted for. While this is a subject that might deserve an entire chapter, for the sake of simplicity we shall assume an elastic model in which the change in the pore volume is proportional to a change in the fluid pressure

$$\frac{\partial \varphi}{\partial t} = \alpha \frac{\partial p}{\partial t}, \tag{2.92}$$

where α is the constant of proportionality. Note that this model includes the situation in which the porosity does not depend upon the fluid pressure as a special case. Substituting Equation (2.92) into Equation (2.91) results in an equation governing the evolution of pressure in the porous medium

$$S_s \frac{\partial p}{\partial t} - \nabla \cdot \left(\frac{k}{\mu} \nabla p \right) - \frac{\beta k}{\mu} \nabla p \cdot \nabla p = 0, \tag{2.93}$$

where $S_s = \varphi\beta + \alpha$ is the specific storage coefficient. This equation is non-linear because of the dependence of the density and the porosity upon the pressure. The pressure-dependence of the density, via Equation (2.88), leads to the right-most term containing the scalar product of the pressure gradient with itself.

There are a few methods by which to simplify, in particular to linearize, Equation (2.93). First, as we are interested in flow in a porous medium, the velocities are typically quite low and hence a quadratic term in the velocity magnitude can usually be safely neglected. Second, we could assume an incompressible fluid in which the fluid density is constant in both space and time. Then those terms containing derivatives of the fluid density vanish and we only need to consider the relationship between the porosity and the fluid pressure. Third, we can assume that the fluid density is variable in time but less so in space (de Marsily, 1986, p. 108). Adopting the first approach, neglecting terms of second order in the pressure gradient magnitude, leads to the following partial differential equation

$$S_s \frac{\partial p}{\partial t} - \nabla \cdot \left(\frac{k}{\mu} \nabla p \right) = 0, \tag{2.94}$$

a diffusion equation governing the evolution of pressure in the porous medium. Note that even with these assumptions, Equation (2.94) may still be non-linear due to the dependence of the storage coefficient (S_s) upon the fluid pressure. Assuming an incompressible or slightly compressible fluid can produce a linear diffusion equation for p. In conclusion, as was done for elastic deformation, combining the conservation of mass and the conservation of momentum (in the form of Darcy's Law) with appropriate constitutive laws produces a governing equation, in this case for fluid pressure. Given sufficient boundary and initial conditions, we can solve this equation for the evolution of the pressure field.

Before moving on, we should note that the pressure equation given above may be also generalized to allow for anisotropic permeability, the presence of a gravitational body force, and pressure-dependent flow properties. We may do this by generalizing Darcy's law (2.90) to

$$\mathbf{q} = -\frac{\mathbf{k}}{\mu}\left(\nabla p - \rho g \mathbf{z}\right), \tag{2.95}$$

where $\mathbf{k}$ is the permeability tensor, accounting for the anisotropy, g is the gravitational constant, and $\mathbf{z}$ is a vector in the direction of the local gravitational field. One final complication is introduced when we allow the flow properties, the porosity, and the absolute permeability to depend on the fluid pressure. We have already noted that the porosity can change as the fluid pressure changes. Furthermore, the permeability is often strongly linked to the porosity. Therefore, there may be situations in which the permeability also depends upon the fluid pressure. This renders the flow Equation (2.94) non-linear, with the non-linearity depending on the particular nature of the pressure-dependence of the permeability.

2.5.2 The flow of two or more fluids

Now consider a multi-component fluid, a fluid composed of distinguishable chemical components. There are two primary situations to treat: miscible and immiscible flow. In the first case the fluid components are able to mix with one another to form a single homogeneous phase, while in the second case there are two fluid phases with a distinct phase boundary and the presence of one phase may impede the flow of the other(s). We end this section with a general formulation that covers both miscible and immiscible flow: the compositional approach. In this formulation phase changes and chemical mixing are allowed between components.

Miscible flow: the advection of passive and active material

In this section, we assume that the flow of one of the chemical components does not interfere with the flow of the others, or of the fluid as a whole. One canonical example of this is a chemical tracer that is transported by advection in a flowing solution. If the tracer concentration is low and of a relatively benign nature, then its presence does not influence the flow of the advecting fluid. Thus, the problem uncouples to some degree, and we can model the flow of the total fluid using the method described previously. Given the details of the flow field of the total fluid, we can then model the evolution of the tracer as the flow progresses. The tracer will certainly be subject to advection as it is 'carried' along by the fluid. In addition, due to molecular processes and variations in concentration, the tracer will be subject to diffusion. Furthermore, small-scale flow variations due to pore-scale heterogeneity in the porous medium will introduce dispersion. Other processes are possible, such as chemical reactions between the advected species and elements of the porous matrix or chemical elements bound in the pore fluids. Initially, we will neglect such reactive tracers and focus on non-reactive, or conservative, tracers.

Non-Reactive Chemical Species

For the moment consider a stationary fluid, such as still water. If the fluid contains a distribution of some chemical species, or if we introduce a drop of some identifiable chemical, the molecules of the chemical will move within the host fluid due to random molecular motion, a process known as **molecular diffusion**. If the distribution of the chemical species was uniform, then molecules moving from one side of an imaginary planar surface within the fluid to the other side would be balanced by the movement of molecules in the opposite direction. However, if there is a variation in the concentration of the chemical species, then there will be a net migration from regions of high chemical concentration to regions of low concentration. A higher-concentration gradient leads to a greater imbalance in the number of molecules on either side of the plane, resulting in a greater flux of molecules from the region of high concentration to the region of low concentration. This idea, known as **Fick's law**, differs little from that underlying Fourier's law for heat flow. For a fluid in motion, this net diffusive migration of material will be superimposed upon the advection of the chemical species due to the flow of the aqueous solution.

In addition to molecular diffusion, there is another micro-scale process associated with a fluid in motion within a porous matrix, acting to redistribute the chemical species. This is **kinematic dispersion**, a mixing of the fluid due to microscopic heterogeneities within the porous medium. The heterogeneities vary from sub-pore scale to greater than pore scale in length, while remaining smaller than the averaging volumes used in the continuum modeling. The overall effect of this random mixing is different from molecular diffusion. That is, there are additional considerations because the porous medium can have a more complicated structure than the fluid. Furthermore, the fluid is in motion, leading to a physical anisotropy with respect to the flow direction. For example, the distribution of porosity can be anisotropic due to layering, fractures and crystalline structure, while the structure of the fluid is almost universally isotropic. It should be noted that the dividing line between kinematic dispersion and advection in a heterogeneous medium is not always clear cut. In fact, some investigators have defined dispersion for an entire formation, lumping all the effects due to heterogeneity into kinematic dispersion. Such global definitions of dispersion can sometimes require more involved models of mixing than those discussed in this section.

Let us now derive an equation governing the evolution of a chemical species in a porous medium. As in the section above, we denote the velocity of the carrier (usually aqueous) fluid by the flow velocity vector $\mathbf{v}$. The total velocity of the chemical species α is denoted by $\mathbf{v}_\alpha$, and a random component of velocity due to diffusion and kinematic dispersion, $\mathbf{v}_d$. The total velocity of the chemical species is given by the sum

$$\mathbf{v}_\alpha = \mathbf{v} + \mathbf{v}_d. \tag{2.96}$$

Thus, one may think of the total velocity of the chemical species α as composed of two components: a component due to advection and a stochastic component due to molecular diffusion and kinematic dispersion. For the quantitative discussion to follow we need to define precisely what is meant by the concentration of a chemical species. The concentration of a chemical species α is the mass of tracer per unit volume of the solution:

$$c = \frac{m_\alpha}{V} = \rho_\alpha \frac{V_\alpha}{V}, \tag{2.97}$$

where ρ_α is the density of the chemical species and V_α is the volume of the chemical contained in the volume V of the fluid. The total mass of the chemical species α is given by the integral

$$m_\alpha = \int\limits_{V_\alpha} \varphi_k \rho_\alpha dV_\alpha \tag{2.98}$$

where φ_k is the kinematic porosity, mentioned in the previous section. Kinematic porosity is needed in the mass calculation because the chemical species will, for the most part, only travel to those locations in the porous medium where the fluid can flow. The integral expression is analogous to the total fluid mass, given above. Using Equation (2.97), we can write this integral in terms of the total fluid volume and the concentration

$$m_\alpha = \int\limits_{V} \varphi_k c dV. \tag{2.99}$$

Now we invoke the principle of the conservation of mass, in this case being the mass of the chemical species of interest. Thus, the mass given by the integral m_α is a constant and

$$\frac{dm_\alpha}{dt} = \frac{d}{dt} \int\limits_{V} \varphi_k c dV = \int\limits_{V} \left[\varphi_k \frac{\partial c}{\partial t} + \nabla \cdot (\varphi_k c \mathbf{v}_\alpha) \right] dV = 0, \tag{2.100}$$

where we have assumed that the porosity does not change with time. This is an acceptable assumption for most tracer transport which tends to occur under steady-state, or time invariant, flow conditions. Note that the the integration volume is to some degree arbitrary. Hence, the integrand must vanish, leading to a differential equation for c. We can re-phrase the last equality in terms of the vector $\mathbf{q}_\alpha = \varphi_k \mathbf{v}_\alpha$ as

$$\varphi_k \frac{\partial c}{\partial t} + \nabla \cdot (c\mathbf{q}_\alpha) = 0, \tag{2.101}$$

and make use of the decomposition (2.96) to write the equation of continuity as

$$\varphi_k \frac{\partial c}{\partial t} + \nabla \cdot (c\mathbf{q} + c\mathbf{q}_d) = 0, \tag{2.102}$$

where the filtration velocity vectors correspond to the partition given in Equation (2.96).

As noted at the beginning of this section, the flow velocity vector for the total fluid, and hence the vector $\mathbf{q}$, is found by solving the pressure Equation (2.93). The vector $\mathbf{q}_d$, which now represents both the diffusive and dispersive velocities, requires additional treatment. As noted above, molecular diffusion induces flow from regions of higher concentration to regions of lower concentration, similar to heat flow. Thus, it should not come as a surprise that the diffusive flux is governed by an equation similar to that of Fourier for heat flow. In particular, Lightfoot and Cussler (1965) have shown that the most general linear relationship between mass movement and concentration gradients is given by Fick's law, in which case the flux is proportional to the concentration gradient. Fick's law was also found to govern the dispersive flux due to sub-scale heterogeneity (Taylor, 1953). Thus, the vector $\mathbf{q}_d$ is related to the concentration via the equation

$$c\mathbf{q}_d = -\mathbf{D}\nabla c, \tag{2.103}$$

where $\mathbf{D}$ is the **dispersion matrix** accounting for both molecular diffusion and small-scale dispersion of the chemical species. Both the idea and nomenclature were introduced by Scheidegger (1954). Even in an isotropic porous medium the dispersion matrix is anisotropic because dispersion in the direction of flow (longitudinal dispersion) generally differs from dispersion in the direction perpendicular to flow (transverse dispersion) (Nikolaevskii, 1959; Bear, 1961; Scheidegger, 1961). The coefficients of the dispersion matrix generally depend upon the flow velocity vector $\mathbf{q}$ (Scheidegger, 1957). Initially, the dispersion matrix was assumed to be symmetric (Scheidegger, 1961), based upon a simple argument involving reversibility. However, the asymmetry of $\mathbf{D}$ was demonstrated by Koch and Brady (1987) and later verified by Auriault et al. (2010). Carbonell and Whitaker (1983) noted that only the symmetric component of the dispersion tensor influences the transport if the medium is uniform, a possible explanation of the earlier suggestions of symmetry.

Substituting (2.103) into the equation of continuity (2.102), we derive a governing equation for the evolution of the tracer

$$\varphi_k \frac{\partial c}{\partial t} - \nabla \cdot (\mathbf{D}\nabla c) + \nabla \cdot (c\mathbf{q}) = 0, \tag{2.104}$$

which is an equation that can represent both diffusive behavior and wave-like behavior, depending upon the values of the coefficients. For example, if the coefficients of the dispersion matrix are sufficiently small that we can neglect the second term in (2.104) which then reduces to

$$\varphi_k \frac{\partial c}{\partial t} + \nabla \cdot (c\mathbf{q}) = 0, \tag{2.105}$$

being a hyperbolic partial differential equation representing non-dispersive tracer propagation, similar in nature to a propagating wave.

Box 2.6 **A concentration-dependent flow field**

As an aside, consider the situation in which the concentration of each chemical species actually influences the total flow field. For example, the amount of a chemical may be sufficiently large so that the fluid density is a function of its concentration. In order to model the movement of the chemical species, we must solve the coupled equations:

$$\varphi \frac{\partial \rho_f}{\partial t} + \rho_f \frac{\partial \varphi}{\partial t} + \rho_f \nabla \cdot \mathbf{q} + \nabla \rho_f \cdot \mathbf{q} = 0, \tag{2.106}$$

and

$$\varphi_k \frac{\partial c}{\partial t} - \nabla \cdot (\mathbf{D}\nabla c) + \nabla \cdot (c\mathbf{q}) = 0,$$

subject to the dependence of the fluid density on the concentration

$$\rho_f = \rho(c). \tag{2.107}$$

The interaction of a saline fluid and fresh water is one example of such coupled flow. Dissolved carbon dioxide in water is another example of coupled flow. However, in such cases the viscosity is also influenced by the concentration of the chemical species.

Reactive Chemical Species

In addition to diffusion and dispersion, there is the possibility that the chemical species will interact with the porous medium itself. Chemical processes such as acid-base, oxidation-reduction, precipitation-dissolution, adsorption-desorption, and ionic reactions are several classes of such interactions. Furthermore, there are also radioactive and biological reactions to consider. The entire suite of reactions may be represented by including a generic term $\mathcal{S}$ in Equation (2.104) representing the addition or removal of the chemical species α,

$$\varphi_k \frac{\partial c}{\partial t} + \mathcal{S} = \nabla \cdot (\mathbf{D} \nabla c) - \nabla \cdot (c\mathbf{q}) . \tag{2.108}$$

Chemical Adsorption and Desorption

Often, some of the chemical species may attach to the minerals making up the solid matrix of the porous medium. We define a mass fraction, representing the mass of the chemical species adsorbed per unit mass of solid matrix. The total mass of solid in a unit volume of the matrix is

$$m_{solid} = (1 - \varphi) \rho_s . \tag{2.109}$$

The source term represents the change in the mass of the chemical over time and is thus given by

$$\mathcal{S} = \frac{\partial [(1 - \varphi) \rho_s F]}{\partial t} = (1 - \varphi) \rho_s \frac{\partial F}{\partial t}, \tag{2.110}$$

where F is the **mass concentration**, representing the mass of substance adsorbed per unit mass of solid. In Equation (2.110), we are assuming that the porosity and solid density do not change with time. This approach must be modified if a sufficient number of molecules attach themselves to the solid matrix so as to modify the porosity or the density of the matrix. Substituting (2.110) into Equation (2.108) produces

$$\varphi_k \frac{\partial c}{\partial t} + (1 - \varphi) \rho_s \frac{\partial F}{\partial t} = \nabla \cdot (\mathbf{D} \nabla c) - \nabla \cdot (c\mathbf{q}) . \tag{2.111}$$

The nature of the adsorption is determined by the relationship between the mass fraction adsorbed per unit mass and the concentration of the chemical species. If we assume that the relationship between these quantities is linear then, at a fixed temperature,

$$F = K_{di} c, \tag{2.112}$$

where the proportionality coefficient, K_{di}, is known as the **distribution coefficient**. The linear relationship assumes that the adsorption is both reversible, and instantaneous. Because the linear relationship holds at a fixed temperature, it is referred to as a linear absorption isotherm or simply as a linear isotherm. Combining the relationship (2.112) with the governing Equation (2.111), we can write

$$[\varphi_k + (1 - \varphi) \rho_s K_{di}] \frac{\partial c}{\partial t} = \nabla \cdot (\mathbf{D} \nabla c) - \nabla \cdot (c\mathbf{q}) , \tag{2.113}$$

or

$$\varphi_k R \frac{\partial c}{\partial t} = \nabla \cdot (\mathbf{D} \nabla c) - \nabla \cdot (c\mathbf{q}), \tag{2.114}$$

where we have defined the retardation factor due to adsorption

$$R = 1 + \frac{1-\varphi}{\varphi_k} \rho_s K_{di}. \tag{2.115}$$

Thus, the chemical adsorption has an effect on the coefficient of the time derivative in Equation (2.114), modifying both the arrival time and the amplitude of a propagating pulse of the chemical species.

As might be expected, in addition to the linear adsorption isotherm, a host of non-linear relationships between F and c have been proposed. We shall just mention some of the more commonly used formulas:

$$F = K_1 c - K_2 c^2$$

known as the isotherm of second degree, as well as Langmuir's isotherm:

$$F = \frac{K_1 c}{1 + K_2 c},$$

Freundlich's isotherm:

$$F = K_1 c^{\frac{1}{n}},$$

and the exponential isotherm:

$$c = K_1 F e^{K_2 F}.$$

These relationships will produce non-linear partial differential equations governing the concentration of the chemical species.

Up to this point we have not distinguished between the mobile pore fluid and the fluid which adheres to the pore walls and is therefore immobile. This distinction can be important, particularly when dealing with fluid recovery such as oil extraction or the clean-up of a contaminant. Furthermore, the 'immobile' fluid fraction can sometimes be induced to move by the introduction of specially designed fluids called surfactants or other types of fluids. To state the problem precisely, let us recall the kinematic porosity φ_k, the pore space occupied by the mobile fluid, and the total porosity φ. Furthermore, let c_i denote the concentration of the immobile fraction. The equation governing the distribution of the chemical species is given by the following modification of Equation (2.111)

$$\varphi_k \frac{\partial c}{\partial t} + (\varphi - \varphi_k) \frac{\partial c_i}{\partial t} + (1-\varphi) \rho_s \frac{\partial F}{\partial t} = \nabla \cdot (\mathbf{D} \nabla c) - \nabla \cdot (c\mathbf{q}), \tag{2.116}$$

which contains an additional term [the second term on the left-hand-side] related to the change in concentration within the immobile fraction of the fluid. Assuming a linear adsorption isotherm, $F = K_{di}c$, and a linear relationship between the concentration in the mobile and immobile phases, $c_i = K_i c$, we can write Equation (2.116) as a single equation for c

$$\varphi_k R_i \frac{\partial c}{\partial t} = \nabla \cdot (\mathbf{D}\nabla c) - \nabla \cdot (c\mathbf{q}) , \tag{2.117}$$

where we have defined the modified retardation factor

$$R_i = 1 + \frac{\varphi - \varphi_k}{\varphi_k} K_i + \frac{1-\varphi}{\varphi_k} \rho_s K_{di} \tag{2.118}$$

that accounts for the change in concentration in the immobile fraction as well as the adsorption onto the solid matrix.

Radioactive decay:

In some instances the chemical species may be subject to radioactive decay. This is particularly true of some types of tracers for which the radioactivity aids in the detection and monitoring. The law governing radioactive decay is well established and the rate of change in the concentration of the substance is proportional to the current concentration

$$\frac{\partial c}{\partial t} = -\lambda c, \tag{2.119}$$

where λ is a decay constant related to the time it takes for half of the material to decay, the half-life, $T_{1/2}$,

$$\lambda = \frac{\ln 2}{T_{1/2}}.$$

Substituting the expression for radioactive decay (2.119), after being multiplied by the kinematic porosity, for the term $\mathcal{S}$ in Equation (2.108) results in the governing equation for a radioactive tracer:

$$\varphi_k \left(\frac{\partial c}{\partial t} + \lambda c \right) = \nabla \cdot (\mathbf{D}\nabla c) - \nabla \cdot (c\mathbf{q}) . \tag{2.120}$$

If there is adsorption as well as radioactivity, then the concentration in the adsorbed phase will also be subject to radioactive decay and hence

$$\frac{\partial F}{\partial t} = -\lambda F. \tag{2.121}$$

Combining this effect with Equation (2.111) gives the radioactive equivalent to Equation (2.111)

$$\varphi\left(\frac{\partial c}{\partial t}+\lambda c\right)+(1-\varphi)\,\rho_s\left(\frac{\partial F}{\partial t}+\lambda F\right)=\nabla\cdot(\mathbf{D}\nabla c)-\nabla\cdot(c\mathbf{q})\,. \tag{2.122}$$

For a linear adsorption isotherm, Equation (2.122) may be written as a linear partial differential equation for the concentration

$$\varphi R\frac{\partial c}{\partial t}+\varphi R\lambda c=\nabla\cdot(\mathbf{D}\nabla c)-\nabla\cdot(c\mathbf{q})\,, \tag{2.123}$$

where R is the retardation factor defined above [Equation (2.118)].

Biological Activity:

Biological reactions can often be modeled in the same fashion as radioactive decay (de Marsily, 1986). However, there is the additional possibility of growth in a biological community.

Multiphase flow of immiscible fluids

Now we consider the simultaneous flow of a number of fluids that are **immiscible**, that is, the constituent fluids are insoluble and remain distinct as they flow. Because the fluids cannot mix and must remain distinct, we encounter a new type of behavior where one fluid can block the flow of another, much like the grains of the solid matrix can prevent flow. This leads to the idea of relative permeability, a permeability multiplier that depends upon the saturation of a particular phase. Such a notion renders the governing equations for multiphase flow non-linear because the presence or absence of a phase can influence the nature of the flow of that phase. If the presence of one fluid phase, such as oil, can block the flow of another, say water, then it is important to keep track of the relative percentages of each phase. Thus, we need to consider the **volumetric saturation** of each fluid:

$$S_i=\frac{\text{pore volume occupied by fluid } i}{\text{total pore volume}}. \tag{2.124}$$

The volumetric saturation of each phase is an important factor, controlling the 'ease' with which each fluid flows through the porous medium. Furthermore, the saturation of liquids and gases of interest (water, various hydrocarbons, carbon dioxide, etc.) can have important economic and environmental implications. Because the fluids must fill the pore space completely, the saturations must sum to unity

$$\sum_i S_i=1,$$

where the summation is over all of the fluid phases.

As in the case of a single fluid, we can consider the various laws governing the flow of each particular phase. The governing principles are the same as those discussed above: the conservation of mass, the conservation of momentum, some type of constitutive relationship or an equation-of-state. In addition, we will need to consider fluid pressure differences between the various phases and the constraint that the total fluid volume cannot exceed the total pore volume. The conservation principles must now reflect variations in the saturation of each phase within the pores. For example, the conservation of mass for the *ith* phase, as expressed by the continuity equation, given by

$$\frac{\partial\left(\varphi\rho_i S_i\right)}{\partial t}+\nabla\cdot\left(\rho_i\mathbf{q}_i\right)=0, \tag{2.125}$$

is similar to Equation (2.85), being the continuity equation for a single fluid. One difference from single phase flow is the presence of the saturation of the *ith* phase, S_i, which reflects the fraction of the phase present in the pore space. Note that the saturation can vary as a function of time as well as a function of space. Another difference between multiphase and single phase flow is implicit in Equation (2.90), contained within the filtration velocity vector $\mathbf{q}_i$. Specifically, the filtration velocity vector is related to the fluid pressure gradient of the *i*th phase via a multiphase generalization of Darcy's law:

$$\mathbf{q}_i=-\frac{k_i(S_i)}{\mu_i}\nabla p_i, \tag{2.126}$$

where $k_i(S_i)$ is the intrinsic permeability for fluid i, a quantity that is proportional to the absolute permeability, k, associated with the porous medium

$$k_i(S_i)=k_{ri}(S_i)\cdot k. \tag{2.127}$$

The proportionality factor $k_{ri}(S_i)$ is known as the **relative permeability**. If we use the more general definition of $\mathbf{q}_i$, given by Equation (2.95), we have

$$\mathbf{q}_i=-\frac{\mathbf{k}_i(S_i)}{\mu_i}\left(\nabla p_i-\rho_i g\mathbf{z}\right), \tag{2.128}$$

where

$$\mathbf{k}_i(S_i)=k_{ri}(S_i)\mathbf{k}. \tag{2.129}$$

We reiterate that the above relationships (2.126) and (2.128), two generalizations of Darcy's law for multiphase flow, are a consequence of the conservation of momentum, reductions of the full Navier–Stokes equations (Whitaker, 1969). For each liquid and/or gas we will have an equation-of-state relating the fluid density of a particular phase to the fluid pressure associated with that phase. For example, the relationship between density and pressure given by

$$\rho_f = \rho_0 e^{\beta(p-p_0)} \tag{2.130}$$

is commonly used for water. Note that the fluid pressure associated with the various phases can differ. In particular, there can be a discontinuity in pressure across the interface separating two immiscible fluids. As noted by Bear (1972), the behavior of the interface between the fluids is similar to that of a stretched membrane and the curvature of the interface is related to the pressure differential p_{ij}^c between phase i and phase j,

$$p_{ij}^c(S) = p_i - p_j, \tag{2.131}$$

known as the **capillary pressure**. Specifically, the radius of curvature of the interface, r, is related to the capillary pressure via the relationship

$$p_{ij}^c = \frac{2\sigma_{ij}}{r}, \tag{2.132}$$

where σ_{ij} is the surface tension between phase i and phase j. Capillary pressure is a function of the saturation of the two phases and the capillary pressure curves are typically determined experimentally. The capillary pressure curve can display hysteresis and the values for an increase in a phase differ from the values associated with a decrease in the phase. Substituting Darcy's law, for example Equation (2.128), into the continuity Equation (2.125) produces

$$\frac{\partial\left(\varphi\rho_i S_i\right)}{\partial t} + \nabla\cdot\left[\rho_i \frac{k_{ri}}{\mu_i}\mathbf{k}\left(\nabla p_i + \rho_i g\mathbf{z}\right)\right] = 0, \tag{2.133}$$

which is subject to the capillary pressure constraint (2.131) and the fact that the saturations of the phases sum to unity

$$\sum S_i = 1. \tag{2.134}$$

The ratio of the relative permeability to the fluid viscosity,

$$\lambda_i = \frac{k_{ri}(S_i)}{\mu_i}, \tag{2.135}$$

is known as the **relative phase mobility**, or simply the mobility of the fluid phase i. Even with all the simplifications involved in producing the system of Equations (2.133), such as the reduction of the Navier–Stokes equations to Darcy's law, these equations can still be rather difficult to solve. One difficulty is the nonlinearity due to the dependence of the relative permeability upon the saturation S_i. Furthermore, as discussed above, for compressible flow the density may also depend upon the pressure, introducing additional non-linearity. The equations for multiphase flow (2.133) are most frequently solved using numerical techniques.

Mass transfer between phases: the compositional approach

In the previous sections we assumed that the phases retained their identities throughout the flow process. However, there are situations in which a chemical component can transform from one phase into another. For example, given particular changes in temperature and/or pressure, water can vaporize to produce a gaseous phase. Or, a gaseous hydrocarbon or carbon dioxide can dissolve into a fluid hydrocarbon oil phase. Thus, each distinct chemical species can exist in one of a number of phases depending upon the pertinent flow conditions. Needless to say, such conversions between phases complicates the bookkeeping associated with mass conservation and increases the number of conservation equations. However, these complications are essential for correctly modeling important processes such as three-phase oil-water-gas flow.

At this stage, because we are not going to consider chemical reactions, the basic quantities being conserved are the distinct chemical species, that we label as chemical components. We assume that the k-th chemical component can exist in N_P possible phases. Note that the term 'phase' does not simply mean the solid, liquid, or gas in the pore space. Rather, it means a quantity that flows as a homogeneous material. For example, there may be a water phase (w) as well as an oil phase (o), both of which are liquids. Furthermore, the liquid phase may also contain dissolved gasses, components that are essentially 'hidden' in a particular phase. Under certain conditions these dissolved components might exsolve and move into one of the gas phases. Consider a system of N_C distinct chemical components that can be distributed over the N_P phases. Let C_i^k denote the mass or mole fraction of component k in phase i. Let S_i denote the saturation of phase i contained in the surrounding pore space. The mass fractions and the saturations are the fundamental dependent variables in the conservation equations. Because the phases are assumed to fill the pore space, the saturations must sum to unity:

$$\sum_{i=1}^{N_P} S_i = 1. \tag{2.136}$$

Similarly, the mole fractions of the components in each phase must add up to unity, introducing the requirement that

$$\sum_{k=1}^{N_C} C_i^k = 1. \tag{2.137}$$

As the compositional formulation contains the miscible and immiscible flow problems as special cases, one would expect it to contain elements of each approach. For example, since the chemical components can diffuse through the various phases we expect that the velocity of each chemical species in each phase, denoted by $\mathbf{v}_i^k$, to be the sum of an average flow velocity of the phase, $\mathbf{v}_i$ and a random component $\mathbf{r}_i^k$ due to the diffusion, as was the case for the flow of a miscible fluid. We considered such a decomposition, in which the total velocity of component k is,

$$\mathbf{v}_i^k = \mathbf{v}_i + \mathbf{r}_i^k, \tag{2.138}$$

as in our treatment of the flow of an miscible fluid [see Equation (2.96)]. There are also elements of immiscible flow that must be taken into account. In particular, we shall require the idea of relative permeability in which the presence or absence of one phase can influence the flow of another [see Equation (2.127)]. We will also need to account for possible differences in fluid pressure in each phase, giving rise to capillary pressure [see Equation (2.131)]. Thus, in addition to the conservation equation derived below, we shall also need constitutive equations in the form of relative permeability curves, capillary pressure curves, and equations-of-state for the densities.

The total mass in a component k is obtained by adding up all the mass fractions from each phase. For the sake of illustration, we shall assume two phases: a gas, denoted by g, and a liquid, denoted by l. The total mass in component k is given by the mole fraction of k in each phase weighted by the mass of each phase (porosity $\times$ density $\times$ saturation)

$$m_k = \int_V \varphi \left[\rho_l S_l C_l^k + \rho_g S_g C_g^k\right] dV,$$

and the conservation of the mass of component k requires that m_k is constant:

$$\frac{dm_k}{dt} = \frac{d}{dt} \int_V \varphi \left[\rho_l S_l C_l^k + \rho_g S_g C_g^k\right] dV = 0. \tag{2.139}$$

When fluid is injected or withdrawn the mass of a particular component k may increase or decrease, at a rate of Q^k. Thus, the rate of change of m_k is given by

$$\frac{dm_k}{dt} = \frac{d}{dt} \int_V \varphi \left[\rho_l S_l C_l^k + \rho_g S_g C_g^k\right] dV = \int_V Q^k dV. \tag{2.140}$$

Making use of the definition of the material derivative of an integral (2.59), keeping in mind that the component k in each phase may move with a distinct velocity $\mathbf{v}_i^k$, given by Equation (2.138), we derive an equation for the conservation of mass of the component k,

$$\int_V \left[\frac{\partial M^k}{\partial t} + \nabla \cdot \left(\mathbf{q}_l^k + \mathbf{q}_g^k\right)\right] dV = \int_V Q^k dV, \tag{2.141}$$

where we have defined the quantities

$$M^k = \varphi\left[\rho_l S_l C_l^k + \rho_g S_g C_g^k\right], \tag{2.142}$$

being the mass per unit volume of component k, and

$$\mathbf{q}_l^k = \varphi \rho_l C_l^k \mathbf{v}_l^k, \tag{2.143}$$

$$\mathbf{q}_g^k = \varphi \rho_g C_g^k \mathbf{v}_g^k, \tag{2.144}$$

being the flux vectors for the liquid and gas phases, respectively. Note that the vectors $\mathbf{q}_l^k$ and $\mathbf{q}_g^k$ are multiphase generalizations of the filtration vector defined previously [see Equation (2.84)]. Deriving Equation (2.141) from Equation (2.140) is left as an exercise. Because we can vary the integration volume arbitrarily, the integrands must vanish, resulting in the differential equation

$$\frac{\partial M^k}{\partial t} + \nabla \cdot \left(\mathbf{q}_l^k + \mathbf{q}_g^k\right) = Q^k, \tag{2.145}$$

a conservation or mass balance equation.

We have yet to invoke the other major conservation law: the conservation of linear momentum. Doing so will provide an expression for the component $\mathbf{v}_i$ of the total velocity vector (2.138). Remember that $\mathbf{v}_i$ denotes the average velocity of phase i. As noted above, for most situations involving fluid flow within a porous medium, the flow velocities of the various phases are slow enough that we can neglect inertial terms. Similarly, product terms in the convective derivative are also small and can be safely ignored. Under these conditions, the Navier–Stokes equation, representing the conservation of linear momentum for phase i, reduces to a multicomponent variant of Darcy's law, producing an expression for the flow velocity of the phase i

$$\mathbf{v}_i = -\frac{\mathbf{k}k_{ri}}{\varphi\mu_i}\left(\nabla p_i - \rho_i g \mathbf{z}\right), \tag{2.146}$$

being similar in form to the generalization of Darcy's law for multiphase flow [see Equations (2.128) and (2.129)]. The other component of $\mathbf{v}_i^k$, that is $\mathbf{r}_i^k$, representing molecular diffusion and small-scale dispersion of the component k within the phase i, is given by Fick's law, as discussed in the section on the flow of miscible flow [see the discussion surrounding Equation (2.103)]. Briefly, the diffusive and dispersive flux is driven by the concentration gradient of the component k within the phase i. The exact expression

$$C_i^k \mathbf{r}_i^k = -\frac{\rho_i}{\varphi} \mathbf{D}_i^k \nabla C_i^k, \tag{2.147}$$

where $\mathbf{D}_i^k$ is the **dispersion tensor**, is similar to the expression (2.103). Combining the various expressions results in an equation for each component. These equations are augmented by the constraints (2.136) and (2.137), as well as the relative permeability curves, the capillary pressure curves, and the equations-of-state for the density. In addition, initial and boundary conditions are incorporated into the approach for solving the resulting governing equations.

2.6 Coupled deformation and fluid flow

2.6.1 General considerations

Now we present a more comprehensive treatment of fluid flow in a deformable matrix. Whereas in the previous section, we assumed that the deformation of the porous matrix played only a minor role in the fluid flow, here both the deformation of the matrix and the flow of a viscous fluid are on an equal footing. In contrast to our earlier constitutive models, heterogeneity is a fundamental aspect of the modeling. That is, rather than a single uniform behavior for the entire material, our model must account for two very different behaviors: that of the fluid and that of the solid (Coussy, 2010). While the constitutive equations for the component materials may be well established, (e.g. a Newtonian fluid) the aggregate behavior of the composite will depend upon other attributes. Specifically, the behavior of the composite material will not only depend on the nature of the individual components but also on the nature of the interaction of these components.

It is best to study the interaction of the individual components at the microscopic scale, modeling the movement of a fluid through the pore space between the solid grains. At that scale, one can use the well-established model of a viscous liquid for the fluid, and an appropriate constitutive model, such as an elastic solid, for the matrix. One must then average over sub-volumes to derive governing equations for the composite material. The importance of this choice has long been recognized, and there are several methods for scaling up from the microscopic to the macroscopic scale, each involving some form of averaging of the microscopic equations.

Currently, three main approaches have been followed in deriving the larger-scale continuum equations. Volume averaging (Bear et al., 1984; Pride et al., 1992; Tuncay and Corapcioglu, 1997) is one straight-forward approach, based upon a theorem for averaging the gradient operator (Slattery, 1968, 1981). However, the averaged equations require an additional **closure relationship** in order to form a complete system. A second approach, based upon the two-space method, sometimes called the method of multiple scales (Levy, 1979; Auriault, 1980; Burridge and Keller, 1981), presents the governing equations in terms of large and small scales and invokes averaging as part of the derivation. Having two scales leads to more complicated intermediate equations, as compared to the averaging method. A third approach, called mixture theory, also known as the theory of interacting continua (Garg, 1971; Morland, 1972; Drumheller, 1978; Truesdell, 1985; Berryman, 1986; Berryman et al., 1988; Santos et al., 1990; Lo et al., 2002; Voyiadjis and Song, 2006), assumes that the various phases (solid, fluid, etc) can be overlapping, that is, existing everywhere in the medium, and also interacting. One problem with mixture theory is that the specific forms of the interaction terms are not always known and must be derived by some other means, such as averaging.

In this exposition we adopt mixture theory as a means to produce governing equations for coupled deformation and flow. In order to keep the arguments short, we will take a somewhat heuristic approach, and we shall present just what is needed to derive the governing equations. In mixture theory it is assumed that each component is present in any given volume of material (Truesdell, 1962, 1985). For example, if we take a one cubic centimeter sample of a water saturated sand, it will contain a large number of sand grains and some significant volume of water. In this approach the exact nature of the microstructural boundary between the solid and the fluid is not treated explicitly. Rather, the microstructure is included implicitly, in the form of interaction terms. One advantage of an approach based on mixture theory is that our derivation of the governing equations can be related to our earlier discussions on elastic deformation and fluid flow, though in an intuitive and non-rigourous manner. In particular, we can base our derivations on the conservation principles for mass and momentum.

In order to illustrate the main ideas, we shall first consider the basic conservation principles for the flow of a fluid through a generic solid, obeying an unspecified constitutive law. Because we are considering two materials interacting across an interface, our previous approach must be modified. Following this, we shall consider two specific materials: an elastic matrix and a viscous fluid.

A single fluid phase

As a first step, consider the flow of a single fluid through a solid matrix. The fluid is assumed to fill the pore space completely. The volume fraction of the material occupied by the pore volume is denoted by the porosity φ. Quantities associated with the fluid are denoted by a subscript f, while those associated with the solid are denoted by a subscript s. For example, we denote the displacement of the solid matrix by the vector $\mathbf{u}_s$ and the displacement of the fluid by a distinct vector $\mathbf{u}_f$. The velocity components of the solid and fluid, the time derivatives of the displacements, are denoted by dots over the respective variables: $\dot{\mathbf{u}}_s$ and $\dot{\mathbf{u}}_f$. In this section $\dot{\mathbf{u}}_f$ is the velocity vector defined in Equation (2.84) written in terms of the time derivative of the fluid displacement.

As in our previous treatments of homogeneous solids and liquids, let us begin by considering the conservation of mass. We stipulate that the total mass (solid plus fluid) in a given volume is conserved if the system is isolated:

$$\frac{d\left(m_s + m_f\right)}{dt} = \frac{d}{dt}\int_V \left[(1-\varphi)\,\rho_s + \varphi\rho_f\right] dV = 0.$$

Because the volume of integration is the total volume, V, we scale the integration by the volume fraction of each material. For the fluid we scale by the porosity, φ, this being the fraction of the volume containing the fluid. Following the procedure outlined by Equation (2.55) to derive the equation of continuity, we can transform the derivative of the total mass integral, and the condition that the total mass is constant becomes

$$\int_V \left[\frac{d\left(\varphi\rho_f\right)}{dt} + \varphi\rho_f \nabla\cdot\dot{\mathbf{u}}_f\right] dV + \int_V \left[\frac{d\,(1-\varphi)\,\rho_s}{dt} + (1-\varphi)\,\rho_s \nabla\cdot\dot{\mathbf{u}}_s\right] dV$$
$$= 0.$$

As noted in the box below, if there is no mass transfer across the interface between the two constituents, then there is no explicit coupling between fluid and solid terms and each of the integrals vanishes separately. Thus, we have a conservation law for the fluid mass (m_f), stated mathematically as [see Equation (2.82), remembering that $\dot{\mathbf{u}}_f$ is just another representation of $\mathbf{v}$],

$$\frac{dm_f}{dt} = \frac{d}{dt}\int_V \varphi\rho_f dV = \int_V \left[\frac{d\left(\varphi\rho_f\right)}{dt} + \varphi\rho_f \nabla\cdot\dot{\mathbf{u}}_f\right] dV = 0, \tag{2.148}$$

and a conservation law for the solid mass (m_s) given by

$$\frac{dm_s}{dt} = \frac{d}{dt}\int_V (1-\varphi)\,\rho_s dV = \int_V \left[\frac{d\,(1-\varphi)\,\rho_s}{dt} + (1-\varphi)\,\rho_s \nabla\cdot\dot{\mathbf{u}}_s\right] dV = 0. \tag{2.149}$$

Because the integration volume can be varied, in particular one can shrink it down to a very small volume around any point of interest, we have equations of continuity for both the fluid and solid components:

$$\frac{\partial\left(\varphi\rho_f\right)}{\partial t}+\nabla\cdot\left(\varphi\rho_f\dot{\mathbf{u}}_f\right)=0, \tag{2.150}$$

$$\frac{\partial\left(1-\varphi\right)\rho_s}{\partial t}+\nabla\cdot\left[\left(1-\varphi\right)\rho_s\dot{\mathbf{u}}_s\right]=0, \tag{2.151}$$

where we have used the definition of the the total derivative as illustrated by Equation (2.49).

Box 2.7 **The making of an interaction term**

In the theory of mixtures each component may be present in a region, however small, about any point in the medium. Let us see what this means for our conservation laws. In particular, let us consider the conservation of linear moment for a porous medium containing a fluid. At any point in the medium the total momentum is related to the sum of the imposed forces according to Newton's law [see Equation (2.25)]. The conservation of total linear momentum is given by

$$\frac{d}{dt}\int\limits_V\left[\left(1-\varphi\right)\rho_s\dot{\mathbf{u}}_s+\varphi\rho_f\dot{\mathbf{u}}_f\right]dV$$

$$=\int\limits_V\left\{\left(1-\varphi\right)\left[\nabla\cdot\sigma_s+\mathbf{f}_s\right]+\varphi\left[\nabla\cdot\sigma_f+\mathbf{f}_f\right]\right\}dV,$$

which can be re-arranged to give

$$\frac{d}{dt}\int\limits_V\left(1-\varphi\right)\rho_s\dot{\mathbf{u}}_s dV-\int\limits_V\left(1-\varphi\right)\left[\nabla\cdot\sigma_s+\mathbf{f}_s\right]dV$$

$$+\frac{d}{dt}\int\limits_V\varphi\rho_f\dot{\mathbf{u}}_f dV-\int\limits_V\varphi\left[\nabla\cdot\sigma_f+\mathbf{f}_f\right]dV=0.$$

While the sum on the left-hand-side of this equation vanishes, the terms associated with the fluid and the solid do not sum to zero separately. Rather, because forces can act across boundaries, the subset of terms associated with each material can be non-zero and we may write

$$\frac{d}{dt}\int\limits_V\left(1-\varphi\right)\rho_s\dot{\mathbf{u}}_s dV-\int\limits_V\left(1-\varphi\right)\left[\nabla\cdot\sigma_s+\mathbf{f}_s\right]dV=\int\limits_V\mathbf{d}_s dV$$

and

$$\frac{d}{dt}\int\limits_V \varphi\rho_f\dot{\mathbf{u}}_f dV - \int\limits_V \varphi\left[\nabla\cdot\sigma_f + \mathbf{f}_f\right]dV = \int\limits_V \mathbf{d}_f dV,$$

where $\mathbf{d}_s = -\mathbf{d}_f$ in order to be consistent with the conservation of total linear momentum . Let us denote $\mathbf{d}_f$ simply as $\mathbf{d}$. The variable $\mathbf{d}$ represents a **momentum transfer** or **interaction term** that we can incorporate into the theory. Note that this quantity is, as yet, unconstrained and we must resort to other means, such as averaging, in order to estimate it. The same principle holds for the conservation of mass. However, because there is no transfer of mass across the internal boundary separating the two constituents, the interaction terms vanish and the sets of terms for each component may be set to zero, leading to Equations (2.148) and (2.149).

Next, consider the conservation of linear momentum for the solid and fluid components, as given by Equation (2.25). As noted in Box 2.7, when forces can act across boundaries, one component can interact with another and we must include momentum transfer terms $\mathbf{d}_s$ and $\mathbf{d}_f$ in the conservation equations. The conservation of linear momentum for the fluid is given by

$$\frac{d}{dt}\int\limits_V \varphi\rho_f\dot{\mathbf{u}}_f dV - \int\limits_V \varphi\left[\nabla\cdot\sigma_f + \mathbf{f}_f\right]dV = \int\limits_V \mathbf{d}_f dV, \tag{2.152}$$

where the subscript f signifies quantities associated with the fluid. A similar equation also holds for the solid phase, but in this case we scale by the solid volume fraction, $1-\varphi$,

$$\frac{d}{dt}\int\limits_V (1-\varphi)\,\rho_s\dot{\mathbf{u}}_s dV - \int\limits_V (1-\varphi)\left[\nabla\cdot\sigma_s + \mathbf{f}_s\right]dV = \int\limits_V \mathbf{d}_s dV \tag{2.153}$$

with the requirement that $\mathbf{d}_f = -\mathbf{d}_s = \mathbf{d}$. Evaluating the derivative of the integrals in Equations (2.152) and (2.153) and making use of the equations of continuity (2.150) and (2.151), as was also done above for Equation (2.63), we arrive at the equations of motion

$$\varphi\rho_f\left\{\frac{\partial\dot{\mathbf{u}}_f}{\partial t} + \dot{\mathbf{u}}_f\cdot\nabla\dot{\mathbf{u}}_f\right\} = \varphi\nabla\cdot\sigma_f + \varphi\mathbf{f}_f + \mathbf{d}, \tag{2.154}$$

$$(1-\varphi)\,\rho_s\left\{\frac{\partial\dot{\mathbf{u}}_s}{\partial t} + \dot{\mathbf{u}}_s\cdot\nabla\dot{\mathbf{u}}_s\right\} = (1-\varphi)\,\nabla\cdot\sigma_s + (1-\varphi)\,\mathbf{f}_s - \mathbf{d}. \tag{2.155}$$

Though we have successfully derived useful conservation laws for mass and momentum, there is still an issue of the coupling between the fluid and the solid.

We do not yet have an expression for the interaction variable $\mathbf{d}$ in terms of the solid and fluid displacements or their derivatives. Because we are not discussing specific material models in this section, only a few general comments are in order. The coupling between the fluid and solid occurs in two places in the conservation of momentum Equations (2.154) and (2.155). There are explicit contributions to $\mathbf{d}_f$ and $\mathbf{d}_s$ that couple the fluid and solid momentum. The primary contribution is due to the viscous dissipation induced by the relative motion of the solid and fluid and the relative acceleration between the solid and the fluid. Considering the momentum balance for an inviscid fluid as it flows through a solid matrix, Garg (1971), Pride et al. (1992), Tuncay and Corapcioglu (1997) and Wilmanski (2006), among others, produced expressions for the coupling force

$$\mathbf{d}_f = -\mathbf{d}_s = \frac{\varphi \mu_f}{k} \left(\dot{\mathbf{u}}_f - \dot{\mathbf{u}}_s \right), \tag{2.156}$$

where μ_f is the fluid viscosity and k is the permeability. The coupling term $\mathbf{d} = \mathbf{d}_f$ is a dissipative force due in part to the relative motion of the solid and fluid and it makes physical sense that it should contain a term depending upon the time derivative of the relative displacement. Equation (2.156) is similar to Darcy's law and is related to the physics of laminar flow that dominates in many applications. The implications of expression (2.156) are actually much more complicated than its simple form suggests. The source of the complexity is the frequency-dependence of the drag forces. In particular, it has been pointed out (Johnson et al., 1987) that there is a distinction between high and low frequency behavior, depending upon the thickness of the viscous skin depth of the flow with respect to pore size. The permeability k is generally taken to be frequency dependent (Johnson et al., 1987; Pride et al., 1992, 1993) and referred to as the dynamic permeability $k(\omega)$.

It is worth spending a little time discussing the nature of the dynamic permeability, particularly its frequency dependence, because it does impact the explicit form of the governing equations and differences in formulation may lead to some confusion. In Box 2.8 we attempt to provide some insight into the nature of the dynamic permeability, following the approach of Johnson et al. (1987). There we describe what is sometimes referred to as a **branching-function** approach (Müller et al., 2010), where exact high- and low-frequency limits are honored, while imposing a causality principle and a simple functional form to connect the limits. This approach has been extended and applied to consider the impact of patchy-saturation and mesoscopic heterogeneity on the frequency-dependence of poroelastic coefficients (Pride et al., 2004). Because of the frequency dependence, the coefficient in the expression (2.156) is actually a integrodifferential operator that is applied to the difference $\left(\dot{\mathbf{u}}_f - \dot{\mathbf{u}}_s\right)$.

Box 2.8 **Dynamic permeability and tortuosity**

The thought experiment that expresses our thinking on this topic was developed in Johnson et al. (1987). The test involves the application of a sinusoidal macroscopic pressure gradient, $\nabla P e^{i\omega t}$, to a large sample of porous material where ∇P is constant. The medium is characterized by its porosity φ and by a measure of the ease of flow, such as the permeability k, while the fluid is characterized by its viscosity (μ_f) and density (ρ_f). The pressure gradient is postulated to give rise to a macroscopically averaged fluid velocity $\mathbf{v} = \dot{\mathbf{u}}_f$. The amplitude of the oscillation is assumed to be small enough that the fluid response is linearly related to the pressure gradient at any given frequency:

$$\mathbf{v} = -\frac{k(\omega)}{\varphi \mu_f} \nabla P e^{i\omega t},$$

which is just Darcy's law, and similarly for the time derivative of $\mathbf{v}$

$$\frac{\partial \mathbf{v}}{\partial t} = -\frac{1}{\rho_f \nu(\omega)} \nabla P e^{i\omega t},$$

where $k(\omega)$ is the 'dynamic permeability' and $\nu(\omega)$ is the 'dynamic tortuosity'. Due to the sinusoidal nature of the pressure gradient and the linear relationship at a given frequency, we can assume that $\mathbf{v}$ also varies in a sinusoidal fashion, and at the same frequency, so that

$$\frac{\partial \mathbf{v}}{\partial t} = i\omega \mathbf{v}$$

and $\nu(\omega)$ and $k(\omega)$ are related:

$$k(\omega) = \frac{1}{\omega} \frac{i\mu_f \varphi}{\rho_f \nu(\omega)}.$$

Based on some reasonable assumptions, Johnson et al. (1987) derive approximate forms for $\nu(\omega)$ and $k(\omega)$. The basic concepts underlying the approach are causality, and the low- and high-frequency limits of the function. To be consistent with causality, $k(\omega)$ must be analytic (have no branch points or poles) for ω in the upper half of the complex plane. Correspondingly, $\nu(\omega)$ has no zeros in this region. The transition from low- to high-frequency behavior is determined by skin depth of the viscous fluid flow, δ, given by

$$\delta = \sqrt{\frac{2\mu_f}{\rho_f \omega}},$$

and its size relative to the characteristic diameter of the pores in the medium.

In the low-frequency regime, as ω approaches zero, the dynamic permeability must approach the static permeability of the medium, k_o:

$$\lim_{\omega \to 0} k(\omega) = k_o$$

and, because of the relationship between $k(\omega)$ and $\nu(\omega)$ given above, it follows that

$$\lim_{\omega\to 0} \nu(\omega) = \frac{1}{\omega}\frac{i\mu_f\varphi}{\rho_f k_o}.$$

In the high-frequency limit the skin depth becomes much smaller than the characteristic pore size and the boundary layer associated with the flow becomes quite small. Outside of the narrow boundary layer, the flow is well approximated by the flow of an ideal fluid and Johnson et al. (1987) argue that

$$\lim_{\omega\to\infty} \nu(\omega) = \nu_\infty + \frac{\beta_\infty}{\sqrt{-i\omega}},$$

where ν_∞ and β_∞ are real-valued and independent of the frequency. For their model of dynamic tortuosity Johnson et al. (1987) consider a function of the general form

$$\nu(\omega) = \nu_\infty + \frac{i\mu_f\varphi}{\omega k_o\rho_f}F(\omega),$$

where $F(\omega)$ is chosen so that the high- and low-frequency limits are honored. In particular, they note that $\lim_{\omega\to 0} F(\omega) = 1$, and

$$\lim_{\omega\to\infty} F(\omega) = \frac{2k_o\nu_\infty}{\Lambda\varphi}\sqrt{\frac{-i\omega\rho_f}{\mu_f}}$$

leading to the following simple model for $F(\omega)$

$$F(\omega) = \sqrt{1 - \frac{4\nu_\infty^2 k_o^2\rho_f}{\mu_f\Lambda^2\varphi^2}i\omega},$$

with the corresponding expression

$$\nu(\omega) = \nu_\infty + \frac{i\mu_f\varphi}{\omega k_o\rho_f}\sqrt{1 - \frac{4\nu_\infty^2 k_o^2\rho_f}{\mu_f\Lambda^2\varphi^2}i\omega},$$

for the dynamic tortuosity. The one parameter that we have not yet discussed is Λ, twice the ratio of the weighted pore volume to the weighted surface area (Johnson et al., 1987). Using the relationship between dynamic tortuosity and dynamic permeability given above, we can develop an expression for $k(\omega)$,

$$\frac{1}{k(\omega)} = \frac{1}{k_o}\left[\sqrt{1 - i\frac{\omega}{\omega_c}\Pi} - i\frac{\omega}{\omega_c}\right].$$

In order to keep the expression compact, we have defined the pore geometry term Π

$$\Pi = 4\frac{\nu_\infty k_o}{\Lambda^2\varphi},$$

and the transition frequency, ω_c, signifying the crossover from viscous dominated flow to flow dominated by inertial forces,

$$\omega_c = \frac{\mu_f \varphi}{\rho_f \nu_\infty k_o}.$$

It was pointed out that the expression generally fits values from both experimental (Charlaix et al., 1988) and numerical modeling (Zhou and Sheng, 1989) except at low frequencies, $\omega < \omega_c$, where the imaginary part generally under-estimates the true value, sometimes by as much as 90 percent. The discrepancy is particularly pronounced when the slope of the channel wall is steep in the throat region. Pride et al. (1993) present a modified expression based upon the physical principle that the drag force is primarily due to two mechanisms:

$$\mathbf{d} = \mathbf{d}^{form} + \mathbf{d}^{friction},$$

including a form factor, $\mathbf{d}^{form}$, due to convergent, divergent, or tortuous flow channels and a frictional effect, $\mathbf{d}^{friction}$, due to friction drag that is most evident in the smallest pores. Pride et al. (1993) show that this model provides an improved fit to values from an analytical solution of a variable-width pore model.

An additional coupling between the fluid and the solid is contained in the expression for the fluid and solid stresses, σ_f and σ_s, respectively. While the exact form of this coupling will depend upon the particular constitutive equations, as well as the nature of the solid material, we can make some general comments: the most basic observation is that, as a fluid-filled porous material deforms, the induce pressure changes can cause the fluid to migrate. This phenomena was elucidated early on by Karl Terzaghi (1943) in relation to the settling of a structure located on a saturated soil. Terzaghi developed the fundamental concept of **effective stress**. For a soil with incompressible grains the effective stress is the difference between the applied stress and the fluid pressure (Wang, 2000, p. 9). Any sudden loading of a fluid-saturated soil is initially borne by both the porous matrix and the fluid itself. As the fluid migrates from the region, due to the corresponding fluid pressure increase, the load is gradually shifted to the solid matrix. In terms of the total stress tensor σ_s, the effective stress tensor, $\bar{\sigma}_s$, is given by

$$\bar{\sigma}_s = \sigma_s + \alpha p_f \mathbf{I}, \tag{2.157}$$

where α is a constant that was introduced by others, p_f is the fluid pressure and $\mathbf{I}$ is the identity matrix. This concept was also adopted by Biot in his insightful analysis of three-dimensional consolidation (Biot, 1941) and in his later study of wave propagation in a poroelastic medium (Biot, 1956a,b), which both still hold up today. We shall have more to say about poroelasticity in the sections that follow.

Within the context of poroelasticity the parameter α is known as the **Biot–Willis** coefficient. α represents the ratio of added fluid volume to the change in bulk volume under the constraint that the pore pressure is constant (Wang, 2000, p. 22).

Multiple fluid phases

Now consider the flow of multiple fluids through a deformable porous solid. For illustration we consider two fluid phases, which we designate as liquid (l) and gas (g). The fluid phases move through a porous, or solid (s), matrix. The most general approach would be a compositional formulation in which we track the various chemical components as they flow though the pore space [see Equations (2.139) to (2.145)]. However, such an approach requires additional bookkeeping that conflicts with our goal of keeping the exposition as simple as possible. Therefore, we illustrate the approach for the flow of just two immiscible fluid phases, in this case a liquid phase and a gas phase. Though there are multiple constituents, the basic principles, the conservation of mass and the conservation of linear momentum, still apply. Furthermore, concepts from mixture theory, such as momentum transfer or interaction terms, can also be incorporated into a formulation involving multiple fluids.

First consider the conservation of mass for each of the phases (solid, liquid, and gas). The fluid phases (liquid and gas) are assumed to occupy the pore space, characterized by the porosity (φ). The remaining fraction of the volume is composed of the solid matrix. The liquid fraction of the fluid is indicated by the saturation S_l. The fraction of gas is given by the gas saturation S_g. The two fluids completely fill the pore space and, as a consequence, we have $S_l + S_g = 1$. If we denote the liquid saturation by $S_l = S$, then the gas saturation is given by $S_g = 1 - S$. Because there are two fluid phases there will be two fluid displacement vectors, denoted by $\mathbf{u}_l$ (liquid) and $\mathbf{u}_g$ (gas). As in the case of a single fluid, we begin with the conservation of the total mass:

$$\frac{d\left(m_l + m_g + m_s\right)}{dt} = \frac{d}{dt}\int_V \left[\varphi S_l \rho_l + \varphi S_g \rho_g + (1-\varphi)\,\rho_s\right] dV = 0,$$

and assume that there is no transfer of mass between the phases (i.e., no phase changes). For multiphase flow this is not always the case and we may actually have mass transfers between fluid phases. In such cases we must modify the approach to include fluid-fluid interaction terms. We will neglect mass transfers across phase boundaries, so that the mass of each fluid in the system is conserved independently, leading to

$$\frac{dm_l}{dt} = \frac{d}{dt}\int_V \varphi S_l \rho_l dV = 0, \tag{2.158}$$

for the liquid and

$$\frac{dm_g}{dt} = \frac{d}{dt}\int_V \varphi S_g \rho_g dV = 0, \tag{2.159}$$

for the gas, if there are no sources or sinks. In the presence of a source or a sink we would have non-zero source terms on the right-hand-sides of Equations (2.158) and (2.159). In order to keep the expressions compact, especially in the presence of multiple fluid phases, we write the conservation equations for the liquid and the gas as a single indexed equation

$$\frac{dm_p}{dt} = \frac{d}{dt}\int_V \varphi S_p \rho_p dV = 0, \tag{2.160}$$

where $p = l, g$ for the liquid and gas, respectively. Now consider the equation for the solid, occupying the remaining $1 - \varphi$ fraction of the volume. The statement of the conservation of the solid mass (m_s), of density ρ_s, is

$$\frac{dm_s}{dt} = \frac{d}{dt}\int_V (1 - \varphi)\, \rho_s dV = 0. \tag{2.161}$$

We can use the expression for the material derivative of an integral, as given by Equation (2.55), and the fact that the volume of integration is arbitrary, to express the conservation of mass for the system (liquid, gas, and solid) as three differential equations

$$\frac{\partial \left(\varphi S_p \rho_p\right)}{\partial t} + \nabla \cdot \left(\varphi S_p \rho_p \dot{\mathbf{u}}_p\right) = 0, \tag{2.162}$$

$$\frac{\partial (1 - \varphi)\, \rho_s}{\partial t} + \nabla \cdot [(1 - \varphi)\, \rho_s \dot{\mathbf{u}}_s] = 0, \tag{2.163}$$

where $p = l, g$.

Another set of conservation equations are those for the linear momentum of each phase. As in the previous section, we must account for the fact that forces can act across phase boundaries, resulting in interactions between constituents. We therefore incorporate momentum transfer or interaction terms that lead to the possibility that, due to such interactions, we may have a source of momentum. The resulting conservation of linear momentum for the two components of the fluid is expressed as

$$\frac{d}{dt}\int_V \varphi \rho_p S_p \dot{\mathbf{u}}_p dV = \int_V \varphi\left[\nabla \cdot \sigma_p + \mathbf{f}_p\right] dV + \int_V \mathbf{d}_p dV, \tag{2.164}$$

where $p = l, g$, while for the solid we have

$$\frac{d}{dt}\int_V (1-\varphi)\,\rho_s \dot{\mathbf{u}}_s dV = \int_V (1-\varphi)\,[\nabla\cdot\sigma_s + \mathbf{f}_s]dV + \int_V \mathbf{d}_s dV, \tag{2.165}$$

where $\mathbf{d}_g + \mathbf{d}_l + \mathbf{d}_s = 0$ [see Box 2.7]. Using the definition of the material derivative of an integral, incorporating the continuity equations [which are Equations (2.162) and (2.163)], and using the fact that the volume of integration is arbitrary, we can derive the set of governing differential equations

$$\varphi S_p \rho_p \left\{\frac{\partial \dot{\mathbf{u}}_p}{\partial t} + \dot{\mathbf{u}}_p\cdot\nabla\dot{\mathbf{u}}_p\right\} = \varphi\nabla\cdot\sigma_p + \varphi\mathbf{f}_p + \mathbf{d}_p, \tag{2.166}$$

$$(1-\varphi)\,\rho_s\left\{\frac{\partial \dot{\mathbf{u}}_s}{\partial t} + \dot{\mathbf{u}}_s\cdot\nabla\dot{\mathbf{u}}_s\right\} = (1-\varphi)\,\nabla\cdot\sigma_s + (1-\varphi)\,\mathbf{f}_s + \mathbf{d}_s, \tag{2.167}$$

the multiphase equivalents of Cauchy's equations of motion [see the derivation associated with Equation (2.64)]. The reader can fill in the remaining steps as an exercise.

The conservation equations provide the basis for a set of governing differential equations that may be solved for solid and fluid displacements. However, the issue of the coupling between the various components, that is among the liquid, the gas, and the solid, still remains. As noted previously, in the discussion associated with Equation (2.156), the coupling occurs in the interaction terms $\mathbf{d}_p$ and $\mathbf{d}_s$ in Equations (2.166) and (2.167). Phase coupling also occurs in the expressions for the stress: σ_p and σ_s. We have already pointed out that one drawback of mixture theory is that the form of the coupling terms must generally be derived using other approaches, such as the method of averaging. For frequencies low enough that Darcy flow dominates, for frequencies below about 10^4 Hz in consolidated sediments, the coupling in the interaction terms $\mathbf{d}_p$, $p = l, g$, and $\mathbf{d}_s$ is primarily due to the viscous drag resulting from the relative motion between the phases (Biot, 1956a,b; Garg and Nayfeh, 1986). Thus, there are terms proportional to the velocity of the fluid and gas relative to the velocity of the solid: $\dot{\mathbf{u}}_l - \dot{\mathbf{u}}_s$ and $\dot{\mathbf{u}}_g - \dot{\mathbf{u}}_s$, and a term proportional to the velocity of the gas relative to the velocity of fluid: $\dot{\mathbf{u}}_g - \dot{\mathbf{u}}_l$. Hence, the coupling terms in the body force vectors are assumed to be of the general form

$$\mathbf{d}_l = -D_{ls}\,(\dot{\mathbf{u}}_l - \dot{\mathbf{u}}_s) + D_{gl}\left(\dot{\mathbf{u}}_g - \dot{\mathbf{u}}_l\right), \tag{2.168}$$

$$\mathbf{d}_g = -D_{gs}\left(\dot{\mathbf{u}}_g - \dot{\mathbf{u}}_s\right) - D_{gl}\left(\dot{\mathbf{u}}_g - \dot{\mathbf{u}}_l\right), \tag{2.169}$$

and

$$\mathbf{d}_s = D_{ls}\,(\dot{\mathbf{u}}_l - \dot{\mathbf{u}}_s) + D_{gs}\left(\dot{\mathbf{u}}_g - \dot{\mathbf{u}}_s\right), \tag{2.170}$$

where, as in the case of a single fluid phase, D_{ls}, D_{gs}, and D_{gl} might be frequency dependent [see the discussion following Equation (2.156)] and also might depend upon capillary pressure in the most general case. The interaction terms given above satisfy the condition $\mathbf{d}_g + \mathbf{d}_l + \mathbf{d}_s = 0$ that is required in order to conserve the total linear momentum of the system. The frequency dependence of these coefficients allows for higher-order derivatives because multiplying (dividing) by ω in the frequency domain is equivalent to differentiating (integrating) with respect to time.

The second source of coupling between the fluids in the pores and the solid matrix is via the stress tensors for the solid, σ_s and the fluids σ_p for $p = l, g$. We can generalize the idea of effective stress, formulated by Terzaghi (1943) for single-phase flow, to the simultaneous flow of multiple fluids. Recall that the effective stress is the stress acting on the solid grains of the porous matrix. The total stress is composed of the sum of the effective stress and pressure in the fluid. When there are two or more fluids within the pores, each fluid may have a distinct pressure due to capillary effects [see Equation (2.131)]. For a given imaginary planar surface cutting through the porous medium, the total pressure will be the pressure transmitted through the grains, the effective pressure, $\bar{\sigma}_{ij}$ plus the pressure transmitted through the fluids. Because, by convention, the stress in a solid is normally taken to be positive in tension, while the fluid pressure is designated positive in compression, there is a sign difference. Hence, the total stress is given by

$$\sigma_s = \bar{\sigma}_s - \alpha p_f \mathbf{I}, \tag{2.171}$$

as given above, in Equation (2.157). The total fluid pressure is the weighted sum of the pressure in each fluid phase. For the liquid and gas mixture under consideration, neither of which can support shear stress, the total stress is given by

$$\sigma_s = \bar{\sigma}_s - \alpha \left(\chi_l p_l + \chi_g p_g\right) \mathbf{I}, \tag{2.172}$$

where χ_p are the weights. Bishop and Blight (1963) have shown that there is a direct linear relationship between the weighting χ_p and the fluid saturation S_p, in fact $\chi_p = S_p$ is one common approximation. Thus, Equation (2.172) may also be written

$$\sigma_s = \bar{\sigma}_s - \alpha \left(S_l p_l + S_g p_g\right) \mathbf{I}. \tag{2.173}$$

This quantity is substituted for the solid stress σ_s in Equation (2.167) to determine the coupling. Equations (2.166) and (2.167), along with the appropriate constitutive equations and equations-of-state, form the elementary equations governing the displacements in the fluid and in the solid matrix.

2.6.2 Coupled elastic deformation and fluid flow: poroelasticity

Now we consider the case in which the solid matrix behaves elastically. The fluid permeating the pores is assumed to behave as a viscous liquid. Some of the earliest treatments of this topic were Kosten and Zwikker (1941), Frenkel (1944), and the well known work of Biot (1956a,b, 1962b,a) [see de Boer (2000) for a comprehensive history]. Their work was concerned with the coupled deformation and flow of an elastic porous matrix saturated with a single fluid phase. We now start with this case and later generalize the formulation to allow for multiple fluids. In an effort to avoid additional complications, we only consider isotropic media. Thus, the stress-strain relationship has a form similar to Equation (2.77) or (2.79). For an anisotropic body one may also generalize the more complicated relationship (2.76) to allow for the presence of one or more fluids.

A single fluid phase

We begin by considering the constitutive equations at play in the poroelastic medium. First, consider the behavior of the fluid saturating the solid. The fluid does not support shear stresses and the stress in the fluid is dominated by the fluid pressure. Therefore, the fluid stress is typically assumed to be of the form

$$\sigma_f = -p_f \mathbf{I} \tag{2.174}$$

where $\mathbf{I}$ is the identity matrix. The interaction of the fluid with the matrix, giving rise to viscous forces, is accounted for via the coupling force $\mathbf{d}_f$ [see Equation (2.156)].

Next, we derive the stress tensor for the fluid-saturated porous matrix. As a prelude, consider the solid matrix in isolation, drained of all fluid. An imaginary plane through a volume of the matrix can be subject to the stresses transmitted through the grains of the matrix. Thus, the stress in the volume will be the effective stress, that is the stress experienced by the matrix. Because we are assuming that the solid matrix deforms as an elastic body, the stress-strain relationship for an isotropic material is given by Equation (2.79), written in tensor or dyadic notation:

$$\bar{\sigma}_s = K_d (\nabla \cdot \mathbf{u}_s) \mathbf{I} + G \left[\nabla \mathbf{u}_s + \nabla \mathbf{u}_s^T - \frac{2}{3} (\nabla \cdot \mathbf{u}_s) \mathbf{I} \right], \tag{2.175}$$

which relates the stress tensor for the solid matrix, being the effective stress tensor $\bar{\sigma}_s$, to the spatial derivatives of the displacement of the solid matrix, $\mathbf{u}_s$. We have switched notation here, replacing the bulk modulus K_b with K_d to denote the drained bulk modulus that we describe shortly. Recall that the trace of the quantity in square brackets vanishes. Also note that we have adopted a linear elastic constitutive model for the porous matrix. This choice is not the only possible model of elasticity, there are also non-linear elastic models as well, but they would take us too far afield.

Now consider the impact of a pore fluid on the deformation. The underlying idea is that of Terzaghi (1943) concerning effective stress and the role of the pore fluid in supporting the load on a porous matrix. Terzaghi formulated his idea for a general, not necessarily elastic, porous medium subject to one-dimensional loading. The idea was extended to three-dimensions by Biot (1941) for an elastic porous medium. From Equation (2.157), we observe that the total stress σ_s is

$$\sigma_s = \bar{\sigma}_s - \alpha p_f \mathbf{I}, \tag{2.176}$$

being a sum of the effective stress and the fluid pressure, the negative sign is due to the different sign convention of compressive stress for the matrix and for the fluid. Thus, if we also incorporate Equation (2.175), the full expression for the total stress is

$$\sigma_s = K_d \left(\nabla \cdot \mathbf{u}_s\right) \mathbf{I} + G \left[\nabla \mathbf{u}_s + \nabla \mathbf{u}_s^T - \frac{2}{3} \left(\nabla \cdot \mathbf{u}_s\right) \mathbf{I} \right] - \alpha p_f \mathbf{I}. \tag{2.177}$$

The constitutive Equations (2.174) and (2.177) state the stress tensors in terms of the solid and fluid displacements and in terms of the fluid pressure p_f. It is advantageous to express the stress tensors entirely in terms of the solid and fluid displacements or their derivatives. In order to do so, we revisit the early work of Biot (1941). Taking the trace (the sum of the diagonal elements) of the matrices comprising each side of Equation (2.177), noting that the trace of the quantity in square brackets vanishes, produces the scalar expression

$$P_c = \alpha p_f - K_d \nabla \cdot \mathbf{u}_s, \tag{2.178}$$

where we have defined the **confining pressure**,

$$P_c = -\frac{\sigma_{11} + \sigma_{22} + \sigma_{33}}{3}, \tag{2.179}$$

representing the average total pressure acting upon a sample [recall that the trace of **I** is 3]. Equation (2.178) is one of the two constitutive equations introduced by Biot (1941)) in his formulation of three-dimensional poroelasticity (Wang, 2000, p. 17). The other constitutive equation postulated by Biot is associated with the increment of fluid content, ζ, being a measure of the relative change in fluid content, and given by

$$\zeta = -\nabla \cdot \mathbf{w}, \tag{2.180}$$

where $\mathbf{w}$ is the motion of the fluid, relative to that of the solid weighted by the porosity of the material

$$\mathbf{w} = \varphi \left(\mathbf{u}_f - \mathbf{u}_s\right). \tag{2.181}$$

Note that

$$\zeta = -\left[\nabla \cdot \varphi \mathbf{u}_f - \nabla \cdot \varphi \mathbf{u}_s\right],$$

represents the difference between the divergence of the vector fields $\varphi \mathbf{u}_f$ and $\varphi \mathbf{u}_s$. Biot postulated that the increment of fluid content was linearly related to the confining pressure (P_c) and the fluid pressure (p_f). The exact relationship is given by (Wang, 2000, p. 35)

$$\zeta = -\frac{\alpha}{K_d} P_c + \frac{\alpha}{K_d B} p_f, \tag{2.182}$$

where B is **Skempton's coefficient** which can be defined as the ratio of pore pressure to confining pressure under undrained conditions, that is:

$$B = \frac{p_f}{P_c} \tag{2.183}$$

when no fluid is allowed to enter or leave a sample ($\zeta = 0$). Rewriting Equations (2.178) and (2.182) as two equations in P_c and p_f, produces the system

$$\nabla \cdot \mathbf{u}_s = -\frac{1}{K_d} P_c + \frac{\alpha}{K_d} p_f, \tag{2.184}$$

$$\nabla \cdot \mathbf{w} = \frac{\alpha}{K_d} P_c - \frac{\alpha}{K_d B} p_f, \tag{2.185}$$

that one can solve for P_c and p_f in terms of $\nabla \cdot \mathbf{u}_s$ and $\nabla \cdot \mathbf{w}$:

$$P_c = -K_u \nabla \cdot \mathbf{u}_s - K_u B \nabla \cdot \mathbf{w}, \tag{2.186}$$

$$p_f = -K_u B \nabla \cdot \mathbf{u}_s - \frac{K_u B}{\alpha} \nabla \cdot \mathbf{w}, \tag{2.187}$$

where K_u is the undrained bulk modulus

$$K_u = \frac{K_d}{1 - \alpha B}. \tag{2.188}$$

Under undrained conditions the fluid content of the sample is constant, no fluid enters or leaves the sample (Rice and Cleary, 1976). This is one of the canonical states of the porous material that is often simulated in experiments in order to estimate some of the parameters that characterize the medium [see Box 2.9]. Drained conditions are also commonly invoked when discussing a porous material. In this state fluid may freely enter or leave the sample, but the fluid pressure is required to remain constant. At various times we may encounter drained or undrained moduli, characterized by these two conditions. Let us take a moment to point out the importance of the product αB in Equation (2.188). As noted by Zimmerman (2000), this product is a poroelastic coupling parameter that indicates if one can safely neglect geomechanical effects when computing fluid pressure variations. In particular, if the product αB is small, then one may neglect the coupling between the deformation of the solid matrix and fluid pressure changes. It is clear from Equation (2.188) that K_u and K_d are virtually identical when $\alpha B \ll 1$.

Equations (2.186) and (2.187) allow us to express both the confining pressure and the fluid pressure in terms of the solid and fluid displacements. Thus, one may formulate a complete set of governing equations solely in terms of the solid and fluid displacements. We are now in a position to combine all the elements discussed above to produce a set of governing equations. The equations of motion for a fluid within a porous matrix, Equations (2.154) and (2.155), together with the expressions for the coupling forces (2.156), give

$$\varphi \rho_f \left\{ \frac{\partial \dot{\mathbf{u}}_f}{\partial t} + \dot{\mathbf{u}}_f \cdot \nabla \dot{\mathbf{u}}_f \right\} - \rho_c \frac{\partial \dot{\mathbf{w}}}{\partial t} = \varphi \nabla \cdot \sigma_f, \tag{2.189}$$

$$(1-\varphi)\,\rho_s \left\{ \frac{\partial \dot{\mathbf{u}}_s}{\partial t} + \dot{\mathbf{u}}_s \cdot \nabla \dot{\mathbf{u}}_s \right\} + \rho_c \frac{\partial \dot{\mathbf{w}}}{\partial t} = (1-\varphi)\,\nabla \cdot \sigma_s,$$

where

$$\rho_c = \rho_f \varphi \frac{\mu_f}{k}$$

is the coefficient for the interaction, or coupling, term and we are neglecting body forces $\mathbf{f}$ in this section.

Box 2.9 The nature of the essential poroelastic coefficients

Up to this point we have encountered a fair number of coefficients and we will introduce a few more before we are through. This bounty of poroelastic constants can be bewildering to those encountering the theory for the first time. Therefore, it is reassuring to know that only three incompressibilities are required to characterize an isotropic porous elastic body. This can be clearly seen in the system of Equations (2.186) and (2.187),

$$P_c = -K_u \nabla \cdot \mathbf{u}_s - K_u B \nabla \cdot \mathbf{w},$$

$$p_f = -K_u B \nabla \cdot \mathbf{u}_s - \frac{K_u B}{\alpha} \nabla \cdot \mathbf{w},$$

relating the system pressures, P_c, p_f, to the divergences the vector fields $\mathbf{u}_s$ and $\mathbf{w}$. Here, the essential coefficients are the undrained bulk modulus K_u, Skempton's coefficient B, and the Biot–Willis coefficient α, introduced after Equation (2.157). However, the three parameters that are perhaps the easiest to define are the undrained bulk modulus, the drained bulk modulus, and Skempton's coefficient. The coefficients K_u and K_d, being the undrained and drained bulk moduli, relate changes in the confining pressure, δP_c, to fractional volume changes, $\delta V/V_o$, under different

boundary conditions: no change in fluid content and no change in fluid pressure, respectively. Skempton's coefficient, B, relates changes in fluid pressure to changes in confining pressure for the case in which the fluid content does not vary.
The three coefficients, K_u, K_d, and B are relatively straight-forward to determine experimentally. For K_u, one measures δP_c and $\delta V/V_o$ when the fluid content of the sample is fixed:

$$K_u = -\left(\frac{\delta P_c}{\delta V/V_o}\right)_{\nabla\cdot\mathbf{w}=0}.$$

To measure K_u one increments the confining pressure by δP_c and measures the change in volume. To estimate K_d, the external or confining pressure P_c is changed and then the fluid content of the sample is adjusted to bring p_f back to its initial value,

$$K_d = -\left(\frac{\delta P_c}{\delta V/V_o}\right)_{\delta p_f=0}.$$

The resulting volume change is then measured and the ratio computed. Skempton's coefficient is determined by incrementing P_c by an amount δP_c and then controlling p_f so that no fluid enters of leaves the sample:

$$B = -\left(\frac{\delta p_f}{\delta P_c}\right)_{\nabla\cdot\mathbf{w}=0}.$$

The change in fluid pressure δp_f required to ensure that the fluid content is constant is used to compute Skempton's coefficient B.

The stresses and pressures may now be written in terms of the spatial derivatives of the solid and fluid displacements. In the last page or so we have taken a few turns in defining various coefficients. In Box 2.9 we follow Pride (2005) and find an essential set of coefficients that can easily be related to both thought experiments and actual laboratory measurements.

Before moving on to the case of multiple fluid phases, we can clean things up a bit, defining a couple of well-known coefficients and presenting an abbreviated form for σ_s. As a start, consider the constitutive equation for the fluid stress σ_f, of the form (2.174)

$$\sigma_f = -p_f\mathbf{I}, \tag{2.190}$$

because an inviscid fluid does not support shear stress. An expression for p_f in terms of the spatial derivatives of $\mathbf{u}_s$ and $\mathbf{w}$ is provided by Equation (2.187)

$$p_f = -C\nabla\cdot\mathbf{u}_s - M\nabla\cdot\mathbf{w}, \tag{2.191}$$

where we have defined the modulus

$$C = K_u B, \tag{2.192}$$

associated with the coupling between the fluid pressure and the elastic deformation of the solid matrix, referred to as **Biot's coupling modulus**, and the **fluid-storage coefficient** (Pride, 2005)

$$M = \frac{C}{\alpha} = \frac{K_u B}{\alpha}, \tag{2.193}$$

representing the amount of fluid that can accumulate in a sample at constant volume (Wang, 2000).

The constitutive expression for the solid component, given by Equation (2.177), can be written in a form that is only slightly more complicated then the constitutive equation for the fluid stress, Equation (2.190):

$$\sigma_s = -P_c \mathbf{I} + \boldsymbol{\tau},$$

where P_c is given by (2.186) which we may rewrite in terms of C

$$P_c = -K_u \nabla \cdot \mathbf{u}_s - C \nabla \cdot \mathbf{w}, \tag{2.194}$$

and

$$\boldsymbol{\tau} = G \left[\nabla \mathbf{u}_s + \nabla \mathbf{u}_s^T - \frac{2}{3} (\nabla \cdot \mathbf{u}_s) \mathbf{I} \right] \tag{2.195}$$

is the contribution to the solid stress due to the (elastic) shear strain. Note that the first term in the expression for σ_s only contains elements on the diagonal of the stress matrix while the second term has a vanishing trace. As a result of the vanishing trace, the quantity $\boldsymbol{\tau}$ is known as the **deviatoric stress tensor**.

Typically, the non-linear terms in Equations (2.189), $\dot{\mathbf{u}}_f \cdot \nabla \dot{\mathbf{u}}_f$ and $\dot{\mathbf{u}}_s \cdot \nabla \dot{\mathbf{u}}_s$, are only significant in a confined region of the medium. For example, such terms may be large near an energetic source, such as an explosion, a vibrational source on soft ground, or near a well subjected to a high flow rate. Alternatively, there may be regions of failure due to inelastic collapse or shear failure in a volume of weakness within the medium. In such regions the assumptions underlying the adoption of an elastic constitutive relation are often violated, for example the use of the infinitesimal strain tensor, and thus the region is best modeled by a more comprehensive approach. However, the non-linear effects tend to decay rapidly with distance from the source and may be neglected outside of a volume immediately surrounding the source or failure zone. While it is quite common to neglect the non-linear terms in Equations (2.189), one can also re-arrange these equations, moving the non-linear terms to the right-hand-side:

$$\begin{aligned} \varphi \rho_f \frac{\partial \dot{\mathbf{u}}_f}{\partial t} - \rho_c \frac{\partial \dot{\mathbf{w}}}{\partial t} &= \varphi \nabla \cdot \sigma_f - \varphi \rho_f \dot{\mathbf{u}}_f \cdot \nabla \dot{\mathbf{u}}_f \\ (1-\varphi)\, \rho_s \frac{\partial \dot{\mathbf{u}}_s}{\partial t} + \rho_c \frac{\partial \dot{\mathbf{w}}}{\partial t} &= (1-\varphi)\, \nabla \cdot \sigma_s - (1-\varphi)\, \rho_s \dot{\mathbf{u}}_s \cdot \nabla \dot{\mathbf{u}}_s \end{aligned} \tag{2.196}$$

and treat them as source terms. In fact it is not unusual to construct an equivalent elastic source by simply 'cutting out' the non-linear region and placing equivalent boundary stresses and/or displacements on the surface of the cut volume. This approach leads to ideas such as source moment tensors and has proven to be powerful in fields such as earthquake seismology (Backus and Mulcahy, 1976). For the remainder of this section, we assume that the source is far away, or that the source may be treated as a boundary condition. In this case the pair of Equations (2.196) may be written as

$$\varphi\rho_f \frac{\partial \dot{\mathbf{u}}_f}{\partial t} - \rho_c \frac{\partial \dot{\mathbf{w}}}{\partial t} = \varphi\nabla \cdot \sigma_f, \tag{2.197}$$

$$(1-\varphi)\,\rho_s \frac{\partial \dot{\mathbf{u}}_s}{\partial t} + \rho_c \frac{\partial \dot{\mathbf{w}}}{\partial t} = (1-\varphi)\,\nabla \cdot \sigma_s. \tag{2.198}$$

These are the basic governing equations for a poroelastic medium containing a single fluid phase, first advanced by Biot (1956a,b). Equations (2.197) and (2.198) may be re-arranged in various ways (Burridge and Keller, 1981; Pride et al., 1992; Lo et al., 2002; Wilmanski, 2006) depending on the desired form. For example, $\phi\rho_f \partial\dot{\mathbf{u}}_s/\partial t$ may be added and subtracted from Equation (2.197) to write it entirely in terms of $\mathbf{w}$ and $\mathbf{u}_s$:

$$\varphi\rho_f \frac{\partial \dot{\mathbf{u}}_s}{\partial t} + \rho_l \frac{\partial \dot{\mathbf{w}}}{\partial t} = \varphi\nabla \cdot \sigma_f, \tag{2.199}$$

where

$$\rho_l = \varphi\rho_f - \rho_c. \tag{2.200}$$

We also obtain a commonly encountered version of the pair of Equations (2.197) and (2.198) by adding Equations (2.197) and (2.198), producing the system

$$\varphi\rho_f \frac{\partial \dot{\mathbf{u}}_s}{\partial t} + \rho_l \frac{\partial \dot{\mathbf{w}}}{\partial t} = \varphi\nabla \cdot \sigma_f, \tag{2.201}$$

$$\rho_w \frac{\partial \dot{\mathbf{u}}_s}{\partial t} + \varphi\rho_f \frac{\partial \dot{\mathbf{w}}}{\partial t} = (1-\varphi)\,\nabla \cdot \sigma_s + \varphi\nabla \cdot \sigma_f, \tag{2.202}$$

where we have defined the weighted density

$$\rho_w = (1-\varphi)\rho_s + \varphi\rho_f. \tag{2.203}$$

While there are situations for which the derivatives of $\dot{\mathbf{w}}$ may be neglected, due to the gradual nature of fluid flow, it is important to retain them for full generality. The derivative terms are crucial for the generation of the Biot fast waves, analogous to an elastic wave. Furthermore, there are situations in which the propagation of a fluid front may lead to inelastic behavior and the rapid collapse of the matrix, generating elastic waves.

Box 2.10 **Gassmann's equations**

We have settled on the essential constants necessary to specify a poroelastic material, such as K_u, B, and α, or the triple K_u, K_d and B. In Box 2.9 we even noted the basic steps required to determine the latter three coefficients in the laboratory. However, laboratory experiments are not always feasible and it can be difficult to acquire an intact sample from a deep formation. Furthermore, we should like to know how the poroelastic coefficients change as the fluid saturation varies or for different lithologies. Therefore, it is useful to relate the poroelastic coefficients directly to the properties of the constituent fluids and the solid matrix. In this area, the work of Gassmann (1951b) stands out as a seminal contribution. Gassmann (1951b) derived an expression for the bulk modulus of a fluid saturated porous material, K_s, and Skemption's coefficient B in terms of the bulk moduli of the constituent solid grains (K_g), the bulk moduli of the fluid (K_f), the bulk modulus of the dry frame (K_d), and the porosity of the material (ϕ):

$$K_s = K_d + \frac{\left(1 - K_d/K_g\right)^2}{\phi/K_f + (1-\phi)/K_g + K_d/K_g^2},$$

and

$$B = \frac{1/K_d - 1/K_s}{1/K_d - 1/K_s + \varphi\left(1/K_f - 1/K_s\right)}.$$

The technique was developed for the study of waves propagating through a packing of spheres, a model for a material such as a gravel or sand, or even a sandstone (Gassmann, 1951a). The result generally holds for disturbances with wavelengths much longer than the size of the pores and for a porous frame composed of a single mineral. Furthermore, in Gassmann's work, the fluid does not influence the shear modulus of the poroelastic material. For the most part, the material constants in these formulas, in particular the bulk modulus of the solid grains and of the fluid, and the porosity are known or can be estimated from available data.

The main difficulty in applying Gassmann's formulas is obtaining the dry frame bulk modulus K_d and shear modulus G. There are numerous theoretical models for estimating these frame moduli, summarized in Berryman (1995) and Mavko et al. (1998). In many practical settings well logs or core samples are used to try and estimate the moduli for formations of interest. The results of Gassmann have been generalized in a number of ways since their publication. For example, Berryman and Milton (1991) have shown how to calculate effective bulk moduli for a porous medium containing two constituents.

Multiple fluid phases

Now we move on to the case in which two or more fluids are present in a deformable elastic porous medium. While the basic conservation principles (mass and momentum) are the same, there are important differences to note; for example, due the movement of the fluid phases and aspects of multiphase flow such as relative permeability and capillary pressure, the governing equations are fundamentally non-linear. This is true even for a poroelastic medium. Many non-numerical studies, focusing on wave propagation, linearize the problem by assuming that the given saturation changes, relative to some background saturation distribution (S_0), are small. This assumption is not generally true in problems involving significant flow, as saturation changes can be large and rather widespread.

The essential conservation laws are the conservation of mass of the fluid and solid components [also see Equation (2.162) and (2.163)]

$$\frac{\partial\left(\varphi S_p \rho_p\right)}{\partial t} + \nabla \cdot \left(\varphi S_p \rho_p \dot{\mathbf{u}}_p\right) = 0, \tag{2.204}$$

$$\frac{\partial\left(1-\varphi\right)\rho_s}{\partial t} + \nabla \cdot \left[\left(1-\varphi\right)\rho_s \dot{\mathbf{u}}_s\right] = 0, \tag{2.205}$$

where $p = l, g$ for two fluid phases and $1, 2, ..., N$ for N fluid phases. The conservation of linear momentum, Equations (2.166) and (2.167) take the form

$$\varphi S_p \rho_p \left\{\frac{\partial \dot{\mathbf{u}}_p}{\partial t} + \dot{\mathbf{u}}_p \cdot \nabla \dot{\mathbf{u}}_p\right\} = \varphi \nabla \cdot \sigma_p + \mathbf{d}_p, \tag{2.206}$$

$$\left(1-\varphi\right)\rho_s \left\{\frac{\partial \dot{\mathbf{u}}_s}{\partial t} + \dot{\mathbf{u}}_s \cdot \nabla \dot{\mathbf{u}}_s\right\} = \left(1-\varphi\right) \nabla \cdot \sigma_s + \mathbf{d}_s. \tag{2.207}$$

The coupling terms in Equations (2.206) and (2.207), denoted by $\mathbf{d}_p$ and $\mathbf{d}_s$ are given above [see Equations (2.168), (2.169), and (2.170)]:

$$\mathbf{d}_p = -\frac{\varphi^2 S_p^2 \mu_p}{k k_{rp}(S_p)}\left(\dot{\mathbf{u}}_p - \dot{\mathbf{u}}_s\right), \tag{2.208}$$

$$\mathbf{d}_s = \sum_{j=1}^{N} \frac{\varphi^2 S_j^2 \mu_j}{k k_{rj}(S_j)}\left(\dot{\mathbf{u}}_j - \dot{\mathbf{u}}_s\right), \tag{2.209}$$

where we have incorporated a multiphase generalization of the coefficients D_{ls} and D_{gs} given by Garg and Nayfeh (1986), that is valid for frequencies below about 10^4 Hz.

The remaining task is to specify the various equations-of-state, constitutive equations, and constraints on the variables and medium parameters such as μ_p the fluid viscosities, φ the porosity, the fluid pressures p_l and p_g, the fluid and solid stresses σ_p and σ_s, the densities ρ_p and ρ_s, the relative permeability and capillary pressure functions. There is also the additional constraint that the saturations must sum to unity, as given by Equation (2.134). Several of these relationships, such as the equations-of-state for the densities and fluid viscosities, the relative permeability curves, and the capillary pressure curves may involve fits to laboratory data.

The coefficients in the previous section and in the multiphase generalization in Chapter 7 are appropriate for Biot's original formulation in which fluid equilibration at the scale of a seismic wavelength, or larger, dominates. As noted in Box 2.11, dispersion and attenuation due to such large-scale processes are not adequate to explain observed field data.

The form of the stresses σ_p and σ_s follows from the constitutive relationships for elastic material and the theory of poroelasticity. In particular, the stresses in the liquid and gas phases are dominated by the pressure, as in Equation (2.174)

$$\sigma_p = -p_p \mathbf{I}, \tag{2.210}$$

where $p = l, g$. For the solid matrix, the stress is that of a drained elastic body, with modifications for the effects of the pore fluids via the concept of effective stress:

$$\sigma_s = K_d (\nabla \cdot \mathbf{u}_s) \mathbf{I} + G \left[\nabla \mathbf{u}_s + \nabla \mathbf{u}_s^T - \frac{2}{3} (\nabla \cdot \mathbf{u}_s) \mathbf{I} \right] - \alpha p_f \mathbf{I}. \tag{2.211}$$

As noted in the discussion surrounding Equation (2.173), the total fluid pressure p_f is given by the weighted sum of the pressure in the liquid and the gas phases, with the weighting given by the fluid saturations:

$$p_f = \sum_{i=1}^{N} S_i p_i. \tag{2.212}$$

There are several paths forward from this point. As was done for a single fluid phase, one can solve for the pressures, such as p_l, p_g, and p_s in terms of the displacement derivatives, $\nabla \cdot \mathbf{u}_l$, $\nabla \cdot \mathbf{u}_g$, and $\nabla \cdot \mathbf{u}_s$ for two fluid phases. This approach was taken in investigations related to wave propagation in a poroelastic medium (Garg and Nayfeh, 1986; Tuncay and Corapcioglu, 1997). Alternatively, we could deal directly with the equations in terms of the displacements, saturations, and pressures (Panday and Corapcioglu, 1989). We shall not pursue either approach here, as there are advantages to both and the best course of action depends upon the particular application.

Box 2.11 **Mesoscopic heterogeneity**

It is generally recognized that, by themselves, Biot's equations do not reproduce the level of intrinsic attenuation that is observed in seismic field data (Pride et al., 2004). To remedy this, additional attenuation mechanisms have been proposed, primarily adding internal structure, such as compliant cracks in the squirt-flow mechanism (Dvorkin et al., 1995) to a poroelastic solid. Most generally, one must add what is known as mesoscopic heterogeneity, spartial variations in properties larger then the pore scale yet smaller then a wavelength, in order to generate sufficient attenuation at the frequencies of observed seismic waves. This solution had been recognized at least since the patchy-saturation model of White (1975) and has evolved into a general model of mesoscopic-scale heterogeneity formalized by Pride et al. (2004), and outlined in this box. Along the way, essential work by Dutta and Ode (1979a,b), Norris (1993), Johnson (2001), and others, provided important advancements, clarifications, and generalizations.

The essential idea is to describe the internal structure or heterogeneity as multiple phases or a dual-porosity/dual-permeability model and then to rewrite the equations as an effective Biot theory (Pride et al., 2004; Pride, 2005). The starting point is the compressibility law for a three-phase medium containing a solid, a liquid, and one other quantity. The additional quantity represents the heterogeneity and may be a fluid phase in the case of patchy-saturation, or a solid phase with differing properties, such as more compliant inclusions. The additional attenuation results from the interaction between this additional phase and the two original phases. The compressibility equations for the three phase system are

$$\begin{pmatrix} \nabla\cdot\mathbf{u}_s \\ \nabla\cdot\mathbf{w}_1 \\ \nabla\cdot\mathbf{w}_2 \end{pmatrix} = -\begin{pmatrix} a_{11} & a_{12} & a_{13} \\ a_{21} & a_{22} & a_{23} \\ a_{31} & a_{32} & a_{33} \end{pmatrix}\begin{pmatrix} P_c \\ p_1 \\ p_2 \end{pmatrix} - \begin{pmatrix} 0 \\ \xi_{int} \\ -\xi_{int} \end{pmatrix}.$$

The quantity

$$\xi_{int} = \frac{i\gamma(\omega)}{\omega}(p_1 - p_2),$$

representing the macroscopic flow, is the average fluid volume transferred from phase 1 to phase 2, normalized by the total averaging volume (Pride, 2005). Note that this system of equations may be written as a generalization of Equations (2.184) and (2.185) to three phases. The compressibility coefficients a_{ij}, given in Pride and Berryman (2003a) and Pride (2005), control the elastic response at the early stages, before the fluid pressure in the two phase equilibrates.

The complex and frequency-dependent coefficient $\gamma(\omega)$ represents the internal transport from one porous phase to another within the sample. Following the branching function approach of Johnson et al. (1987) discussed in Box 2.8, Pride and Berryman (2003b) determine that $\gamma(\omega)$ has the approximate form

$$\gamma(\omega) = \gamma_o\sqrt{1 - i\frac{\omega}{\omega_o}}$$

where ω_o is a transition frequency related to the fluid-pressure diffusion and γ_o is an internal transport coefficient (Pride and Berryman, 2003a,b).

The next step is to write the equations for the three-phase system as an effective Biot theory with a solid and a fluid phase. The presence of the internal structure, due to the third phase, leads to complex and frequency-dependent coefficients. These coefficients are in addition to the dynamic permeability $k(\omega)$ already present in conventional Biot theory, as discussed in Box 2.8.

The approach taken by Pride et al. (2004) and Pride (2005) is to assume that the new third phase is entirely enveloped by the other phases so that its flux into and out of the volume is zero. Placing $\nabla \cdot \mathbf{w}_2$ equal to zero in the compressibility equations given above allows us to eliminate p_2 from the equations, resulting in equations relating $\nabla \cdot \mathbf{u}_s$ and $\nabla \cdot \mathbf{w}_1$ to P_c and p_1. The resulting 2×2 system can be inverted for P_c and p_1, as in the Biot Equations (2.186) and (2.187) for a single fluid. However, now the coefficients K_u, K_d, and B depend upon the compressibilities a_{ij}, $\gamma(\omega)$, and ω according to

$$\frac{1}{K_d(\omega)} = a_{11} - \frac{a_{13}^2}{a_{33} - \gamma/i\omega}$$

$$B(\omega) = \frac{-a_{12}\left(a_{33} - \gamma/i\omega\right) + a_{13}\left(a_{23} + \gamma/i\omega\right)}{\left(a_{22} - \gamma/i\omega\right)\left(a_{33} - \gamma/i\omega\right) - \left(a_{23} + \gamma/i\omega\right)^2}$$

$$\frac{1}{K_u(\omega)} = \frac{1}{K_d(\omega)} + B(\omega)\left[a_{12} - \frac{a_{13}\left(a_{23} + \gamma/i\omega\right)}{a_{33} - \gamma/i\omega}\right].$$

The additional frequency-dependence provided by the mesoscopic heterogeneity allows for a level of dispersion (frequency-dependent velocity) and attenuation that matches observed field data (Pride et al., 2004). The formulation is sufficiently general to cover a variety of situations. Pride et al. (2004) consider three specific cases in their study: a patchy-saturation model, a heterogeneous matrix model, and a model representing squirt-flow.

The approach described in the box, invoking heterogeneity on a scale smaller then a seismic wavelength, so-called mesoscopic heterogeneity, has been found to lead to the correct level of attenuation and dispersion. This approach leads to complex and frequency-dependent coefficients that can be easily incorporated into our formulation.

2.7 Summary

We have reviewed the basic governing equations controlling fluid flow and deformation. We will invoke these equations repeatedly in the chapters that follow. Although these equations seem to be mathematically disparate, covering a broad range of processes ranging from pressure diffusion to poroelastic wave propagation, in the next few chapters we will explore a common framework for their solution. Several practical applications will be presented to illustrate the power and utility of these trajectory-based methods in obtaining solutions that are not only computationally efficient but also visual, intuitive, and easy to interpret.

3
Trajectory-based modeling

3.1 Introduction

In this chapter, we develop asymptotic and trajectory-based methods for the solution of partial differential equations that arise in flow and transport through permeable media. After some background, we discuss specific forms of asymptotic solutions applicable to a wide range of problems in hydrology, geophysics, and petroleum engineering. We describe the construction of a trajectory-based solution and illustrate the procedure by an application to the wave equation. A major advantage of the asymptotic method is the partitioning of the modeling into a travel time calculation and an amplitude calculation. For example, the asymptotic method leads to the eikonal equation that governs the geometry of a propagating fluid front and the travel time of the front. Such travel times form the basis for many of the imaging and inversion techniques discussed in subsequent chapters. We conclude the chapter with an overview of a multiple scale asymptotic technique that is well-suited to problems involving spatial heterogeneity and non-linearity, two features that appear frequently in the modeling of multiphase flow through natural porous media.

3.1.1 Transients and propagating disturbances

As we noted in Chapter 1, our focus will be on using observations related to fluid flow and transport to better characterize the properties of a porous medium. The observations might be from an external experiment, such as a geophysical survey, attempting to image flow-related attribute changes. Alternatively, the observations might be data from tests that involve flow directly, such as the injection of a trace chemical or the monitoring of transient fluid pressure. The important point is that we will be dealing with transient disturbances that propagate through the medium. Thus, there will be a source of the disturbance, such as an injection interval. The transient will travel from the source outward into the medium, ultimately reaching an observation point where one or more instruments record the passing disturbance.

Based upon causality, there is a region 'ahead' of the propagating front, representing the undisturbed volume and a region 'behind' the front through which the disturbance has already propagated. We assume that the surface separating the undisturbed and disturbed volumes may be characterized by an equation of the form

$$S(\mathbf{x}, t) = 0, \tag{3.1}$$

defining its spatial position at time t (Whitham, 1974, p. 236). As long as the front varies with time, so that

$$\frac{\partial S}{\partial t} \neq 0,$$

we can solve the equation $S(\mathbf{x}, t) = 0$ for t as a function of the spatial coordinates, at least locally. We write this function as

$$t = \sigma(\mathbf{x}),$$

and the equation for the hypersurface [a surface in space and time coordinates] becomes

$$S(\mathbf{x}, t) = t - \sigma(\mathbf{x}) = 0. \tag{3.2}$$

We illustrate this in Figure 3.1 with an example from tracer transport. The physical field of interest, representing a quantity such as a jump in pressure, an elastic compression, a tracer concentration, a change in fluid saturation, and so on, is denoted by $u(\mathbf{x}, t)$. In the undisturbed region ahead of the propagating front, $u(\mathbf{x}, t)$ is assumed to be zero. With the passage of the front $u(\mathbf{x}, t)$ takes on non-zero values, either in a continuous or in a discontinuous fashion, as determined by the physical processes at play. At any given instant in time t, the field value $u(\mathbf{x}, t)$ depends upon the distance from the front and the position on the front:

$$u(\mathbf{x}, t) = \begin{cases} \Upsilon(\mathbf{x}, S(\mathbf{x}, t)) & S(\mathbf{x}, t) > 0 \\ 0 & S(\mathbf{x}, t) < 0 \end{cases}. \tag{3.3}$$

Note that Υ depends upon $\mathbf{x}$ both explicitly and implicitly, through the surface $S(\mathbf{x}, t)$. It is possible to relax the requirement that the field directly ahead of the front is zero. That is, due to dissipation and dispersion, the changes associated with the front may have 'diffused' a bit ahead of the surface defined by the vanishing of S. Conceptually, one can think of the front as a moving boundary layer separating the two regions (undisturbed and disturbed). As in a boundary layer, the interface between the two regions might be expected to widen into a zone of finite width.

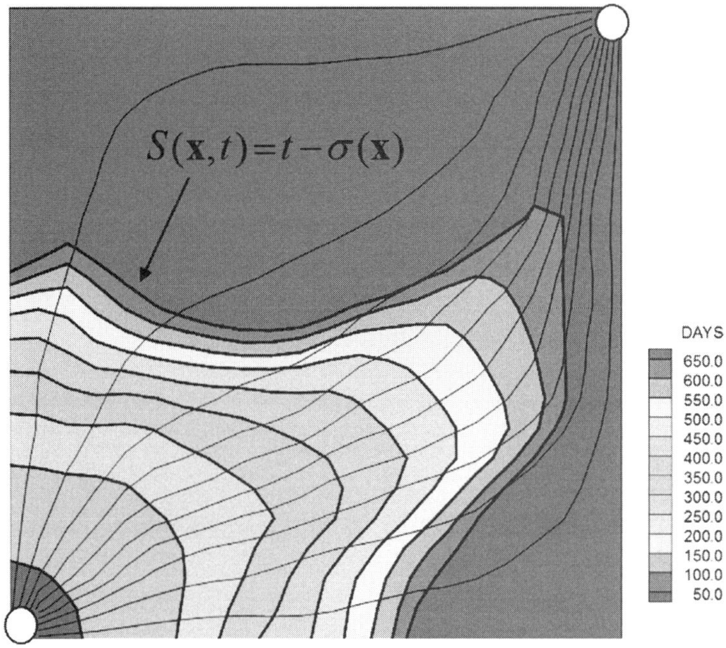

Figure 3.1 An illustration of a propagating tracer front for an injector (bottom left)-producer (top right) pair. Here, $\sigma(\mathbf{x})$represents the tracer travel time along trajectories (streamlines). The fronts are defined by the isochrones (equal travel time) and $S(\mathbf{x}, t)$ identifies the front location at a specific time. For the colour version, please refer to the plate section.

3.1.2 An illustrative example

Physical systems are most frequently modeled using differential equations. We have already encountered examples in Chapter 2 and we shall discuss some of them in more detail in the remainder of the book. Because of their importance, we will attempt to illustrate the ideas introduced in this chapter by way of an example differential equation. A fairly commonly encountered form, encompassing many of the equations found here, is

$$\frac{\partial}{\partial x_i}\left(c_{ij}\frac{\partial u}{\partial x_j}\right) + Bu = 0, \tag{3.4}$$

where we have adopted the Einstein convention of summing over repeated indices. Equation (3.4) is a second-order, scalar, partial differential equation for the function $u(\mathbf{x})$. The wave equation (Whitham, 1974, p. 3)

$$\nabla \cdot \left(c^2 \nabla u\right) - \frac{\partial^2 u}{\partial t^2} = 0, \tag{3.5}$$

provides a well-studied example. In order to put it in the form (3.4), we can extend the definition of $\mathbf{x}$ to the 4-dimensional vector $(x_0, x_1, x_2, x_3) = (t, x, y, z)$. Alternatively, we could apply the Fourier transform, discussed below, producing

$$\nabla \cdot \left(c^2 \nabla U\right) + \omega^2 U = 0, \tag{3.6}$$

where U is the Fourier transform of u.

To extract a solution in the form of a propagating front, with minimal complications, let us further simplify Equation (3.5) by considering a medium with constant properties, c, and one spatial dimension:

$$c^2 \frac{\partial^2 u}{\partial x^2} - \frac{\partial^2 u}{\partial t^2} = 0. \tag{3.7}$$

This equation may be written in a factored form

$$\left(c\frac{\partial}{\partial x} - \frac{\partial}{\partial t}\right)\left(c\frac{\partial}{\partial x} + \frac{\partial}{\partial t}\right) u = 0 \tag{3.8}$$

and we may treat each factor separately (Whitham, 1974, p. 3). Thus, the first factor

$$c\frac{\partial u}{\partial x} + \frac{\partial u}{\partial t} = 0, \tag{3.9}$$

represents the simple one-way wave equation. In Chapter 1 we showed that this equation admits a solution

$$u(x,t) = f\left(t - \frac{x}{c}\right) \tag{3.10}$$

where f is an arbitrary function. If f is in the form of a sharp jump or step, then our solution represents a propagating front, such as the one described by Equation (3.3). The other factor in Equation (3.8) introduces a front propagating in the opposite direction. Initial and boundary conditions dictate the solutions that are active at a given location at a given time.

3.2 Series representation of a moving front

The form (3.3) of a moving front is too general and does not provide enough insight, flexibility, or efficiency to be useful for treating the inverse problem, where we image the internal properties of an object. For efficiency, we would like to represent a propagating disturbance in terms of a succession of relatively simple functions. For insight, the first few terms of the series should pertain to the most important aspects of our solution, serving as a basis for modeling and imaging. These points should become clearer as we proceed through this chapter.

Historically, the most important expansion has been in the form of powers of a variable, in this case powers of S, such as a Taylor series expansion. Power series

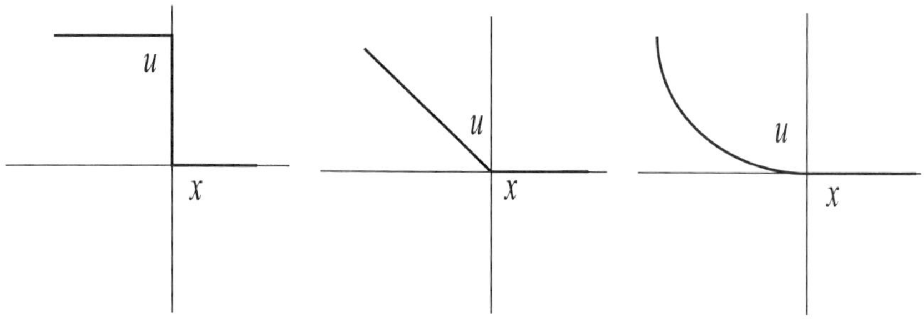

(a) 0th order discontinuity. (b) 1st order discontinuity. (c) 2nd order discontinuity.

Figure 3.2 Examples of discontinuous functions of different orders.

are one of the most general forms for solutions of linear and non-linear partial differential equations (Bender and Orszag, 1978). Furthermore, if we consider a complex valued $S = r \exp i\theta$, where r is the amplitude and θ is the phase, then the series solution comprises a Fourier series representation of u (Needham, 2000, p. 77). Because of the discontinuity in the representation (3.3) at $S = 0$, we must modify the power series representation somewhat. For example, we could generalize the modeling, allowing for weak solutions containing discontinuities (Whitham, 1974), and develop a formal mathematical apparatus for that approach. Instead we adopt a more informal approach, simply requiring continuity in the solutions up to a particular order m. That is, we will allow the front to contain discontinuities in derivatives beyond order m. Physically, this allows for fronts characterized by different degrees of smoothness (Figure 3.2). One can expand the function $\Upsilon(\mathbf{x}, S)$ as a Taylor series in S for points immediately behind the front

$$u(\mathbf{x}, t) = \begin{cases} \Upsilon_0(\mathbf{x}) S^m/m! + \Upsilon_1(\mathbf{x}) S^{m+1}/(m+1)! + \cdots & S > 0 \\ 0 & S < 0 \end{cases} \tag{3.11}$$

where m might even involve fractional powers, for example to model cylindrical waves (Whitham, 1974). The coefficients, such as Υ_0, Υ_1, represent spatial variations in the magnitude of the discontinuity. In fact, the n-th coefficient represents the $m + n$-th one-sided derivative with respect to S

$$\Upsilon_n = \frac{\partial^{m+n} \Upsilon}{\partial S^{m+n}} \tag{3.12}$$

where the derivatives are taken from behind the front towards the leading edge of the front, from positive S towards zero. The derivatives do not exist in the other direction due to the discontinuity at $S = 0$. One can see that the first discontinuity is in the m-th derivative. Specifically, derivatives with respect to S of lower order than m vanish on the surface defined by $S = 0$. Thus, the lower-order derivatives vanish on both sides of the interface and are thereby continuous, equal to the constant value zero. However, the m-th derivative with respect to S gives a non-zero value of $\Upsilon_0(\mathbf{x})$ at, or immediately behind the front, and vanishes ahead of the front. Thus the derivative jumps by Υ_0 across the front.

Recalling our expression for the propagating disturbance,

$$S(\mathbf{x}, t) = t - \sigma(\mathbf{x}) = 0,$$

we can represent the propagating front (3.11) as the sum

$$u(\mathbf{x}, t) = \sum_{n=0}^{\infty} \Upsilon_n(\mathbf{x}) f_n[t - \sigma(\mathbf{x})] \tag{3.13}$$

where $f_n[t - \sigma(\mathbf{x})]$ is the discontinuous function

$$f_n[t - \sigma(\mathbf{x})] = \begin{cases} [t - \sigma(\mathbf{x})]^{n+m}/(n+m)! & t > \sigma(\mathbf{x}) \\ 0 & t < \sigma(\mathbf{x}) \end{cases}, \tag{3.14}$$

and m denotes the lowest-order discontinuity across the hypersurface. This sum is over successively higher-order discontinuous functions, with the discontinuity located at the hypersurface defined by $S(\mathbf{x}, t)$. We can use the Heaviside function $H(S)$, defined as a step-like change from a value of zero to a value of one, with the jump occurring at the front where $S = 0$:

$$H(S) = \begin{cases} 1 & S > 0 \\ 0 & S < 0 \end{cases}, \tag{3.15}$$

to write the series (3.13) as

$$u(\mathbf{x}, t) = \sum_{n=0}^{\infty} \Upsilon_n(\mathbf{x}) \frac{[t - \sigma(\mathbf{x})]^{n+m}}{(n+m)!} H[t - \sigma(\mathbf{x})]. \tag{3.16}$$

A power series expansion in terms of higher-order discontinuous functions is known as a **ray series**. The form of the solution given by the sum (3.13) is really quite general and encompasses many useful series solutions to partial differential equations used in modeling physical processes.

3.3 The frequency domain and high-frequency approximations

The frequency domain, entered via the Fourier transform, provides another perspective on a propagating discontinuity, one that has a long and fruitful history. The **Fourier integral transformation** ($\mathcal{F}$), of a function f in the space-time domain to a function F in the space-frequency (ω) domain is

$$F(\mathbf{x}, \omega) = \mathcal{F}[f(\mathbf{x}, t)] = \frac{1}{\sqrt{2\pi}} \int_{-\infty}^{\infty} f(\mathbf{x}, t) e^{i\omega t} dt, \tag{3.17}$$

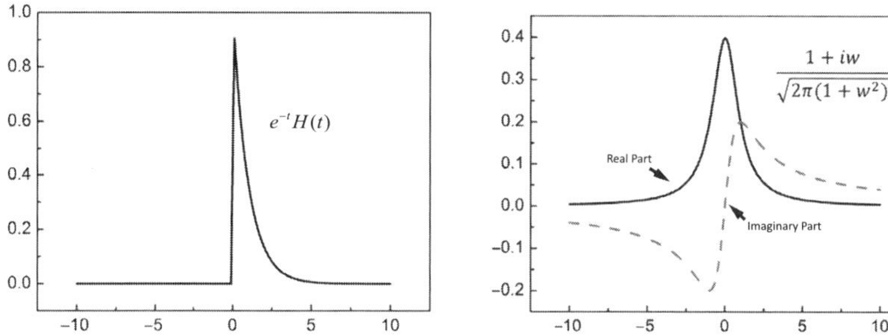

(a) Input function in the time domain. (b) Output Fourier transform.

Figure 3.3 Example of the Fourier transform pair for an impulsive function.

(Bracewell, 2000). There are a several definitions of the Fourier transform, with different sign conventions and normalizations to ensure that the transformation back to the time domain, the inverse Fourier transform, has a very similar form. Here, we are adopting the form that is commonly used in physics, particularly when spatial variables are involved. In Figure 3.3 we plot an example of a Fourier transform pair (Bracewell, 2000). Note that the Fourier transform is a complex quantity. For our purposes a key property for transforming the series representation (3.16) into a series in the frequency domain is the Fourier transform of the product of the step function, $H(t)$, and a power of t,

$$\mathcal{F}\left[H(t)t^n\right] \sim (-i\omega)^{-(n+1)} \tag{3.18}$$

where $\omega > 0$ in order to avoid a singularity at the origin (Bracewell, 2000; Gradshteyn and Ryzhik, 1965). Applying the Fourier transform we can map the function $u(\mathbf{x}, t)$, given by Equation (3.16), into the frequency domain:

$$U(\mathbf{x}, \omega) = e^{-i\omega\sigma(\mathbf{x})} \sum_{n=0}^{\infty} \frac{\Upsilon_n(\mathbf{x})}{(-i\omega)^{(m+n+1)}}, \tag{3.19}$$

for $\omega > 0$. The requirement that $\omega > 0$ is not an issue because we are interested in a high-frequency approximation, where ω is large. In deriving the expression (3.19) we have used both the linearity of the Fourier transform and the fact that a translation or shift in the time domain by the factor $\sigma(\mathbf{x})$ is equivalent to multiplication by $e^{-i\omega\sigma(\mathbf{x})}$ in the frequency domain (Bracewell, 2000). This series in inverse powers of the frequency ω is the basis for the **high-frequency approximation** that has been used to solve many important problems. When ω is a large number, only the first few terms of the series will be significant. For processes governed by linear differential equations with coefficients that are independent of time, the frequency-domain approach works remarkably well and has a somewhat intuitive appeal. Much of the earliest rigorous work on high-frequency asymptotic solutions arose from efforts to reconcile existing geometrical optics approximations with Maxwells equations governing the propagation of electromagnetic waves (Sommerfeld, 1964; Kline and Kay, 1965; Luneburg, 1966). Why have we chosen the Fourier transform? There is no fundamental reason why we could not use another integral transform such as the Laplace transform. One advantage of the Fourier transform is the physical interpretation associated with the frequency domain solution.

3.4 Asymptotic series and solutions

3.4.1 General ideas

As we have seen, transient disturbances are generally represented by functions that may be discontinuous across a propagating front. Non-linear processes can lead to discontinuous solutions, such as shocks, even when the initial state and the governing equation are smooth (Whitham, 1974; Drumheller, 1998; Debnath, 2005). This brings us into the realm of generalized functions or distributions (Lighthill, 1958; Bracewell, 2000). For generalized functions the classical notions of continuity and differentiability will not always apply. We shall need analytical and numerical methods to handle such functions. One useful technique is the use of asymptotic series (Erdélyi, 1956) to represent functions that may not have sufficient differentiability to allow for a straight-forward Taylor series expansion. While the theory of asymptotic series was initially developed by Poincaré (1886), their use dates back at least to Euler (Euler, 1754).

At the beginning of this chapter we constructed a representation of a transient disturbance using a Taylor series expansion that was only valid on one side of the propagating front. Due to the discontinuity, it is not possible to use a single Taylor series to represent the function over its entire range of interest. In particular, the Taylor series fails at the edge of the front, a region of some significance, due to a discontinuous change in some of its derivatives. Thus, a single Taylor series expansion cannot represent the transient disturbance over the range of interest.

We would like to construct a representation of a propagating disturbance that, as in a Taylor's series, provides an expansion in terms of a succession of relatively simple functions. Ideally, the first few terms of the series would provide the most important elements of the approximation, particularly in the limit of some desired condition. An example of such a series is provided by the Fourier transformation of the discontinuous series (3.16), given by the expansion (3.19) in inverse powers of $-i\omega$. The series solution (3.19) is a representation in terms of simple functions and the first few terms provide an increasingly accurate representation of $U(\mathbf{x}, \omega)$ as ω takes on larger and larger values. Note that, given enough terms, the series may diverge for a fixed value of ω, depending upon the values of the successive expansion coefficients $\Upsilon_n(\mathbf{x})$. This follows from the fact that, unlike a Taylors series expansion, we cannot assume that all derivatives exist for all values of $\mathbf{x}$. Following Poincaré's lead, we can abstract these ideas and use such **asymptotic series** to represent approximate solutions to physical problems. The conditions that we shall invoke will typically involve a high-frequency assumption or the idea of a medium with smoothly varying properties.

While the early use of asymptotic series, with their divergent behavior, was initially met with some skepticism, their spectacular success in solving physical problems won them respect in time. Formalism quickly followed and asymptotic methods now comprise an important segment of applied mathematics and mathematical physics (Bender and Orszag, 1978). Consider a function, $f(\mathbf{x}, \varepsilon)$, that depends upon a set of independent variables such as position, $\mathbf{x}$, and an expansion parameter, ε, representing quantities of interest in a physical problem, such as the inverse of the frequency $(1/\omega)$, a ratio of length scales, or the non-dimensionalized width of a boundary layer. The function could be expanded in the vicinity of $\mathbf{x}$ using a Taylor series

$$f(\mathbf{x}, \varepsilon) = f(\mathbf{x}) + f'(\mathbf{x})\varepsilon + \cdots + \frac{f^n}{n!}(\mathbf{x})\varepsilon^n + \cdots$$

However, such an expansion is limited to functions that are differentiable in the region around $\mathbf{x}$. In the asymptotic approach we generalize the series expansion in terms of a sequence of elementary functions of ε, denoted by $\Gamma_i(\varepsilon)$:

$$f(\mathbf{x}, \varepsilon) = \gamma_0(\mathbf{x}) + \gamma_1(\mathbf{x})\Gamma_1(\varepsilon) + \cdots + \gamma_n(\mathbf{x})\Gamma_n(\varepsilon) + \cdots \tag{3.20}$$

The basis functions of the expansion, $\Gamma_i(\varepsilon)$, are referred to as **gauge functions**. The expansion is unique for any given set of gauge functions, but one has considerable latitude in their choice. Commonly, successive powers of $\varepsilon(1, \varepsilon, \varepsilon^2, \ldots, \varepsilon^n, \ldots)$ are used. In most cases the gauge functions are dictated by the physics of the problem. For example, as we shall see in Chapter 4, an expansion in terms of powers of $\sqrt{\varepsilon}$, where $\varepsilon = 1/\omega$, is needed in order to honor the physics of diffusion. It is certainly worth spending some effort finding a good set of gauge functions.

A key feature of asymptotic expansions is that as ε approaches a limiting value, that we will take to be 0, the expansion is increasingly dominated by the earliest terms. That is, we require that all the gauge functions of an asymptotic sequence satisfy

$$\lim_{\varepsilon \to 0} \frac{\Gamma_{n+1}(\varepsilon)}{\Gamma_n(\varepsilon)} = 0 \tag{3.21}$$

for all n. We can write the relationship (3.21) between the two functions Γ_n and Γ_{n+1} succinctly through the use of the **order symbol** small o (Panton, 2005, p. 351).

The statement that the function $\Gamma_{n+1}(\varepsilon)$ is of smaller order than the function $\Gamma_n(\varepsilon)$ is written

$$\Gamma_{n+1}(\varepsilon) = o\left[\Gamma_n(\varepsilon)\right] \tag{3.22}$$

and is equivalent to Equation (3.21). One can then define an equivalence relationship between functions in terms of the order symbol o. For example, we say that a function is **asymptotically equal** to the first three terms of its expansion, which we write as:

$$f(\mathbf{x}, \varepsilon) \sim \gamma_0(\mathbf{x}) + \gamma_1(\mathbf{x})\Gamma_1(\varepsilon) + \gamma_2(\mathbf{x})\Gamma_2(\varepsilon), \tag{3.23}$$

if they differ by a term of order $o\left[\Gamma_2(\varepsilon)\right]$.

There is the practical issue of calculating the coefficients of the asymptotic expansion. The utility of a Taylor series expansion rests on the expression of the coefficients in terms of the derivatives of the function, as in Equation (3.12). The coefficients of an asymptotic expansion can be defined in terms of the limit as $\varepsilon \to 0$:

$$\begin{aligned}
\gamma_0(\mathbf{x}) &= \lim_{\varepsilon \to 0} f(\mathbf{x}, \varepsilon) \\
\gamma_1(\mathbf{x}) &= \lim_{\varepsilon \to 0} \frac{f(\mathbf{x}, \varepsilon) - \gamma_0(\mathbf{x})}{\Gamma_1(\varepsilon)} \\
\gamma_2(\mathbf{x}) &= \lim_{\varepsilon \to 0} \frac{f(\mathbf{x}, \varepsilon) - \gamma_0(\mathbf{x}) - \gamma_1(\mathbf{x})\Gamma_1(\varepsilon)}{\Gamma_2(\varepsilon)}
\end{aligned} \tag{3.24}$$

and similarly for the coefficients of the higher-order terms. In most applications the coefficients are determined by substituting the asymptotic expansion into the governing equations and considering terms of the same order. We will illustrate this a number of times in the chapters that follow.

To some degree we have had to give up on our hope that the asymptotic series converges as we sum a large number of terms. Part of the problem is that we are trying to represent functions that may contain discontinuities in one or more of its derivatives. Thus, the function possibly contains singularities, hence we might expect a divergent series. However, because we are most interested in an accurate approximation in a few terms, summing a large number of terms is somewhat burdensome, the divergence of the series is usually not an issue. The ultimate divergence or convergence of a series is a property of its tail (Dingle, 1973, p. 4), the terms far out in the summation, and is not related to the usefulness of the initial terms in approximating a function (Panton, 2005, p. 352).

To motivate the use of an asymptotic expansion, consider the ways in which one might represent a special function, the modified Bessel function of order zero, $I_0(x)$

(Abramowitz and Stegun, 1972, p. 374). A straight-forward power series expansion can be written in the form

$$I_0(x) = 1 + \frac{\frac{1}{4}x^2}{(1!)^2} + \frac{(\frac{1}{4}x^2)^2}{(2!)^2} + \frac{(\frac{1}{4}x^2)^3}{(3!)^2} + \frac{(\frac{1}{4}x^2)^4}{(4!)^2} + \ldots \tag{3.25}$$

and is absolutely convergent. For large values of x we will need many terms in this series in order to approximate $I_0(x)$ to sufficient accuracy. This is because the convergence of the power series involves a race between the large terms in the numerator, powers of x, and the large terms in the denominator, the factorials that depend upon the number of terms used. The terms are not negligible until the denominator becomes much larger than the numerator. As noted by Panton (2005, p. 352), the situation becomes even more difficult when the terms in the expansion alternate in sign, as in a Taylor series expansion of the Bessel function $J_0(x)$. Such a long series can also be difficult to evaluate numerically because one must operate with some very large numbers. The representation (3.25) provides very little insight into the behavior of $I_0(x)$ for large values of x. In contrast, the leading term of asymptotic expansion of $I_0(x)$, in the limit of large values of x, is given by (Abramowitz and Stegun, 1972, p. 378)

$$I_0(x) \sim \frac{e^x}{\sqrt{2\pi x}}. \tag{3.26}$$

From the asymptotic representation (3.26) we can get an accurate estimate of $I_0(x)$ for large values of x in an efficient manner. Furthermore, the expression provides some insight into the behavior of $I_0(x)$ for large values of x. In the applications presented in this book the computation of successive terms in the asymptotic series will require some effort, thus it is important to obtain an accurate solution with just a few terms. The applicability of asymptotic techniques to a wide range of problems, involving dispersion, dissipation, heterogeneity, and non-linearity is also important.

Exercise 3.1. Compare the results of the power series (Equation 3.25) and the asymptotic solution (Equation 3.26) representation of the modified Bessel function, $I_0(x)$ for $x = 10$. Notice that you need up to 9 terms in the power series to obtain equivalent results.

Exercise 3.2. The asymptotic approach can be used to identify the dominant behavior of the local solution of a differential equation. Recall that the asymptotic solution is a representation of the behavior of the solution of a differential equation near a point (possibly singular) as an infinite series. The first term of the series is called the *Leading Behavior* and the most rapidly changing component of the first term is known as the *Controlling Factor* (Bender and Orszag, 1978). Let us find the controlling factor of the following differential equation.

$$y' = \sqrt{x}\,y, x \to 0^+.$$

Solution: Start with a test function, usually in the form of an exponential (Bender and Orszag, 1978), $y(x) = e^{S(x)}$. Then, $y' = S'e^S$ and after substituting into the differential equation, $S'e^S = \sqrt{x}e^S$ and $S' = \sqrt{x}$ as $e^S \neq 0$. After integration, $S(x) \sim \frac{2}{3}x^{3/2}$. Hence the controlling factor of the solution will be $e^{\frac{2}{3}x^{3/2}}$.

Notice that in the above quadrature we did not include the constant of integration, hence the use of $\sim$ rather than the equal sign. In order to obtain the leading behavior, we need to include the constant of integration and substitute back into the differential equation and carry out additional steps. We will not dwell upon the details here as there are excellent references that specifically address the asymptotic solution of differential equations (Bender and Orszag, 1978). Instead, in the section that follows we focus on two asymptotic solutions of special interest to fluid flow in porous media.

3.4.2 Examples from the frequency domain

The discussion in Section 3.4.1 on asymptotic series and solutions provides some idea of the flexibility that is available to us and the broad outlines of what is needed for a useful power series expansion. However, that discussion is too general to provide a practical tool for solving problems. For example, the asymptotic series was in terms of a generic set of gauge functions and we require a little more guidance concerning useful choices. We noted the importance of the physics of the problem in determining the specific form of the gauge functions. For a propagating front, the frequency domain representation (3.19) follows from a Taylor's series expansion (3.11) and the application of the Fourier transform. For physical problems specified by one or more linear partial differential equations with spatially varying coefficients, the slightly more general asymptotic power series

$$U(\mathbf{x}, \omega) = e^{-\sigma(\mathbf{x},\omega)} \frac{1}{(-i\omega)^{\alpha}} \sum_{n=0}^{\infty} \frac{A_n(\mathbf{x})}{(-i\omega)^{\beta n}}, \tag{3.27}$$

where α and β are real constants, is sufficiently flexible to cover a wide range of applications. Note that the phase function σ now has a general frequency dependence, and the numerical values of α and β are determined by the particular set of governing equations. These values may be determined, as in the Chapter 4 on diffusive processes, by an analysis of the equations associated with the simplest possible version of the medium. For example, we may utilize an analytic solution for a homogeneous, isotropic, and infinite medium and consider its asymptotic representation in order to determine α and β.

Some appreciation of the generality of the series (3.27) follows from the consideration of two specific cases. First, consider the case in which $\sigma(\mathbf{x}, \omega) = i\omega\tau(\mathbf{x})$, $\alpha = 1$, and $\beta = 1$, resulting in the asymptotic expansion

$$U(\mathbf{x}, \omega) = -\frac{e^{-i\omega\tau(\mathbf{x})}}{i\omega} \sum_{n=0}^{\infty} \frac{A_n(\mathbf{x})}{(-i\omega)^n},$$

where $\tau(\mathbf{x})$ is an arbitrary function of $\mathbf{x}$. In the limit as $\omega \to \infty$ the first term is the most significant and we have

$$U(\mathbf{x}, \omega) \sim -\frac{e^{-i\omega\tau}}{i\omega} A_0(\mathbf{x})$$

and upon applying the inverse Fourier transform to return to the time-domain, we have the solution

$$u(\mathbf{x}, t) = A_0(\mathbf{x}) H[t - \tau(\mathbf{x})].$$

The solution represents a propagating step with an amplitude that varies with $\mathbf{x}$ and a propagation time of $\tau(\mathbf{x})$. This functional form is characteristic of solutions to the wave equation (3.9), that is the function given in (3.10).

A very different solution results if we consider the asymptotic series specified by $\sigma(\mathbf{x}, \omega) = -\sqrt{-i\omega}\tau(\mathbf{x})$, $\alpha = 0$, and $\beta = 1/2$,

$$U(\mathbf{x}, \omega) = e^{-\sqrt{-i\omega}\tau(\mathbf{x})} \sum_{n=0}^{\infty} \frac{A_n(\mathbf{x})}{(\sqrt{-i\omega})^n},$$

encountered in Chapter 4. In the limit of high-frequency, $\omega \to \infty$, the asymptotic solution is again dominated by the zeroth-order ($n = 0$) term

$$U(\mathbf{x}, \omega) \sim e^{-\sqrt{-i\omega}\tau} A_0(\mathbf{x}).$$

This approximation transforms back to the function

$$u(\mathbf{x}, t) = \mathcal{F}^{-1}\left[e^{-\sqrt{-i\omega}\tau(\mathbf{x})} A_0(\mathbf{x})\right] = A_0(\mathbf{x}) \frac{\tau(\mathbf{x})}{2\sqrt{\pi t^3}} e^{-\tau^2(\mathbf{x})/4t} H(t)$$

in the time-domain, a form similar to the analytical solution of the diffusion equation (Crank, 1975) for a homogeneous medium. Thus, the series (3.27) can encompass many different functional forms, representing a wide range of physical processes.

3.5 Characteristics and trajectories

The central topic of this book is the development of trajectory-based solutions for modeling of fluid flow and transport and the use of these solutions for imaging flow properties. Here we will describe an algorithm for treating general first-order partial differential equations that results in such a trajectory-based solution. Note that the technique will not be applied directly to the governing equations introduced in Chapter 2 and discussed in more detail in later chapters. Such equations are typically second-order partial differential equations. Rather, we will use this method to solve equations emerging from the asymptotic approach. An application to the wave equation should illustrate how it all works.

Consider a first-order partial differential equation written in the form

$$F(\mathbf{x}, t, \mathbf{p}, q, u) = 0, \tag{3.28}$$

where $u(\mathbf{x}, t)$ is a function of the three spatial coordinates, $\mathbf{x}$, and time, t. The additional variables $\mathbf{p}$ and q are associated with the derivatives of u. Specifically, $\mathbf{p}$ is a vector of the spatial derivatives of u, given by $\mathbf{p} = \nabla u$, while q is the time derivative of $u(\mathbf{x}, t)$

$$q = \frac{\partial u}{\partial t}.$$

We have chosen to distinguish the space and time variables in Equation (3.28) and in our derivation. We could just as easily extend the vector of independent variables, $\mathbf{x}$, to include time, simplifying the presentation a bit. Note that there is no restriction on the nature of the partial differential equation (3.28), it may represent a linear, a quasi-linear, or a fully non-linear equation.

In this section we will show that the first-order partial differential equation (3.28) is equivalent to a system of ordinary differential equations, the **characteristic equations**. The ordinary differential equations will define a solution along a trajectory, that is along a curve in both space,

$$\mathbf{x} = \mathbf{x}(s) \tag{3.29}$$

and time

$$t = t(s) \tag{3.30}$$

parameterized by the variable s. The nature of the parameter s is left open at this point, it may represent distance along the trajectory, or position with respect to a reference point, or time itself.

There are at least two ways to derive the system of ordinary differential equations defining the trajectory and the solution $u(\mathbf{x}, t)$. First, there is a geometrical approach that arose early in the study of partial differential equations (Courant and Hilbert, 1962). Second, there is a rather straight-forward analytic derivation (Logan, 2008). We present the second approach, as it should be easier for the reader to follow and also easier to generalize. In our derivation we invoke Einstein's summation convention, introduced in Box 2.2 of Chapter 2, where repeated indices indicate summation from 1 to 3.

The differential equations specifying the path and the solution along the path must be related to Equation (3.28) in some fundamental fashion. In particular, the equations will most likely be determined by the explicit dependence of F on its arguments, as characterized by its partial derivatives with respect to these quantities. In order to have a complete system of ordinary differential equations defining the trajectory, we need to consider the derivatives of all of the independent and dependent variables $\mathbf{x}$, t, $\mathbf{p}$, q, and u, along the trajectory parameterized by s. While it might seem obvious that we should start with the derivatives of $\mathbf{x}(s)$ and $t(s)$, let us begin with the derivatives of $\mathbf{p}(s)$ and $q(s)$ instead. As we shall see very shortly, this choice will quickly pay off in terms of a useful set of differential equations defining the trajectory. If we compute these derivatives explicitly and write them in terms of the derivatives of $\mathbf{x}$ and t with respect to s, we have

$$\frac{dp_i}{ds} = \frac{d}{ds}\left(\frac{\partial u}{\partial x_i}\right) = \frac{dx_j}{ds}\frac{\partial^2 u}{\partial x_j \partial x_i} + \frac{dt}{ds}\frac{\partial^2 u}{\partial t \partial x_i} \tag{3.31}$$

$$\frac{dq}{ds} = \frac{d}{ds}\left(\frac{\partial u}{\partial t}\right) = \frac{dx_j}{ds}\frac{\partial^2 u}{\partial x_j \partial t} + \frac{dt}{ds}.\frac{\partial^2 u}{\partial t^2} \tag{3.32}$$

In order to relate these derivatives to the dependence of $F(\mathbf{x}, t, \mathbf{p}, q, u)$ on its arguments, we take the partial derivatives of Equation (3.28) with respect to the spatial variables and time,

$$\frac{\partial F}{\partial x_i} + \frac{\partial F}{\partial u}\frac{\partial u}{\partial x_i} + \frac{\partial F}{\partial p_j}\frac{\partial^2 u}{\partial x_j \partial x_i} + \frac{\partial F}{\partial q}\frac{\partial^2 u}{\partial t \partial x_i} = 0 \tag{3.33}$$

$$\frac{\partial F}{\partial t} + \frac{\partial F}{\partial u}\frac{\partial u}{\partial t} + \frac{\partial F}{\partial p_j}\frac{\partial^2 u}{\partial x_j \partial t} + \frac{\partial F}{\partial q}\frac{\partial^2 u}{\partial t^2} = 0, \tag{3.34}$$

where $i = 1, 2, 3$ signifies the component of $\mathbf{x}$. Comparing the set of Equations (3.31) and (3.32) with Equations (3.33) and (3.34) we note similarities in the sets of mixed partial derivatives. In fact, if we equate their coefficients:

$$\frac{dx_j}{ds} = \frac{\partial F}{\partial p_j} \tag{3.35}$$

$$\frac{dt}{ds} = \frac{\partial F}{\partial q} \tag{3.36}$$

then the two right-most terms in Equations (3.33) and (3.34) are identical to the right-hand-sides of Equations (3.31) and (3.32). Then we can replace the two right-most terms in Equations (3.33) and (3.34) by dp_i/ds and dq/ds, respectively. Solving the two resulting equations for these derivatives, we have

$$\frac{dp_i}{ds} = -\frac{\partial F}{\partial x_i} - \frac{\partial F}{\partial u}\frac{\partial u}{\partial x_i} \tag{3.37}$$

$$\frac{dq}{ds} = -\frac{\partial F}{\partial t} - \frac{\partial F}{\partial u}\frac{\partial u}{\partial t}. \tag{3.38}$$

Equations (3.35), (3.36), (3.37), and (3.38) constitute eight ordinary differential equations relating the derivatives of the independent variables along the trajectory to derivatives of F. Counting the number of independent and dependent variables that are functions of s in Equation (3.28), $\mathbf{x}$, t, $\mathbf{p}$, q, and u, we see that we shall need nine ordinary differential equations to determine all relevant quantities. If we consider the expression for the total derivative of the dependent variable u, we obtain one additional equation

$$\begin{aligned}\frac{du}{ds} &= \nabla u \cdot \frac{d\mathbf{x}}{ds} + \frac{\partial u}{\partial t}\frac{dt}{ds} \\ &= \mathbf{p} \cdot \frac{d\mathbf{x}}{ds} + q\frac{dt}{ds}.\end{aligned} \tag{3.39}$$

Replacing the derivatives of $\mathbf{x}$ and t with respect to s by the partial derivative of F given in Equations (3.35) and (3.36), provides the ninth and final ordinary differential equation,

$$\frac{du}{ds} = p_j\frac{\partial F}{\partial p_j} + q\frac{\partial F}{\partial q}. \tag{3.40}$$

Taken together, Equations (3.35), (3.36), (3.37), (3.38), and (3.40) define a trajectory-based solution for any first-order partial differential Equation (3.28), be it linear, quasi-linear, or non-linear. These equations constitute the characteristic equations associated with Equation (3.28) and the trajectories are known as the **characteristic curves** (Logan, 2008, p. 87). In Box 3.1 we summarize this set of equations, writing them as vector equations, for future reference.

Box 3.1 **The ray equations.**

We summarize the equations for the trajectory, $\mathbf{x}(s)$, $t(s)$, the derivatives of the dependent variables, $\mathbf{p}(s)$, $q(s)$, and the dependent variable $u(\mathbf{x}, t)$ along the trajectory.

$$\frac{d\mathbf{x}}{ds} = \nabla_p F$$
$$\frac{dt}{ds} = \frac{\partial F}{\partial q}$$
$$\frac{d\mathbf{p}}{ds} = -\nabla F - \mathbf{p}\frac{\partial F}{\partial u}.$$
$$\frac{dq}{ds} = -\frac{\partial F}{\partial t} - q\frac{\partial F}{\partial u}.$$
$$\frac{du}{ds} = \mathbf{p} \cdot \nabla_p F + q\frac{\partial F}{\partial q},$$

where ∇_p signifies that the partial derivatives are with respect to the components of the vector $\mathbf{p}$ rather than $\mathbf{x}$. Later in the book we shall need these results.

3.6 Trajectory-based modeling: the wave equation

Let us apply the approaches from the last two sections to the wave equation (3.5) transformed into the frequency domain [see Equation (3.6)],

$$c^2\nabla \cdot \nabla U + \nabla c^2 \cdot \nabla U + \omega^2 U = 0, \tag{3.41}$$

where $U(\mathbf{x}, \omega)$ is the Fourier transform of $u(\mathbf{x}, t)$. Our goal is to derive a semi-analytic solution to Equation (3.41) that is valid at high frequencies, an asymptotic solution. To this end, we seek a solution in the form of the asymptotic power series (3.27). Our first task is to determine the values of the exponents α and β to use in the expansion (3.27). At this stage, rather than pursuing an analysis of the wave equation in an infinite, isotropic, and homogeneous medium, we shall simply take advantage of previous work along those lines (Aki and Richards, 1980a; Chapman, 2004). An expansion of the form

$$U(\mathbf{x}, \omega) = e^{-i\omega\sigma(\mathbf{x})} \sum_{n=0}^{\infty} \frac{A_n(\mathbf{x})}{(-i\omega)^n} \tag{3.42}$$

has been found to produce a useful asymptotic solution of the wave equation in the limit of high frequency (Kline and Kay, 1965).

Upon substituting the high-frequency asymptotic representation (3.42) into Equation (3.41), we can apply the vector differential operators to each term. For example, ∇ applied to $U(\mathbf{x}, \omega)$ gives

$$\nabla U(\mathbf{x}, \omega) = \left[-i\omega\nabla\sigma \sum_{n=0}^{\infty} \frac{A_n(\mathbf{x})}{(-i\omega)^n} + \sum_{n=0}^{\infty} \frac{\nabla A_n(\mathbf{x})}{(-i\omega)^n} \right] e^{-i\omega\sigma(\mathbf{x})}. \tag{3.43}$$

The operator $\nabla\cdot$ may be applied to the expression (3.43) to obtain $\nabla \cdot \nabla U$, but we will leave that computation as an exercise. We simply present the results of the substitution of these operators, and the asymptotic representation of U, into the governing equation,

$$
\begin{aligned}
&(-i\omega)^2 \left[c^2\nabla\sigma \cdot \nabla\sigma - 1\right] \sum_{n=0}^{\infty} \frac{A_n}{(-i\omega)^n} \\
&+(-i\omega) \left[\nabla \cdot \left(c^2\nabla\sigma\right) \sum_{n=0}^{\infty} \frac{A_n}{(-i\omega)^n} + 2c^2 \sum_{n=0}^{\infty} \frac{\nabla\sigma \cdot \nabla A_n}{(-i\omega)^n}\right] \\
&+ \sum_{n=0}^{\infty} \frac{\nabla \cdot (c^2\nabla A_n)}{(-i\omega)^n} = 0
\end{aligned}
\tag{3.44}
$$

where we have grouped the terms according to the power of the coefficient $-i\omega$. In Equation (3.44) we have divided out the common factor $e^{-i\omega\sigma}$ with the understanding that it does not vanish.

For large values of ω we are interested in the terms of highest order in ω, or conversely, terms of the lowest order in $1/\omega$. Considering terms of order $(-i\omega)^2$, contained in the top line of Equation (3.44), we obtain the equation

$$\left[c^2\nabla\sigma \cdot \nabla\sigma - 1\right] A_0 = 0. \tag{3.45}$$

Assuming that A_0, the zeroth-order amplitude of the propagating disturbance does not vanish, Equation (3.45) implies that

$$\nabla\sigma \cdot \nabla\sigma - \frac{1}{c^2} = 0. \tag{3.46}$$

This is the eikonal equation, a differential equation for the phase function $\sigma(\mathbf{x})$. As we shall see, there are many roads to the eikonal equation, and it appears in several contexts when modeling propagating disturbances. The eikonal equation is a first order, non-linear, partial differential equation of the form (3.28),

$$F(\mathbf{x}, \mathbf{p}) = \mathbf{p} \cdot \mathbf{p} - \frac{1}{c^2(\mathbf{x})} = 0, \tag{3.47}$$

where $\mathbf{p} = \nabla\sigma$. We may use the method described in Section 3.5 to derive a trajectory-based solution for $\sigma(\mathbf{x})$. To apply that approach we first write $\mathbf{x}$ and $\mathbf{p}$ as functions of position, s, along the trajectory. Then, Equations (3.36) and (3.38) in Section 3.5 [also the first and third equations written in Box 3.1] give a self-contained system of ordinary differential equations for the trajectory $\mathbf{x}(s)$ and the gradient of σ, $\mathbf{p}(s)$,

$$\frac{d\mathbf{x}}{ds} = 2\mathbf{p}$$
$$\frac{d\mathbf{p}}{ds} = -\nabla\left(\frac{1}{c^2}\right).$$

These ordinary differential equation may be integrated using available numerical methods (Press et al., 1992). The differential equations indicate that the trajectory $\mathbf{x}(s)$ is determined by the spatial variation of $1/c^2$, along with the boundary conditions, primarily the source and receiver locations and the angle at which the trajectory leaves the source, the **take-off angle**.

The phase σ is determined by the characteristic equations in Section 3.5, and may be found by a one-dimensional integration along the trajectory $\mathbf{x}(s)$. In particular, the final ray equation in Box 3.1, for $F(\mathbf{x}, \mathbf{p})$ given by Equation (3.47) produces an equation for σ

$$\frac{d\sigma}{ds} = 2\mathbf{p}\cdot\mathbf{p} = \frac{2}{c^2(\mathbf{x})},$$

where we have used the eikonal equation (3.47) to substitute for $\mathbf{p}\cdot\mathbf{p}$. The first ray equation allows us to substitute for $\mathbf{p}$ in Equation (3.47) giving

$$\frac{d\mathbf{x}}{ds}\cdot\frac{d\mathbf{x}}{ds} = \left(\frac{dx}{ds}\right)^2 = \frac{4}{c^2}$$

Upon integrating the expression for the derivative of σ along the trajectory, we arrive at

$$\sigma(\mathbf{x}) = \int \frac{2}{c^2(\mathbf{x})} ds = \int \frac{dx}{c(\mathbf{x})},$$

an explicit expression for $\sigma(\mathbf{x})$, defined along the trajectory.

The high-frequency asymptotic approach works well for equations containing coefficients that vary spatially but do not depend upon time. However, when the coefficients are time-dependent then one must contend with products of time-varying functions. The Fourier transform of such product terms becomes a convolution in the frequency domain, coupling the various frequencies (Bracewell, 2000). This coupling also occurs for non-linear problems, when the coefficients are functions of the dependent variables. Thus, the high-frequency asymptotic approach, while still applicable, can become cumbersome in the face of non-linearity and in the presence of time-dependent coefficients. For this reason, in Section 3.7 we will discuss a class of techniques that, while certainly applicable in the frequency domain, work just as well in the time domain.

Exercise 3.3. The transport equation

$$\frac{\partial c(\mathbf{x},t)}{\partial t} + \mathbf{v}(\mathbf{x}) \cdot \nabla c(\mathbf{x},t) = 0$$

describes the motion of a passive tracer, with concentration denoted by c(x,t), advected in a steady velocity field $\mathbf{v}(\mathbf{x})$ with no diffusion or dispersion. Show that a high-frequency asymptotic solution leads to the equation

$$\mathbf{v}(\mathbf{x}) \cdot \nabla \tau(\mathbf{x}) = 1,$$

for the phase $\tau(\mathbf{x})$.

Hint: Consider a frequency domain solution of the form

$$C(\mathbf{x},\omega) = -\frac{e^{-i\omega\tau(\mathbf{x})}}{i\omega} \sum_{k=0}^{\infty} \frac{A_k(\mathbf{x})}{(i\omega)^k}.$$

and insert the solution into the governing differential equation. Then equate the coefficient of the highest power of ω [lowest power of $1/\omega$] to zero to obtain the phase equation. The phase equation describes the travel time of the tracer particle in the flow field. We elaborate on this in Chapter 5.

Exercise 3.4. The diffusion equation

$$\phi(\mathbf{x})\mu c_t \frac{\partial p(\mathbf{x},t)}{\partial t} - \nabla \cdot [k(\mathbf{x})\nabla p(\mathbf{x},t)] = 0$$

governs the evolution of pressure in a permeable medium. Show that a high-frequency asymptotic solution leads to the equation

$$\nabla\tau(\mathbf{x}) \cdot \nabla\tau(\mathbf{x}) = \frac{\phi(\mathbf{x})\mu c_t}{k(\mathbf{x})}$$

for the phase,

Hint: Consider a frequency domain solution of the form

$$P(\mathbf{x},\omega) = -e^{\sqrt{-i\omega}\,\tau(\mathbf{x})} \sum_{k=0}^{\infty} \frac{A_k(\mathbf{x})}{(\sqrt{-i\omega})^k}.$$

Insert this solution into the governing equation and equate the coefficient of the highest power of $\sqrt{-i\omega}$ to zero, as discussed in the previous exercise. The equation for the phase describes the travel time of the peak of a pressure pulse for an impulse source. We discuss this further in Chapter 4.

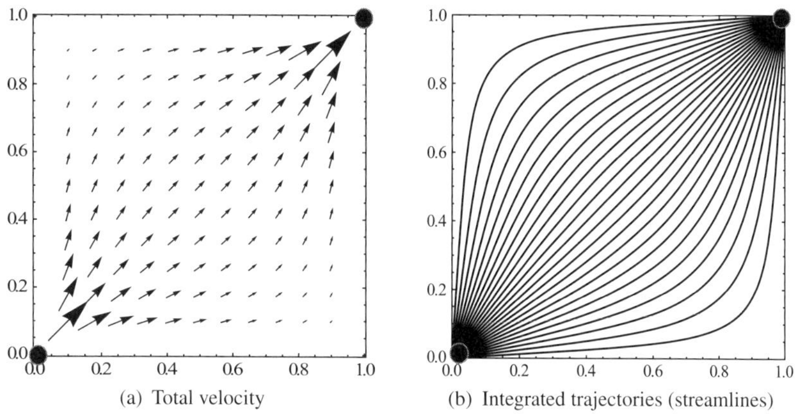

Figure 3.4 The velocity field and the resulting characteristics (streamlines) for an injector (bottom left) and producer (top right) configuration with no-flow boundaries.

Exercise 3.5. Consider the convective transport equation describing the motion of a passive tracer in a steady velocity field $\mathbf{v(x)} = (v_x, v_y, v_z)$, with no diffusion or dispersion,

$$\frac{\partial c(\mathbf{x}, t)}{\partial t} + \mathbf{v}(\mathbf{x}) \cdot \nabla c(\mathbf{x}, t) = 0.$$

Show that the characteristics are given by integrating the velocity field,

$$\frac{dx}{v_x} = \frac{dy}{v_y} = \frac{dz}{v_z} = dt.$$

Also, show that the concentration $c(\mathbf{x}, t)$ does not change along the characteristics.

Hint: The differential equation can be written in the form

$$F(\mathbf{x}, t, \mathbf{p}, q, c) = q + \mathbf{v} \cdot \mathbf{p} = 0.$$

Use the results in Box 3.1 to derive the characteristics equations. The integration of the velocity field leads to characteristics or streamlines along which tracer particles travel. We discuss the streamline-based transport calculations in detail in Chapter 5. Figure 3.4 is an example of the velocity distribution $\mathbf{v}(\mathbf{x})$ in a two-dimensional flow field and the resulting streamlines, locally tangent to the velocity field.

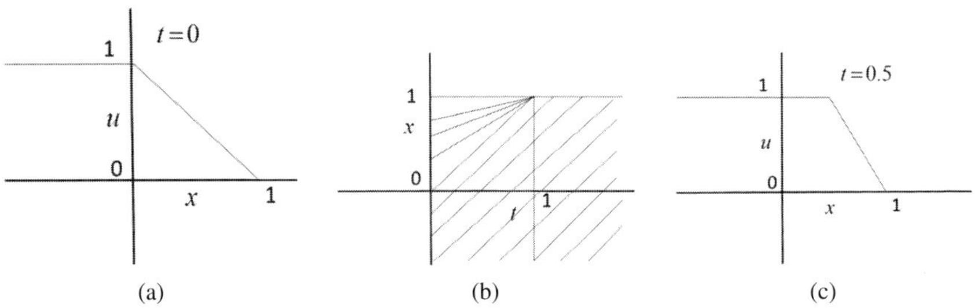

Figure 3.5 Inviscid Burger equation: (a) Initial solution at $t = 0$. (b) Characteristics associated with the inviscid Burger equation. (c) Solution at $t = 0.5$.

Exercise 3.6. Consider the inviscid Burger equation

$$\frac{\partial u}{\partial t} + u\frac{\partial u}{\partial x} = 0.$$

where $u(x,t)$ represents the density along a one-dimensional path. With the initial data given as follows (Trangenstein, 1988)

$$u(x,0) = \begin{cases} 1, & x \leq 0 \\ 1-x, & 0 < x < 1 \\ 0, & 1 \leq x \end{cases}$$

derive the equation for the characteristics (trajectories) and show that along the characteristics u is constant.

Hint: Write the equation as

$$F(x,t,p,q,u), = q + up$$

where $p = \partial p/\partial x$ and utilize the ray equations in Box 3.1. The initial conditions and the characteristics are shown in the Figure 3.5. Notice that the characteristics are straight lines that intersect whenever $1 < x < t$. Because u is constant along the characteristics, the solution $u(x,t)$ can be obtained by tracing back along the characteristics to the initial data, until the characteristics intersect. The intersection of characteristics leads to the formation of shocks and discontinuities as discussed in Chapter 6 in the context of multiphase flow problems.

Exercise 3.7. Compute the operator $\nabla \cdot \nabla U$ where ∇U is given by the asymptotic expression (3.43).

3.7 Multiple scale asymptotics

3.7.1 Overview

One of our goals is the imaging of the internal structure of an object using external observations. While we might know the major features, such as large-scale faults and/or interfaces, the body is assumed to be heterogeneous, with variations in properties between these boundaries. This is particularly true for flow properties, such as permeability, because they can vary significantly within a known reservoir. Typically, given remote and sparse observations one cannot expect to resolve small-scale heterogeneity. Rather, the observations only constrain large-scale variations in internal properties. Therefore, it makes sense to develop efficient modeling techniques, tailored to a medium containing smoothly varying heterogeneity between known boundaries. In this section we shall develop an asymptotic approach for this very case. In the class of techniques described here, we introduce the idea of transforming the independent variables, $\mathbf{x}$ and t, in conjunction with an asymptotic series representation of the dependent variable(s). Such an approach has a long history, starting with the early work of Poincaré (1886) [republished in 1992] and Lindstedt (1883). The significant advantage of the approach, as it relates to the subject of this book, is that it is applicable to problems involving both heterogeneity and nonlinearity.

Because we will illustrate this approach with applications in almost every chapter, our discussion here will be brief and expository in nature. However, we will end this section with an application of the technique to the wave equation, providing a concrete illustration of the procedure. Finally, our primary applications will involve the calculation and interpretation of arrival times related to flow, transport, and associated geophysical monitoring. Therefore, we will not spend much time discussing the calculation of amplitudes here, or in any of the other chapters.

We are interested in disturbances propagating in a medium with smoothly varying heterogeneity. Hence, we can assume that, away from boundaries, the coefficients in the equation(s) governing the transient behavior depend upon slowly varying coordinates. Put another way, if the disturbance is characterized by a length scale l, then the heterogeneity is assumed to vary over a larger length scale $L \gg l$. The ratio of these two length scales defines a parameter

$$\varepsilon = \frac{l}{L}.$$

For a smoothly varying medium or a relatively sharp front, ε is a small number, much smaller than 1. Many non-linear processes, such as multiphase fluid flow, give rise to self-sharpening fronts, similar to shock waves in gas dynamics (Courant and Friedrichs, 1948; Chorin and Marsden, 1993). Our assumptions are reasonable when modeling the propagation of such fronts in a medium with smoothly varying attributes. For sharp boundaries, such as interfaces between layers, the assumptions will no longer hold. However, such discontinuities are typically treated as boundary conditions in the modeling. Similar considerations enter into ray-based modeling and geometrical optics (Kline and Kay, 1965). The presence of boundaries between regions with smoothly varying properties leads to conditions on trajectories crossing the interfaces, such as Snell's law (Chapman, 2004, p. 201).

Many of the properties of the propagating front, such as its amplitude, are modulated by the heterogeneity. Therefore, it makes sense to rewrite the governing equations in terms of the new variables

$$\mathbf{X} = \varepsilon \mathbf{x} \tag{3.48}$$

characterized by the scale parameter $\varepsilon \ll 1$. A similar expression holds for the time t and the rescaled time $T = \varepsilon t$. For non-linear processes, $\mathbf{X}$ and T might signify the length and time scales over which the effects of the non-linearity become significant. Non-linear processes may require a more flexible relationship between the two scales, in its most general form we may write this relationship as

$$\mathbf{X} = \chi\left(\mathbf{x}, \varepsilon\right) \tag{3.49}$$

where χ is a specified function. For example, for quasi-linear problems involving dispersion and dissipation the relationship between the independent space and time variables is:

$$X = \varepsilon^{a}\left(x - \lambda t\right)$$

and

$$T = \varepsilon^{a+1} t$$

where a is determined by the length scale of the non-linearity and the length scale of the dispersion and λ is determined by the propagation velocity associated with the linearized system (Debnath, 2005, p. 604). Such a transformation was developed for solitary wave propagation in a plasma and is known as the **Gardner–Morikawa transformation** (Gardner and Morikawa, 1960). In the discussion that follows we shall work exclusively with the relationship (3.48). However, its possible generalization to (3.49) should be kept in mind.

The underlying idea is the existence of a localized faster variation, in space and/or time, associated with a transient propagating disturbance, contrasting with much slower variations in the surrounding medium (Maslov and Omel'yanov, 2001, p. 3). If we adopt a coordinate system advancing with the wave front, then we can relate the approach to that of boundary layer theory (Cole and Kevorkian, 1963; Kevorkian and Cole, 1996). In order to position ourselves on the front as it propagates, we shall need a measure of the front position, such as the surface $S(\mathbf{x}, t)$ [Equation (3.1)] from Section 3.1, or its spatial component $\sigma(\mathbf{x})$ [Equation (3.2)]. We will generally represent the location and propagation of the front by the **phase function** $\theta(\mathbf{x}, t)$. The phase function will be a critical component of our treatment of a propagating disturbance. To highlight its importance, the phase is explicitly included in the expression for the disturbance:

$$u(\mathbf{x}, t) = U(\mathbf{X}, T, \theta) \tag{3.50}$$

where $\mathbf{X}$ depends upon $\mathbf{x}$ and ε through the definition (3.48) or (3.49). Thus, ultimately u depends upon $\mathbf{x}$ and t. However, during intermediate steps in our application of the method of multiple scales, we shall treat U as a function of the variables $\mathbf{X}$, T, and θ.

Note that the introduction of a distinct phase function is similar to writing the solution in a complex or polar form

$$u(\mathbf{x}, t, \theta) = A_0(\mathbf{x}, t) \exp i\theta(\mathbf{x}, t) \tag{3.51}$$

with an amplitude function A_0 and a distinct phase function θ. Such a representation is assumed in the linear **WKB method** (Bender and Orszag, 1978) and in the high-frequency asymptotic method given previously [see Equation (3.19)]. However, we cannot always assume the simple exponential form (3.51) because, as noted by Luke (1966) and Miura and Kruskal (1974), it does not usually suffice for solutions of non-linear equations. Even linear problems may require a more general form for zeroth-order solution, for example in an area where there is intense focusing (Ludwig, 1966). We need to allow for a more general dependence on the phase, as given in (3.50), particularly when we consider non-linear problems. The function θ determines the variation of $U(\mathbf{X}, T, \theta)$ across the front, which is assumed to occur over a short distance and over a small time span. As such, it is a rapidly changing quantity that depends upon $\mathbf{x}$ and t (Whitham, 1974, p. 494; Jeffrey and Kawahara, 1982, p. 146). As noted by Luke (1966), the idea of the representation (3.50) is to include the relatively fast local oscillations by including the variable θ while the dependence upon $\mathbf{X}$ and T take care of the slower variations.

Due to the dependence of $U(\mathbf{X}, T, \theta)$ on $\theta(\mathbf{x}, t)$ there is both an explicit and an implicit dependence upon the independent variables. The derivatives of $u(\mathbf{x}, t)$, reflecting this dual dependence are given by

$$\begin{aligned}\frac{\partial u}{\partial x_i} &= \frac{\partial X_i}{\partial x_i}\frac{\partial U}{\partial X_i} + \frac{\partial \theta}{\partial x_i}\frac{\partial U}{\partial \theta} \\ &= \varepsilon\frac{\partial U}{\partial X_i} + \frac{\partial \theta}{\partial x_i}\frac{\partial U}{\partial \theta},\end{aligned} \tag{3.52}$$

where we have used the relationship (3.48) between the two coordinate systems. A similar expression exists for the derivative with respect to t.

The final task is to represent the dependence of $U(\mathbf{X}, T, \theta)$ upon ε. Most commonly the relationship is expanded in a power series in ε, that we write in the somewhat generalized form,

$$U(\mathbf{X}, T, \theta) = \sum_{n=0}^{\infty} \varepsilon^{\beta n+\alpha} A_n(\mathbf{X}, T, \theta). \tag{3.53}$$

In fact, this form is almost universally adopted in applications of the technique to non-linear, dispersive, diffusive, and heterogeneous media. Just as in the high-frequency asymptotic expansion (3.27), the constants α and β are determined by the physics of the problem and $A_n(\mathbf{X}, T, \theta)$ are the successive amplitude corrections. The amplitude corrections account for such complications as spatial variations in material properties and/or non-linearity. Note the similarity of this expansion to the high-frequency asymptotic approximation (3.27) if we take

$$\varepsilon = -\frac{1}{i\omega}. \tag{3.54}$$

3.7.2 *Illustration via an application to the wave equation*

As a simple illustration, consider the wave equation 3.5 introduced at the beginning of this chapter

$$c^2\nabla \cdot \nabla u + \nabla c^2 \cdot \nabla u - \frac{\partial^2 u}{\partial t^2} = 0. \tag{3.55}$$

The first order of business involves writing the differential operators in Equation (3.55) in terms of the slow variables and the phase, then rewriting the derivatives, as in Equation (3.52). For example, the gradient of $u(\mathbf{x}, t) = U(\mathbf{X}, T, \theta)$ is given by

$$\nabla U = \varepsilon \overline{\nabla} U + \nabla\theta \frac{\partial U}{\partial \theta} \tag{3.56}$$

where the overbar indicates that the partial derivatives in the gradient operator are with respect to the slow variables $\mathbf{X}$. Expanding all differential operators in Equation (3.55) in this fashion, we have the expression

$$\begin{aligned} &\varepsilon^0 \left[c^2\nabla\theta \cdot \nabla\theta \frac{\partial^2 U}{\partial \theta^2} - \left(\frac{\partial \theta}{\partial t}\right)^2 \frac{\partial^2 U}{\partial \theta^2} \right] \\ &+ \varepsilon^1 \left[c^2\overline{\nabla} \cdot \left(\nabla\theta \frac{\partial U}{\partial \theta}\right) + c^2\nabla\theta \cdot \overline{\nabla}\left(\frac{\partial U}{\partial \theta}\right) + \overline{\nabla} c^2 \cdot \nabla\theta \frac{\partial U}{\partial \theta} - 2\frac{\partial \theta}{\partial t}\frac{\partial^2 U}{\partial \theta \partial T} \right] \\ &+ \varepsilon^2 \left[c^2\overline{\nabla} \cdot \overline{\nabla} U + \overline{\nabla} c^2 \cdot \overline{\nabla} U - \frac{\partial^2 U}{\partial T^2} \right] = 0. \end{aligned} \tag{3.57}$$

In this equation we have grouped terms according to the power of ε that results from rewriting the derivatives. Now we substitute the asymptotic power series expansion (3.53),

$$U(\mathbf{X}, T, \theta) = \sum_{n=0}^{\infty} \varepsilon^n A_n(\mathbf{X}, T, \theta). \tag{3.58}$$

where we have taken $\alpha = 0$, and $\beta = 1$, into Equation (3.57) and consider terms of various orders in ε. Because $\varepsilon \ll 1$ only the first few terms are important. In fact, our interest in the trajectories and the travel times means that we can concentrate on terms of order zero.

Terms of order $\varepsilon^0 \sim 1$

Substituting the power series expansion (3.58) for $U(\mathbf{X}, T, \theta)$ into Equation (3.57) and considering terms of order $\varepsilon^0 \sim 1$, gives

$$\left[c^2 \nabla\theta \cdot \nabla\theta - \left(\frac{\partial\theta}{\partial t}\right)^2\right] \frac{\partial^2 A_0}{\partial\theta^2} = 0. \tag{3.59}$$

Equation (3.59) is equivalent to the highest-order term of the high-frequency asymptotic expansion (3.44). Assuming that the second derivative of A_0 with respect to θ does not vanish, we have a non-linear partial differential equation for $\theta(\mathbf{x}, t)$

$$c^2 \nabla\theta \cdot \nabla\theta - \left(\frac{\partial\theta}{\partial t}\right)^2 = 0. \tag{3.60}$$

of the same form as Equation (3.46).

Because Equation (3.60) is a first order, non-linear, partial differential equation for $\theta(\mathbf{x}, t)$, we may use the techniques from Section 3.5 to derive a trajectory-based solution. However, we can simplify our task if we take advantage of the mathematical structure of Equation (3.60). Specifically, the terms are homogeneous in the sense that they are both products of first derivatives. Additionally, the equation is separable in the sense that all the spatially dependent terms and derivatives can be isolated from the time derivatives. The symmetry and separability of the equation suggests that the solution should also share those properties. One form of the solution that comes to mind is similar to Equation (3.2) for the propagating front

$$\theta(\mathbf{x}, t) = t - \sigma(\mathbf{x}). \tag{3.61}$$

Substituting this for θ in Equation (3.60) produces a reduced equation for $\sigma(\mathbf{x})$

$$\nabla\sigma \cdot \nabla\sigma - \frac{1}{c^2} = 0. \tag{3.62}$$

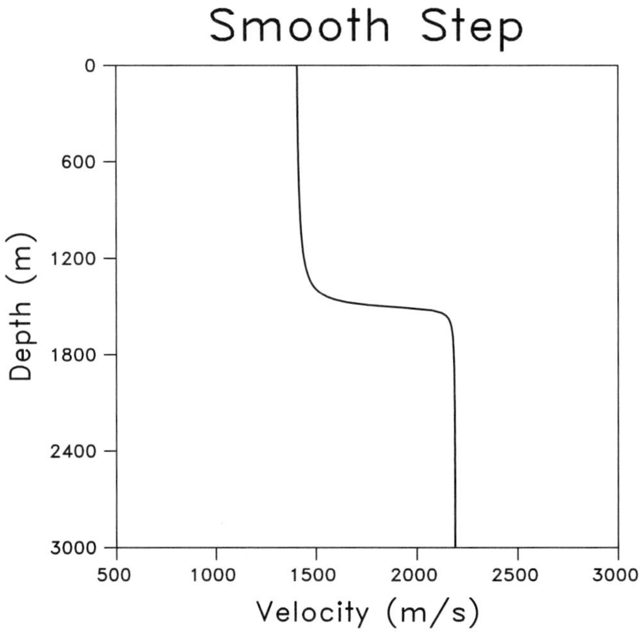

Figure 3.6 The velocity function $c(\mathbf{x})$ used for an example calculation of the ray trajectories and for the solution of the eikonal equation. The velocity only varies as a function of the depth (z). The steep jump in velocity simulates an interface at $z = 1500m$.

This is the eikonal equation, first encountered in Section 3.6. So now we are in familiar territory, faced with a non-linear, first-order, partial differential equation that may be written in the form of Equation (3.28)

$$F(\mathbf{x}, \mathbf{p}) = \mathbf{p} \cdot \mathbf{p} - \frac{1}{c^2} = 0 \tag{3.63}$$

where $\mathbf{p} = \nabla\sigma(\mathbf{x})$. We may therefore follow the procedures outlined in Section 3.5 and derive the ray equations equivalent to Equation (3.63). Since we have already done the work in our treatment of the wave equation using a high-frequency asymptotic expansion (Section 3.6), we merely write down the appropriate ray equations

$$\frac{d\mathbf{x}}{ds} = 2\mathbf{p} \tag{3.64}$$

$$\frac{d\mathbf{p}}{ds} = -\nabla\left(\frac{1}{c^2}\right) \tag{3.65}$$

$$\frac{d\sigma}{ds} = \frac{2}{c^2}. \tag{3.66}$$

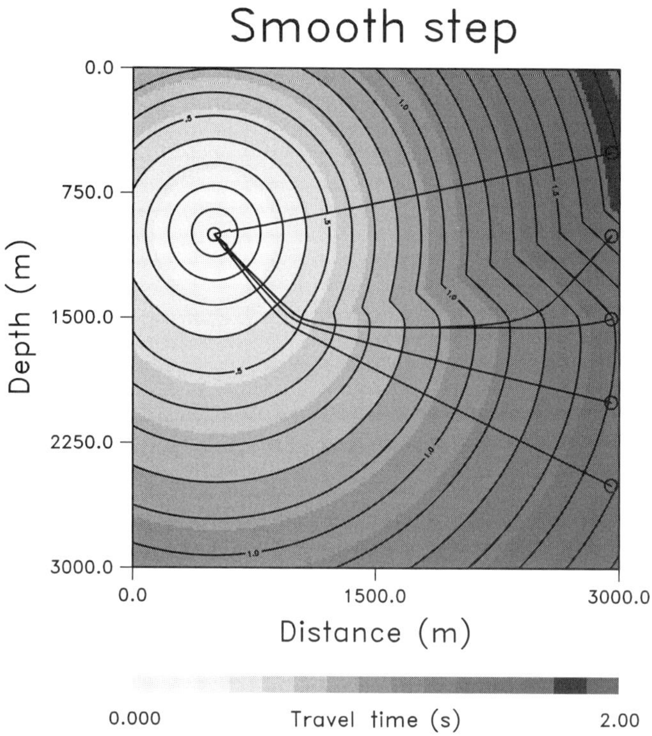

Figure 3.7 Raypaths from a seismic source to five receivers at the rightmost end of the simulation grid. The velocity model for the computations is plotted in Figure 3.6. Travel time contours, obtained from a finite-difference solution of the eikonal equation, are also shown.

We can also derive an explicit expression for the spatial component of the phase function as an integral along the trajectory $\mathbf{x}(s)$

$$\sigma(\mathbf{x}_o) = \int_{\mathbf{x}_s}^{\mathbf{x}_o} \frac{dx}{c(\mathbf{x})} \tag{3.67}$$

using the first and final ray equations, along with the eikonal equation (3.63). The ray equations and the expression for $\sigma(\mathbf{x})$ provide one method for solving the eikonal equation. There are also efficient numerical techniques, primarily finite-differences, for the direct solution of the eikonal equation 3.62. For illustration purposes, in Figure 3.7, we plot a finite-difference solution of the eikonal equation, for the velocity model in Figure 3.6. We will briefly touch upon the finite difference solution of the eikonal equation in Chapter 4. However, this is a subject worthy of an extended discussion. In fact there are excellent texts available, describing the various numerical approaches (Sethian, 1999; Osher and Fedkiw, 2003).

Terms of order ε

The terms of order ε, contained in the first two lines of Equation (3.57), can be treated in a similar fashion. Because we are mainly interested in calculating the trajectories and travel times we will leave this treatment as an exercise. The result is a transport equation governing the evolution of the amplitude $A_0(\mathbf{X}, T, \theta)$. In some of the applications that follow, particularly in Chapter 4, we illustrate the treatment of the terms of order ε.

Exercise 3.8. Derive an expression for the zeroth-order amplitude function $A_0(\mathbf{X}, \omega)$ by considering terms of order ε in Equation (3.57).

Construction of the zeroth-order solution

The lowest-level approximation is provided by the zeroth-order term in the asymptotic series (3.58)

$$U(\mathbf{X}, T, \theta) = A_0(\mathbf{X}, T, t - \sigma(\mathbf{x})), \tag{3.68}$$

where we have inserted the separable form (3.61) for the phase. This is a propagating disturbance, similar in form to the moving fronts encountered early on in this chapter.

3.7.3 Summary

We have presented an asymptotic and trajectory-based method for solving the governing equations related to flow and transport through porous media. The asymptotic method leads to the eikonal equation, an equation for the phase or the travel time of the front. The approach presented here is general and encompasses wave propagation, convective transport, and also diffusive processes such as transient pressure transmission in permeable media. The asymptotic method is particularly well-suited for partial differential equations with spatially varying coefficients, a situation typically encountered in Earth sciences. A major advantage of the method is that it allows partitioning of the modeling into a phase or propagation time calculation and amplitude calculations. Much of the geologic heterogeneity is embedded in the phase. This fact can be used to our advantage in simplifying the amplitude calculations. Thus, semi-analytic methods can significantly speed up computations. The visual and intuitive nature of the calculations greatly facilitates analysis and interpretation of field data. We will illustrate the power and utility of the approach using a variety of applications in the chapters that follow.

4

Equations in diffusion form

4.1 Introduction

Diffusive processes are often assumed to have very little in common with solutions of the hyperbolic partial differential equations governing wave-like propagation. Indeed, solutions of the diffusion equation decay in an exponential fashion with distance from the source of a disturbance. Thus, to propagate an observable disturbance away from a source it is often necessary to continuously supply mass or energy. For example, in order to propagate an observable pressure disturbance from an injection point to a distant location it is typically necessary to introduce fluid at a constant rate into the porous medium for a sustained period of time. However, as indicated in Chapter 1, transient solutions to the diffusion equation do indeed propagate, and, as will be shown in this chapter, one may derive an expression governing the phase of such a propagating disturbance. The governing equation for the phase is a Hamilton–Jacobi equation, identical in form to the eikonal equation encountered earlier. However, the interpretation and the effect of the phase function is significantly different from that of a propagating wave, in keeping with the very different nature of diffusion. In this chapter we shall consider one approach for developing a trajectory-based solution for diffusion equation, a high-frequency asymptotic expansion. We shall illustrate the approach and its utility with a few applications, including diffusion tomography.

4.2 A high-frequency asymptotic solution

Here we derive a high-frequency asymptotic solution for the diffusion equation. The phrase 'high-frequency' is not commonly associated with diffusive processes. However, a high-frequency component is present in the early rapid variations of pressure due to the onset of fluid injection. Also, the notion of what constitutes a high frequency is relative to the temporal variation of the background field, which can be gradual. Admittedly, the approximation does not have the long reach associated with processes such as elastic wave propagation and is often restricted to a region surrounding the source. Still, the range over which we can utilize a transient pressure disturbance is far enough to provide a powerful tool for imaging flow properties between a given source and receiver. Furthermore, our asymptotic analysis allows us to draw a connection between the well developed modeling of wave propagation and transient solutions to the diffusion equation. Such a connection is useful in understanding coupled processes and equations of a mixed nature: those that are neither entirely wave-like nor completely diffusive in character.

4.2.1 Motivation and derivation of the asymptotic expression

Let us begin with an equation in diffusion form containing coefficients that vary with position

$$\nabla \cdot \lambda(\mathbf{x}) \nabla p = S(\mathbf{x}) \frac{\partial p}{\partial t}. \tag{4.1}$$

Such an equation governs the evolution of fluid pressure, $p(\mathbf{x}, t)$, in a porous medium. In fact, Equation (4.1) is a variation of Equation (2.94) for a heterogeneous medium characterized by spatially varying properties, $\lambda(\mathbf{x})$ and $S(\mathbf{x})$. The fluid mobility $\lambda(\mathbf{x})$ is the ratio of medium permeability $k(\mathbf{x})$ and fluid viscosity μ,

$$\lambda(\mathbf{x}) = \frac{k(\mathbf{x})}{\mu}.$$

The storage coefficient, $S(\mathbf{x})$ is the product of porosity, $\varphi(\mathbf{x})$ and total system compressibility (rock plus fluid) c_t,

$$S(\mathbf{x}) = \varphi(\mathbf{x}) c_t.$$

Because we are interested in a high-frequency solution we first apply the Fourier transform

$$P(\mathbf{x}, \omega) = \frac{1}{\sqrt{2\pi}} \int_{-\infty}^{\infty} p(\mathbf{x}, t) e^{i\omega t} dt \tag{4.2}$$

(Bracewell, 2000; Debnath, 2005), introduced earlier in Chapter 3. The Fourier transform will take us into the frequency domain and makes the dependence of the pressure $P(\mathbf{x}, \omega)$ upon frequency, ω, explicit. It is the natural setting in which to derive a series solution for P in terms of powers of ω. Note that we are adopting a sign convention that is often used in wave propagation within the Earth (Aki and Richards, 1980a, p. 130). Applying the integral transform (4.2) to the diffusion Equation (4.1) results in

$$\nabla \cdot \lambda(\mathbf{x}) \nabla P + i\omega S(\mathbf{x}) P = 0. \tag{4.3}$$

To extract the high-frequency component we now consider a power series representation of the pressure. As noted in the introduction to the high-frequency asymptotic approximation in Chapter 3, we motivate the form of our power series solution by first considering solutions for a homogeneous medium. This was the approach taken by Virieux et al. (1994) in their study of diffusive electromagnetic imaging. Thus, consider Equation (4.3) when $\lambda(\mathbf{x})$ and $S(\mathbf{x})$ are constants λ and S, respectively

$$\lambda \nabla \cdot \nabla P + i\omega SP = 0. \tag{4.4}$$

For a homogeneous medium Equation (4.4) may display some degree of symmetry depending upon the initial or boundary conditions. The source of the disturbance may be treated both as an initial condition and as a boundary condition. Thus, the symmetry of the source will determine the overall symmetry of the solution. For example, when we consider a line source, such as injection in a long well, in a homogeneous medium, the problem and hence the solution has cylindrical symmetry. In that case it makes sense to write the divergence operator $\nabla \cdot \nabla$ in cylindrical coordinates (r, θ, z) (Davis, 1967, p. 193), where r is the radial distance from line of symmetry, θ is the angle about the line of symmetry, and z is the distance along the line of symmetry. In cylindrical coordinates Equation (4.4) becomes

$$\frac{\lambda}{r}\frac{\partial}{\partial r}\left(r\frac{\partial P}{\partial r}\right) + \frac{\lambda}{r^2}\frac{\partial^2 P}{\partial \theta^2} + \lambda\frac{\partial^2 P}{\partial z^2} + i\omega SP = 0. \tag{4.5}$$

Due to the symmetry of the problem, we expect the pressure field to only depend upon the radial (r) coordinate, and Equation (4.5) is an ordinary differential equation for $P(r, \omega)$:

$$\frac{\lambda}{r}\frac{d}{dr}\left(r\frac{dP}{dr}\right) + i\omega SP = 0, \tag{4.6}$$

where the frequency ω is treated as a parameter and not as a variable. Multiplying this equation by r^2 and expanding the derivative gives

$$r^2\frac{d^2P}{dr^2} + r\frac{dP}{dr} + i\omega r^2\frac{S}{\lambda}P = 0, \tag{4.7}$$

a modified form of Bessel's equation (Dingle, 1973, p. 281).

Box 4.1 **A series solution of the modified Bessel's equation**

Our derivation of an asymptotic representation of the modified Bessel function (4.23) begins with the governing Equation (4.7). After dividing through by r and defining

$$\beta = -i\omega\frac{S}{\lambda}, \tag{4.8}$$

the result is the equation

$$r\frac{d^2P}{dr^2} + \frac{dP}{dr} - r\beta P = 0. \tag{4.9}$$

The coefficient β can be removed by writing Equation (4.9) in terms of the new variable

$$s = \sqrt{\beta} r \tag{4.10}$$

with the corresponding derivative

$$\frac{ds}{dr} = \sqrt{\beta}. \tag{4.11}$$

Writing the governing equation in terms of s gives

$$\frac{d^2P}{ds^2} + \frac{1}{s}\frac{dP}{ds} - P = 0. \tag{4.12}$$

We shall be interested in a power series solution of Equation (4.12). However, this equation has singularities when s approaches zero and ∞. We shall assume that we are a sufficient distance away from the origin that r is not small relative to $\sqrt{\beta}$. Thus, we only need to consider the singularity at ∞. In order to capture the behavior in the limit of large s, we first consider the Equation (4.12) when ω and hence β and s are large [see (4.10)]. For large values of s, Equation (4.12) is well approximated by

$$\frac{d^2P}{ds^2} - P = 0, \tag{4.13}$$

which has the stable solution

$$P(s,\omega) \sim e^{-s}. \tag{4.14}$$

Now consider a solution of the full Equation (4.12), of the form

$$P(s,\omega) = e^{-s}p^{-}(s). \tag{4.15}$$

Substituting the form of the solution (4.15) into Equation (4.12) gives an equation for $p^{-}(s)$

$$s\frac{d^2p^{-}}{ds^2} + (1-2s)\frac{dp^{-}}{ds} - p^{-} = 0. \tag{4.16}$$

We shall consider the function $p^{-}(s)$ in the form of a power series with a leading order term s^{σ} present, representing the singular behavior as s approaches ∞:

$$p^{-}(s) = s^{\sigma}\sum_{n=0}^{\infty} A_n s^{-n}. \tag{4.17}$$

In order to discover the leading order behavior of the solution, that is the value of σ, we substitute a trial solution of the form

$$p^{-}(s) = s^{\sigma}, \tag{4.18}$$

which is just the s component of the zeroth term in the series (4.17), into Equation (4.16) which produces the equation

$$\frac{\sigma^2}{s} - (2\sigma + 1) = 0. \tag{4.19}$$

For large values of s, Equation (4.19) implies that

$$\sigma = -\frac{1}{2} \tag{4.20}$$

and so the full solution is of the form

$$P(s, \omega) = e^{-s} s^{-1/2} \sum_{n=0}^{\infty} A_n s^{-n}. \tag{4.21}$$

We can determine the coefficients A_i by substituting the full solution (4.21) into the governing Equation (4.12). Because

$$s = \sqrt{\beta} r = \sqrt{-i\omega}\sqrt{\frac{S}{\lambda}} r,$$

the asymptotic power series solution of the original problem, the modified Bessel Equation (4.9), is of the form

$$P(r, \omega) = \frac{1}{\sqrt{\sqrt{-i\omega}}} e^{-\sqrt{-i\omega}\sqrt{S/\lambda} r} \sum_{n=0}^{\infty} A_n \left(\frac{\sqrt{\lambda/S}}{r}\right)^{n+1/2} \left(\sqrt{-i\omega}\right)^{-n} \tag{4.22}$$

As noted by Virieux et al. (1994) for an impulsive source, that is a source given by a scaled delta function $\delta(t)$ located on the source line, Equation (4.7) has the solution

$$P(r, \omega) = 2K_0 \left[r\sqrt{-i\omega}\sqrt{\frac{S}{\lambda}} \right] \tag{4.23}$$

where $K_0(x)$ is the modified Bessel function of order zero (Press et al., 1992). In Box 4.1 we apply a power series expansion method directly to the modified Bessel equation to derive an asymptotic representation of (4.23).

As pointed out by Virieux et al. (1994), and demonstrated in Box 4.1, the solution of the modified Bessel equation, specifically the modified Bessel function of order zero, has the asymptotic expansion:

$$P(r, \omega) = e^{-\sqrt{-i\omega}\sqrt{S/\lambda} r} \sum_{n=0}^{\infty} A_n \left(\frac{\sqrt{\lambda/S}}{r}\right)^{n+1/2} \left(\sqrt{-i\omega}\right)^{-n} \tag{4.24}$$

for an impulsive line source. In the expression (4.24) we have suppressed the $1/\sqrt{\sqrt{-i\omega}}$ term because, as noted by Virieux et al. (1994), this term depends upon the spatial distribution of the source and thus depends upon the type of source. Typically, the solution associated with an impulse is convolved with a source function so such factors appear in this later step. We should note that solutions of the form (4.24) also hold for planar (one-dimensional) and point (three-dimensional) sources (Virieux et al., 1994).

For a medium with smoothly varying properties $\lambda(\mathbf{x})$ and $S(\mathbf{x})$ are slowly varying functions of $\mathbf{x}$. If we let λ and S in Equation (4.24) depend upon position, then we can write the asymptotic expansion in the generalized form

$$P(\mathbf{x}, \omega) = e^{-\sqrt{-i\omega}\sigma(\mathbf{x})} \sum_{n=0}^{\infty} \frac{A_n(\mathbf{x})}{\left(\sqrt{-i\omega}\right)^n} \tag{4.25}$$

as in Virieux et al. (1994) and Vasco et al. (2000). In this expression we have grouped all terms depending upon the spatial variables into the two functions $\sigma(\mathbf{x})$ and $A_n(\mathbf{x})$. The functions $A_n(\mathbf{x})$ represent amplitude terms associated with the various orders of $1/\sqrt{-i\omega}$. The quantity $\sigma(\mathbf{x})$, known as the **phase** or **pseudo-phase**, shares characteristics with the phase of a propagating wave. However, there are important differences because the frequency-dependent multiplier is $\sqrt{-i\omega}$. The $\sqrt{\omega}$ frequency-dependence leads to a very different temporal variation in the solution, once the inverse Fourier transform is applied and we return to the time-domain. To see this, consider the inverse Fourier transform applied to the zeroth-order term of the series (4.25)

$$p(\mathbf{x}, t) = \mathcal{F}^{-1}\left[e^{-\sqrt{-i\omega}\sigma(\mathbf{x})} A_0(\mathbf{x})\right] = A_0(\mathbf{x}) \frac{\sigma(\mathbf{x})}{2\sqrt{\pi t^3}} e^{-\sigma^2(\mathbf{x})/4t} H(t), \tag{4.26}$$

(Virieux et al., 1994), similar in form to the solution of the diffusion equation in a homogeneous medium. Note that the phase or pseudo-phase, $\sigma(\mathbf{x})$, appears both as an argument of the exponential, and as a multiplier of the amplitude. Thus, the solution decays in an exponential fashion as a function of the square of the phase. This decay will likely dominate the variation of the amplitude $A_0(\mathbf{x})$ with distance from the source.

To summarize the results of this section, an asymptotic power series representation of the solution for a homogeneous medium suggests a frequency-domain solution for a heterogeneous medium of the form (4.25). At high-frequencies, the series should be dominated by terms of the lowest order in $1/\sqrt{-i\omega}$. For the zeroth-order ($n = 0$) expression there are two unknown functions $\sigma(\mathbf{x})$ and $A_0(\mathbf{x})$. In the section that follows, we shall determine these functions and explore some of their properties.

4.2.2 Determination of the phase and amplitude functions

The functions $\sigma(\mathbf{x})$ and $A_n(\mathbf{x})$ are found in a rather straight-forward fashion. We simply substitute the asymptotic series (4.25)

$$P(\mathbf{x},\omega) = e^{-\sqrt{-i\omega}\sigma(\mathbf{x})} \sum_{n=0}^{\infty} \frac{A_n(\mathbf{x})}{\left(\sqrt{-i\omega}\right)^n}$$

into the frequency-domain version of the governing equation, Equation (4.3), written in the expanded form

$$\lambda(\mathbf{x})\nabla \cdot \nabla P + \nabla\lambda(\mathbf{x}) \cdot \nabla P + i\omega S(\mathbf{x})P = 0, \tag{4.27}$$

and consider terms of various orders in $1/\sqrt{-i\omega}$. Most of our effort is spent in computing the spatial derivatives of the asymptotic series. For example, the gradient of $P(\mathbf{x},\omega)$ is given by

$$\nabla P(\mathbf{x},\omega) = e^{-\sqrt{-i\omega}\sigma(\mathbf{x})} \left[-\sqrt{-i\omega}\nabla\sigma(\mathbf{x}) \sum_{n=0}^{\infty} \frac{A_n(\mathbf{x})}{\left(\sqrt{-i\omega}\right)^n} + \sum_{n=0}^{\infty} \frac{\nabla A_n(\mathbf{x})}{\left(\sqrt{-i\omega}\right)^n} \right]. \tag{4.28}$$

We can gain some insight into the phase function, $\sigma(\mathbf{x})$, by considering the expression (4.28) for high frequencies, when it is accurately represented by the zeroth-order term. For large values of ω expression (4.28) may be approximated as

$$\nabla P(\mathbf{x},\omega) = -e^{-\sqrt{-i\omega}\sigma(\mathbf{x})}\sqrt{-i\omega}A_0(\mathbf{x})\nabla\sigma(\mathbf{x}) \tag{4.29}$$

and the phase gradient is oriented parallel to the pressure gradient but pointing in the direction of decreasing pressure. For lower frequencies the spatial variation of the amplitude function $A_0(\mathbf{x})$ has an influence on the pressure gradient.

Returning to the governing Equation (4.27), the term $\nabla \cdot \nabla P$ is obtained by applying the divergence operator to the expression (4.28). We shall not show this step but leave it as an exercise for the reader. The complete expression for Equation (4.27) is given by

$$\begin{aligned} &-(i\omega)\ [\lambda\nabla\sigma \cdot \nabla\sigma - S] \sum_{n=0}^{\infty} \frac{A_n}{\sqrt{-i\omega^n}} \\ &-(\sqrt{-i\omega}) \left[\nabla \cdot (\lambda\nabla\sigma) \sum_{n=0}^{\infty} \frac{A_n}{\sqrt{-i\omega^n}} + 2\lambda \sum_{n=0}^{\infty} \frac{\nabla\sigma \cdot \nabla A_n}{\sqrt{-i\omega^n}} \right] \\ &+ \sum_{n=0}^{\infty} \frac{\nabla \cdot (\lambda\nabla A_n)}{\sqrt{-i\omega^n}} = 0 \end{aligned} \tag{4.30}$$

where we have factored out the exponential multiplier, grouped terms according to their order in $\sqrt{-i\omega}$ and suppressed the dependence of λ, S, σ, and A_n upon $\mathbf{x}$ in order to streamline the expression. This equation contains an infinite number of terms of various orders in $\sqrt{-i\omega}$. Since we are interested in high-frequency solutions, we shall assume that the frequency ω is large and that terms of the highest order in $\sqrt{-i\omega}$, or conversely, of the lowest order in $1/\sqrt{-i\omega}$ are the most significant. In the sub-sections that follow we shall consider terms of order two and one in $\sqrt{-i\omega}$, the two highest orders in Equation (4.30).

Terms of order $(\sqrt{-i\omega})^2 \sim -i\omega$: an expression for the phase

Collecting terms of the highest order, those containing $(\sqrt{-i\omega})^2 \sim -i\omega$, leads to the equation

$$[\lambda \nabla\sigma \cdot \nabla\sigma - S] A_0 = 0. \tag{4.31}$$

Assuming that the amplitude $A_0(\mathbf{x})$ does not vanish, Equation (4.31) provides a partial differential equation for the phase $\sigma(\mathbf{x})$:

$$\nabla\sigma \cdot \nabla\sigma - \frac{1}{\kappa(\mathbf{x})} = 0, \tag{4.32}$$

known as the **eikonal equation**. In Equation (4.32) we have defined the quantity

$$\kappa(\mathbf{x}) = \frac{\lambda(\mathbf{x})}{S(\mathbf{x})} \tag{4.33}$$

which is the diffusivity (de Marsily, 1986, p. 162) associated with the original diffusion Equation (4.1). From Equation (4.32) we see that the phase depends upon the diffusivity of the medium. Also, from Equation (4.29) we have seen that for high frequencies the phase gradient is in the direction of the pressure gradient. We will make use of this property in Section 4.3.2 to solve for pressure distribution. Greater physical insight about the phase can be achieved once we construct a zeroth-order approximation to the solution of the diffusion equation as discussed in the next section. For now, it suffices to say that the phase is related to the propagation of the peak of pressure disturbance corresponding to an impulse source.

The eikonal Equation (4.32) is a non-linear, scalar, first-order partial differential equation, a particular example of a Hamilton–Jacobi equation discussed in Chapter 3. It has two attractive features that will help in finding a solution $\sigma(\mathbf{x})$. First, it does not depend upon time or frequency, only upon spatial coordinates. Thus, through the form (4.25) of our asymptotic solution, we have successfully peeled off the frequency dependence. That is, we have isolated a function related to the propagation that does not depend upon frequency and hence is independent of time. Second, the eikonal equation is a first-order scalar differential equation and is therefore amenable to the various methods for solving such equations (Courant and Hilbert, 1962; Debnath, 2005). We shall have more to say about that in a moment. There are also some difficulties associated with the eikonal equation. It is non-linear, so that linear techniques may not be applied. A consequence of this, well known in elastic wave propagation, is the possibility of multi-valued solutions. Also, the coefficient $\kappa(\mathbf{x})$ may be any given function of $\mathbf{x}$ and, in general, we may not take advantage of any particular symmetry.

Because the eikonal Equation (4.32) is a scalar, first-order, partial differential equation we can solve it using numerical methods such as finite differences. In fact there are techniques suited specifically for Hamilton–Jacobi equations such as the eikonal equation (Crandall and Lions, 1983; Crandall et al., 1984; Osher and Fedkiw, 2003). These methods introduce a viscosity term to suppress the multiplicity of solutions, producing a unique solution corresponding to the smallest phase values. The approach was introduced into seismology by Vidale (1988). The technique has advanced further, through a combination of a narrow band methodology, in which changes are made to a set of points in a narrow region around an outwardly propagating front, and a fast heapsort algorithm which finds the grid point with the smallest phase value. The technique, known as the Fast Marching Method, was introduced by Sethian (1996) and is described in greater detail later in this chapter. Such viscosity solutions do not recover the full multi-valued solution. However, one can expend additional effort, such as formulating and solving escape equations (Formal and Sethian, 2002) or by combining the finite difference approach with ray tracing (Benamou, 1996), and compute multi-valued solutions. If we choose to use an eikonal solver such as Fast Marching Method to compute the phase on a numerical grid, we can use the phase as a spatial coordinate to simplify the pressure equation. Specifically, we can define a reduced equation for the pressure along the trajectory and solve it using standard finite difference methods. We discuss this approach in detail in the application section.

An alternative approach is provided by the method of characteristics (Courant and Hilbert, 1962, p. 75) introduced in Chapter 3. The technique starts with the eikonal Equation (4.32) written in the form

$$F(\mathbf{x}, \mathbf{p}) = \mathbf{p} \cdot \mathbf{p} - \frac{1}{\kappa(\mathbf{x})} = 0, \tag{4.34}$$

where we have defined the conjugate momentum vector

$$\mathbf{p} = \nabla\sigma. \tag{4.35}$$

In the method of characteristics we derive a system of ordinary differential equations equivalent to the eikonal Equation (4.32), The solutions are defined upon trajectories, or paths, through the medium, and points along the path are parameterized by s. The trajectory itself is denoted by the position vector $\mathbf{x}(s)$. The characteristic equations associated with a first-order partial differential equation, and given in Box 3.1 in Chapter 3, are

$$\frac{d\mathbf{x}}{ds} = \nabla_{\mathbf{p}} F = 2\mathbf{p} \tag{4.36}$$

$$\frac{d\mathbf{p}}{ds} = -\nabla_{\mathbf{x}} F - \mathbf{p}\frac{\partial F}{\partial \sigma} = \nabla \left(\frac{1}{\kappa(\mathbf{x})} \right), \tag{4.37}$$

where $\nabla_{\mathbf{p}} F$ is the gradient of F with respect to the components of $\mathbf{p}$ and $\nabla_{\mathbf{x}}$ is the standard gradient with respect to the spatial variables. Because there is no explicit dependence upon σ, the partial derivative $\partial F(\mathbf{x}, \mathbf{p})/\partial \sigma$ vanishes in Equation (4.37). These equations comprise a system of ordinary differential equations for the raypath $\mathbf{x}(s)$ and the conjugate momenta along the raypath. Note that the trajectories are completely determined by the function $\kappa(\mathbf{x})$.

In order to obtain a unique solution, the differential Equations (4.36) and (4.37) must be accompanied by a set of boundary conditions. Because the variables are $\mathbf{x}(s)$ and $\mathbf{p}(s)$, the boundary conditions can involve either or both of these quantities at either one or both endpoints of the trajectory. Thus, for example, we can specify the starting point $\mathbf{x}(0)$ and the conjugate momentum vector $\mathbf{p}(0)$ at the initial point of the trajectory, the source point. This brings up the various approaches for solving the coupled system of Equations (4.36) and (4.37). Two conditions, prescribed at the starting point of the trajectory, constitute an **initial value problem** that may be integrated numerically (Press et al., 1992, p. 701). Alternatively, when the boundary conditions are specified at the initial and final points, one has a **two point boundary value problem**. Two point boundary value problems generally require more effort to solve (Press et al., 1992, p. 745). For example, one technique, the shooting method, requires solving a sequence of initial value problems. Two point boundary value problems can have multiple solutions, as trajectories leaving a source at different angles can bend around due to heterogeneity and intersect at the target point.

A complete wavefield is obtained when one has determined all contributing trajectories between a source point and a receiver point. When a single trajectory connects a source and receiver the phase field is single valued. A multivalued phase function results when two more trajectories meet at a point. The advantage of the trajectory-based approach is that it can handle multiple arrivals quite naturally because it develops the solution trajectory by trajectory and does not rely on an underlying regular grid. The disadvantage is that the approach requires more effort to implement numerically than does a straight-forward finite difference solution of the eikonal Equation (4.32). Furthermore, it can be difficult to map the results into a regular grid. One property of a diffusive disturbance is the rapid exponential decay with propagation distance. In practical terms this means that interference from multiple arrivals is likely to be less of a problem due to the much smaller amplitude of contributions from later arrivals that have traveled greater distances.

The third ray equation derived in Section 3.5, see Box 3.1, gives an expression for $\sigma(\mathbf{x})$ along the trajectory $\mathbf{x}(s)$:

$$\frac{d\sigma}{ds} = \mathbf{p} \cdot \nabla_p F$$

or

$$\frac{d\sigma}{ds} = 2\mathbf{p} \cdot \mathbf{p} = \frac{2}{\kappa(\mathbf{x})}. \tag{4.38}$$

In the right-most equality we have used the eikonal Equation (4.34). Using Equation (4.36) to substitute for $\mathbf{p}$, we find that

$$\left(\frac{dx}{ds}\right)^2 = \frac{4}{\kappa(\mathbf{x})} \tag{4.39}$$

or

$$\frac{dx}{ds} = \frac{2}{\sqrt{\kappa(\mathbf{x})}}. \tag{4.40}$$

Integrating Equation (4.38) from a source location $\mathbf{x}_s$ to an observation point $\mathbf{x}_o$ gives

$$\sigma(\mathbf{x}_o) = \int_{\mathbf{x}_s}^{\mathbf{x}_o} \frac{2}{\kappa(\mathbf{x})} ds = \int_{\mathbf{x}_s}^{\mathbf{x}_o} \frac{1}{\sqrt{\kappa(\mathbf{x})}} dx \tag{4.41}$$

upon using Equation (4.40). Thus, by solving the coupled ray Equations (4.36) and (4.37) we obtain the trajectory, $\mathbf{x}(s)$, and by an integration of (4.41) along the trajectory we obtain the phase. As mentioned before, the phase function has an important physical interpretation. The phase is associated with the propagation time of the peak of an impulsive pressure disturbance. We will sometimes refer to the phase function as the 'diffusive time of flight' of a pressure pulse. In Figure 4.1 we illustrate the trajectories for a heterogeneous permeability field. The phase variation or diffusive time of flight along the trajectories are also shown. It is clear that the effects of permeability heterogeneity are embedded in the phase variations. Note that in Equation (4.41), the phase has unit of square root of time which is consistent with the scaling behavior of diffusive processes.

It is interesting to draw a comparison at this point with convective trajectories or streamlines which we will discuss in detail in Chapter 5. Unlike the diffusive trajectories that correspond to the propagation of a pressure pulse, streamlines are a depiction of the velocity field and represent the flowpaths of a passive tracer. Analogous to the diffusive time of flight, we can define a 'convective time of flight' along the streamlines. For comparison purposes, we have also shown the streamlines and time of flight in Figure 4.1. Both the diffusive and the convective time of flight are very useful for characterizing of the flow properties in heterogeneous permeable media, as we will see.

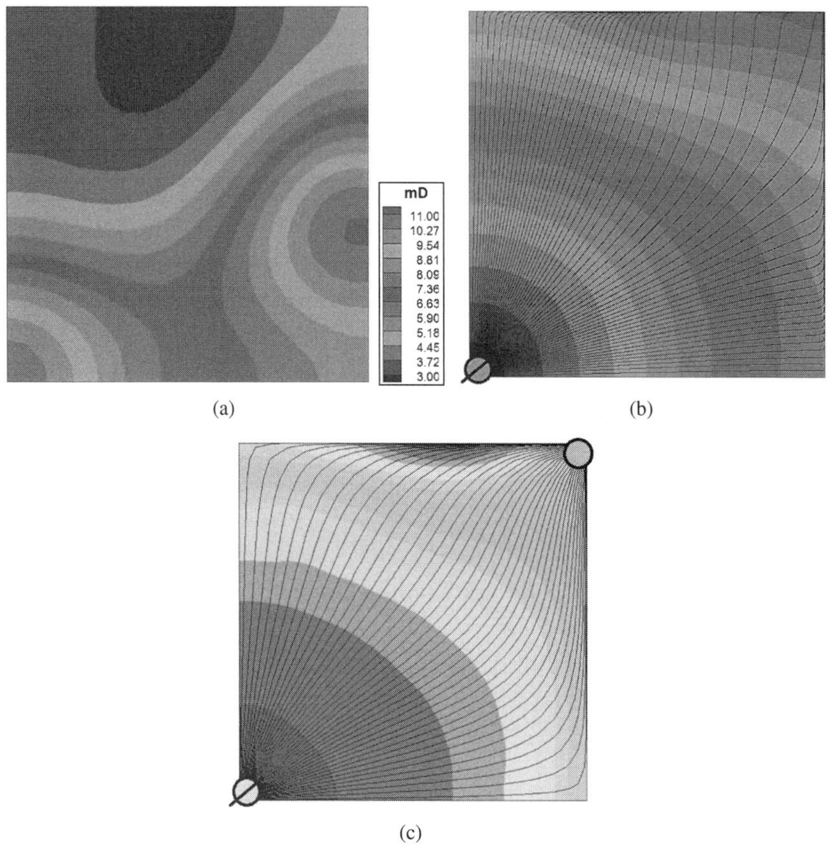

Figure 4.1 An illustration of the spatial variation of the phase function $\sigma(\mathbf{x})$ along trajectories for a heterogeneous permeable medium. (a) Permeability field (b) Trajectories and phase (c) Streamlines and convective time of flight for an injector (bottom left)-producer (top right) configuration. For the colour version, please refer to the plate section.

Terms of order $\sqrt{-i\omega}$: an expression for the amplitude

A complete specification of the zeroth-order approximation, the first term in the series (4.25), requires the amplitude function $A_0(\mathbf{x})$. Isolating the terms of the next highest order in Equation (4.30), those containing $\sqrt{-i\omega}$:

$$(\lambda \nabla \sigma \cdot \nabla \sigma - S) A_1 - \nabla \cdot (\lambda \nabla \sigma) A_0 - 2\lambda \nabla \sigma \cdot \nabla A_0 = 0, \tag{4.42}$$

we arrive at an equation containing the zeroth-order amplitude $A_0(\mathbf{x})$ and its derivative. Because the phase function σ solves the eikonal Equation (4.32), the coefficient of the $A_1(\mathbf{x})$ term vanishes and Equation (4.42) reduces to

$$\nabla \cdot (\lambda \nabla \sigma) A_0 + 2\lambda \nabla \sigma \cdot \nabla A_0 = 0, \tag{4.43}$$

a partial differential equation for $A_0(\mathbf{x})$ known as the **transport equation**. Note, if we multiply Equation (4.43) by A_0 and collect terms, we can write the transport equation in the form of a divergence

$$\nabla \cdot \left(\lambda A_0^2 \nabla \sigma\right) = 0.$$

Assuming that the phase is known, one may solve the transport equation for the amplitude function. Dividing Equation (4.43) by $A_0(\mathbf{x})$, and expanding the first term and dividing by $\lambda(\mathbf{x})$ gives

$$2\nabla\sigma \cdot \nabla \ln A_0 + \nabla\sigma \cdot \nabla \ln \lambda + \nabla \cdot \nabla\sigma = 0, \tag{4.44}$$

a first-order, linear partial differential equation for $\ln A_0(\mathbf{x})$. The coefficients of this differential equation depend upon the components of $\nabla\sigma$ and $\nabla \ln \lambda(\mathbf{x})$. Thus, the amplitude is determined by the gradient of the phase and the gradient of the logarithm of the function $\lambda(\mathbf{x})$.

One may solve for the amplitude $A_0(\mathbf{x})$ using one of several approaches. First, one may discretize and solve the transport Equation (4.44), a linear partial differential equation for the scalar amplitude function, directly using numerical methods such as finite differences (Press et al., 1992). Alternatively, one may integrate the linear differential equation for $\ln A_0(\mathbf{x})$ along the trajectory using methods for ordinary differential equations. Such an integration will require calculating the divergence of the vector field $\nabla\sigma$. This quantity is usually obtained by tracing a suitable number of trajectories adjacent to the desired ray. Alternatively, one can devise equations for the direct calculation of the divergence of $\nabla\sigma$ along the ray. Finally, because the amplitude occurs in only one term in Equation (4.44), one can manipulate the differential equation further and obtain an explicit expression for its value [see Kline and Kay (1965, p. 156), Kravtsov and Orlov (1990, p. 22)]. The exact equation describing the evolution of the amplitude along the trajectory is

$$A_0(r) = A_0(r_0)\sqrt{\frac{\lambda(r_0)}{\lambda(r)}\frac{\sqrt{\sqrt{\kappa(r)}J(r_0)}}{\sqrt{\sqrt{\kappa(r_0)}J(r)}}}, \tag{4.45}$$

where $J(r)$ is the Jacobian, measuring the expansion of the surface area of the front with distance r along the trajectory. This term is related to the divergence of the vector field $\mathbf{p}$ and is obtained in the same fashion, by tracing several neighboring rays and calculating their divergence. This procedure is described in more detail elsewhere (Kravtsov and Orlov, 1990; Chapman, 2004).

4.2.3 Construction and interpretation of the zeroth-order solution

Having obtained the phase, $\sigma(\mathbf{x})$, and amplitude, $A_0(\mathbf{x})$ functions, as described above, we may construct the lowest-order approximation of the solution to the diffusion Equation (4.1). The zeroth-order term in the asymptotic series (4.25) is

$$P_0(\mathbf{x}, \omega) = A_0(\mathbf{x}) \, e^{-\sqrt{-i\omega}\sigma(\mathbf{x})}. \tag{4.46}$$

This expression is useful if we wish to transform data into the frequency domain and interpret quantities in that realm. However, it is much more common to work in the time-domain, requiring the application of the inverse Fourier transform to (4.46). As noted earlier [see Equation (4.26)], the inverse Fourier transform of the expression (4.46) is

$$p(\mathbf{x}, t) = \mathcal{F}^{-1}\left[e^{-\sqrt{-i\omega}\sigma(\mathbf{x})} A_0(\mathbf{x})\right] = A_0(\mathbf{x}) \frac{\sigma(\mathbf{x})}{2\sqrt{\pi t^3}} e^{-\sigma^2(\mathbf{x})/4t} H(t). \tag{4.47}$$

This expression is similar to the analytical solution of the diffusion equation for a homogeneous medium with an impulsive source (Crank, 1975). The main difference is the presence of the spatially varying amplitude and phase functions, $A_0(\mathbf{x})$ and $\sigma(\mathbf{x})$, respectively.

We have a rough understanding of the role of the function $A_0(\mathbf{x})$. It is the amplitude of the pressure change that propagates along the trajectories and is a function of position along the path $\mathbf{x}(s)$. From the differential Equation (4.44) we see that changes in the amplitude $A_0(\mathbf{x})$ are primarily due to the gradient of $\lambda(\mathbf{x})$ and the divergence of the trajectories, as measured by $\nabla \cdot \mathbf{p} = \nabla \cdot \nabla\sigma$. The role of the phase function, $\sigma(\mathbf{x})$, in the evolution of the pressure pulse $p(\mathbf{x}, t)$ is decidedly more complicated. It occurs in two locations in expression (4.47), both as an amplitude term, and as a term in the argument of the exponential.

In an effort to gain some insight into the nature of the phase function and its role in the evolution of the pressure pulse, consider the zeroth-order, semi-analytic expression (4.47). Plotting the evolving pulse for varying distances from the source, as in Figure 1.5 of Chapter 1, suggests that the function (4.47) has a single peak when considered as a function of time. One can show that the phase is directly related to the arrival time of this peak. At this peak or pressure extremum, the derivative of $p(\mathbf{x}, t)$ with respect to time vanishes. If we consider only positive times the derivative is given by the quantity

$$\frac{\partial p(\mathbf{x}, t)}{\partial t} = A_0(\mathbf{x}) e^{-\sigma^2(\mathbf{x})/4t} \left(-\frac{3}{2\sqrt{t^5}} + \frac{\sigma^2(\mathbf{x})}{4\sqrt{t^7}} \right). \tag{4.48}$$

For finite values of the phase function and non-zero times, the exponential term does not vanish and hence the condition for an extremum in pressure is given by

$$-\frac{3}{2\sqrt{t^5}} + \frac{\sigma^2(\mathbf{x})}{4\sqrt{t^7}} = 0. \tag{4.49}$$

Equation (4.49) is satisfied when either t approaches infinity or when $6t = \sigma^2(\mathbf{x})$. Thus, the phase function $\sigma(\mathbf{x})$ may be interpreted in terms of the time at which the peak pressure is observed at a particular point $\mathbf{x}$, denoted by T_{max}

$$\sigma(\mathbf{x}) = \sqrt{6T_{max}}. \tag{4.50}$$

Using this relationship and the explicit expression (4.41) for the phase function, we may relate T_{max} to the properties of the medium along the trajectory connecting the observation point to the source point:

$$\sqrt{T_{max}} = \frac{1}{\sqrt{6}} \int_{\mathbf{x}_s}^{\mathbf{x}_o} \frac{1}{\sqrt{\kappa(\mathbf{x})}} dx. \tag{4.51}$$

Because of the relationship (4.50) between the phase function and the propagation time of the pressure peak, $\sigma(\mathbf{x})$ has been refered to as the **diffusive time of flight**, even though it is actually proportional to the square root of the arrival time. Equation (4.51) serves as the basis for a form of diffusion equation tomography, in which 'arrival times' are used to image variations in the diffusivity $\kappa(\mathbf{x})$ (Vasco et al., 2000). We shall demonstrate this using an example given at the end of this chapter.

In closing this section, we will return to the example in Figure 4.1. In Figure 4.2 we have converted the 'diffusive time of flight' in Figure 4.1 to physical time using the 2D equivalent of Equation 4.50 (see Exercise 4.1). For comparison purposes, we have also shown the convective tracer travel time. Notice that the diffusive 'pressure front' propagates much faster compared to the convective 'tracer front'. This is consistent with field observations where pressure response can be observed much sooner than the tracer arrivals.

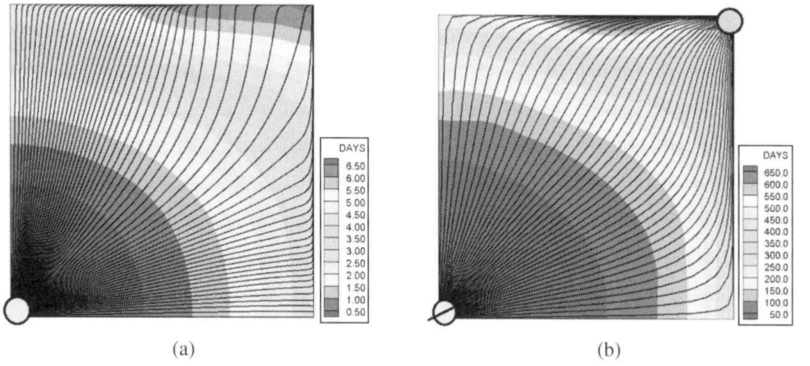

Figure 4.2 The propagation of the diffusive pressure front (a) versus convective tracer front (b). For the colour version, please refer to the plate section.

Exercise 4.1. Show that the relationship between the phase and the arrival time of the 'peak' pressure for an impulse source as given in Equation (4.50) takes the following forms for one- and two-dimensional flow.

1D:$\sigma(\mathbf{x}) = \sqrt{2T_{max}}$,

2D:$\sigma(\mathbf{x}) = \sqrt{4T_{max}}$.

Hint: The inverse Fourier transform of Equation (4.46) for two-dimensional flow is given by

$$p(\mathbf{x}, c) = A_0(\mathbf{x}) \frac{\sigma(\mathbf{x})}{2\sqrt{\pi t^2}} e^{-\sigma^2(\mathbf{x})/4t} H(t).$$

4.2.4 An alternative method for calculating the phase and the trajectories

One approach for calculating the trajectory, $\mathbf{x}(s)$, and the phase $\sigma(\mathbf{x})$ is to solve the ray Equations (4.36), (4.37), and then integrating (4.41). Numerical routines are available for solving such systems of differential equations (Press et al., 1992). However, because these ray-tracing techniques are distinct from numerical methods such as finite differences typically used to model fluid flow, their implementation and testing can be a barrier to their use. Fortunately, one can calculate the phase and trajectories directly from the output of a numerical flow simulation and without solving the ray equations (Vasco and Finsterle, 2004). We briefly outline the approach in this section.

One may ask, why it is even necessary to calculate the phase or trajectories, given that the simulator has already calculated the pressure variation throughout the reservoir. Besides being visual and physically intuitive, the simple answer is that the phase and trajectories may be used for a tomographic-style inversion of pressure arrival times, as discussed in the Applications section below. Such arrival time inversions, a by-product of the partitioning of the trajectory-based modeling into distinct travel time and amplitude computations, has some advantages in comparison to the direct inversion of the pressure variation. First, the relationship between arrival times and reservoir parameters is quasi-linear (Cheng et al., 2005), in contrast to the decidedly non-linear relationship associated with pressure amplitudes. Practically, this means that an inversion of travel times is much less sensitive to the initial reservoir model and less likely to converge to a local minimum. Therefore, a travel time inversion is a good method for constructing an initial reservoir model, prior to inverting the pressure amplitude observations. Second, as shown below, the trajectory-based expression (4.51) for the travel time provides a semi-analytic sensitivity, allowing for an efficient inversion algorithm. Sensitivities are partial derivatives of the arrival time with respect to subsurface properties such as porosity and permeability. Calculating the sensitivities efficiently typically constitutes an integral part of an inversion algorithm. Finally, the travel time inversion makes use of the early, rapidly varying portion of the transient pressure curve. Therefore, one does not have to continue a pressure test until a steady-state is reached, resulting in rapid field experiments. For example, arrival time inversion is useful in the analysis of crosswell slug tests (Brauchler et al., 2010).

The procedure for constructing the phase and trajectories is straight-forward and follows from the results of the previous section. One first conducts a numerical simulation, computing the pressure variation throughout the region of interest. Next, extract the time at which $\partial p/\partial t$ attains its maximum, based upon the simulated pressure histories for each grid block. Equation (4.50), stating that

$$\sigma = \sqrt{6T_{max}},$$

is then used to map the peak pressure arrival time into a phase value. This procedure is followed for each grid block, resulting in phase estimates throughout the reservoir volume. With the phase defined over the simulation grid one can compute

$$\mathbf{p} = \nabla\sigma$$

using a numerical differencing algorithm. The trajectory is found by stepping down the gradient of the phase from an observation point $\mathbf{x}$ to the source at $\mathbf{x}_s$. An iterative procedure, such as a second-order Runge-Kutta technique, may be used to calculate the trajectory. For example, Heun's method is one approach that is easy to implement. In this algorithm one improves upon a step by computing the gradient at an intermediate point $\hat{\mathbf{x}}_i$, given by

$$\hat{\mathbf{x}}_i = \mathbf{x}_i + \mathbf{p}(\mathbf{x}_i)\delta s.$$

where δs is the intermediate step length. The actual step taken to the next point along the path $\mathbf{x}_{i+1}$ is computed using the average of the gradient at the current point, $\mathbf{x}_i$, and the intermediate point

$$\mathbf{x}_{i+1} = \mathbf{x}_i + \frac{1}{2}\left[\mathbf{p}(\mathbf{x}_i) + \mathbf{p}(\hat{\mathbf{x}}_i)\right]\delta s.$$

Note that calculating the trajectory in this fashion is far simpler than solving the full set of ray Equations (4.36), (4.37), and (4.38). The numerical integration of these differential equations requires two-point raytracing, and more complicated computer code then Heun's method.

4.3 Applications

The advantages of the trajectory-based approaches described in this book are most apparent when trying to image the internal structure of an object using remote observations, the so-called inverse problem. The utility springs not simply from the efficiency of the methods, there is also an important conceptual advancement. Specifically, the partitioning of the modeling into phase (travel time) and amplitude computations introduces additional flexibility into both the forward and inverse problems. This flexibility has been used to some advantage in such fields as medical imaging, geophysical tomography, and petroleum reservoir characterization. We shall examine the differences between travel time and amplitude computations below, in our discussion on model parameter sensitivities. Before that, we will touch upon the numerical solution of the eikonal equation and its use in flow visualization and well drainage calculations. The phase or 'diffusive time of flight' has two primary applications. First, it can be utilized to visualize the interaction between heterogeneity and fluid flow. Second, the phase may be used as a spatial coordinate to decouple the effects of heterogeneity from fluid flow calculations.

4.3.1 Fast Marching Solutions of the eikonal equation

We have seen that a trajectory-based solution to the eikonal equation involves the construction of a ray path, or trajectory, by solving Equations (4.36) and (4.37) and then calculation of the line integral in Equation (4.41) to compute the phase. An alternative approach to computing the phase function is the Fast Marching Method which does not require explicit construction of the trajectories (Sethian, 1999). Here we briefly review the solution of the eikonal equation using the Fast Marching Method.

Fast Marching Method

In this section we illustrate the solution of the eikonal equation, Equation (4.32),

$$\nabla\sigma \cdot \nabla\sigma - \frac{1}{\kappa(\mathbf{x})} = 0,$$

using the Fast Marching Method. This technique can be used to solve the eikonal equation. It is a single-pass method utilizing the fact that $\sigma(\mathbf{x})$ only depends upon the value of σ along the characteristic(s) passing through the point $\mathbf{x}$ (Sethian, 1996). Thus, the solution σ can be constructed in an orderly one-pass fashion from smaller values of σ to larger values.

The basic framework of Fast Marching Method comprises the following steps (Sethian, 1999):

(1) Label all grid nodes as *unknown*;

(2) Assign σ values (usually zero) to the nodes corresponding to the initial position of the propagating front and label them as *accepted*;

(3) For each node that is *accepted*, locate its immediate neighboring nodes that are *unknown* and label them as *considered*;

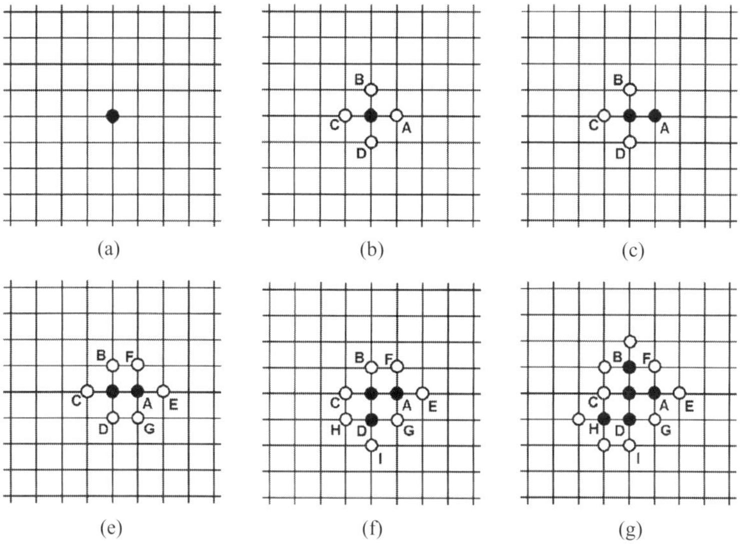

Figure 4.3 Illustration of FMM in a 5-stencil Cartesian grid.

(4) For each node labeled *considered*, update its value based on its *accepted* neighbors using the minimum of the local solutions of Equation (4.32) as illustrated below [see Equation (4.52)];

(5) Once all nodes labeled *considered* have been locally updated, we pick the node which has the minimum value among them and label it as *accepted*;

(6) Go to step (3) until all nodes are *accepted*.

In a 5-stencil Cartesian grid, these steps can be illustrated in Figure 4.3. We put one point as the initial position of the propagating front and label it as *accepted* (solid) as shown in (a). Then its immediate neighbors A, B, C, and D are marked as *considered* (open circles) as shown in (b). After the σ values of A, B, C, and D have been updated, we pick the smallest one (suppose it is A) and mark it as *accepted* as shown in (c). Then its neighbors E, G, and F are added into the *considered* as shown in (d). These steps will repeat for the next *accepted* point (suppose it is D) as shown in (e) and continued as shown in (f) until all nodes are visited. Local updates of σ, using Equation (4.32) for the 5-stencil Cartesian grid, can be written with the standard finite difference notation as (Sethian, 1996):

$$max(D_{ij}^{-x}\sigma, -D_{ij}^{+x}\sigma, 0)^2 + max(D_{ij}^{-y}\sigma, -D_{ij}^{+y}\sigma, 0)^2 = \frac{1}{\kappa(\mathbf{x})}. \tag{4.52}$$

Here the standard finite difference operator D for $\pm x$ directions can be written as $D_{ij}^{-x}\sigma = (\sigma_{i,j} - \sigma_{i-1,j}/\Delta x)$ and $D_{ij}^{+x}\sigma = (\sigma_{i+1,j} - \sigma_{i,j}/\Delta x)$. Similar equations hold for $\pm y$ directions. In Equation (4.52), the value of σ at *unknown* points is regarded as infinity and the 'max' function is used to guarantee the 'upwind' criteria.

Equation (4.52) leads to a quadratic equation and its minimum positive root gives us the value at point (i, j). We calculate σ values from each of the four quadrants (bottom-left, bottom-right, top-left, and top-right) by the similar finite difference formulation and then take the minimum σ value. The generalization of the Fast Marching Method to complex grids and anisotropic medium has been discussed by Sethian (1999). In particular, in the presense of anisotropy the characteristics may not be aligned with the computed σ-gradient resulting in erroneous front propagation. Under such conditions the σ-gradient must be computed carefully, by expanding the stencil for finite-difference calculations. More details can be found in Sethian (1999).

Exercise 4.2. Use the software 'FRONTS3D' to solve the eikonal equation and visualize the diffusive time of flight (phase) distribution for a homogeneous reservoir with a single central well (data file 'Homogeneous_GRAT.fip'). Also, visualize the evolution of the pressure front in physical time by thresholding at various times. Repeat the exercise for the heterogeneous example (data file 'Heterogeneous_GRAT.fip').

Well drainage: horizontal wells with multistage fractures

We have seen that the phase function $\sigma(\mathbf{x})$ is related to the propagation time of the peak pressure disturbance caused by an impulse source. In the reservoir engineering literature, the propagation distance of the pressure disturbance is known as the **depth of investigation** and the associated volume defines the **drainage volume**, a measure of the well performance (Lee, 1982). The drainage volume is that region of the reservoir contributing to the oil and gas production of the well at any given time. The quantification of well drainage volume is useful in well performance predictions, identifying new well locations and planning, designing and optimizing hydraulic fractures in unconventional hydrocarbon reservoirs such as shale oil and gas.

The asymptotic approach provides a relatively straightforward technique for computing and visualizing well drainage under very general conditions. We can solve Equation (4.32) for the phase $\sigma(\mathbf{x})$ using the Fast Marching Method and compute the propagation time for the pressure front using the relationship in Equation 4.50. By thresholding the propagation time at selected intervals, we can visualize the evolution of the well drainage. Most importantly, the approach allows us to account for the spatial variations in reservoir properties, such as porosity and permeability, understand the interaction between hydraulic and natural fractures and incorporate complex well geometry. Unconventional resources, such as shale gas, have taken a significant share in the energy supply in the United States

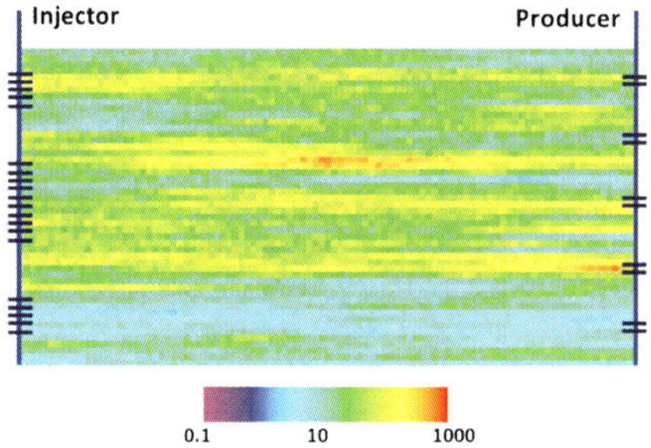

(a) Heterogenous permeability field.

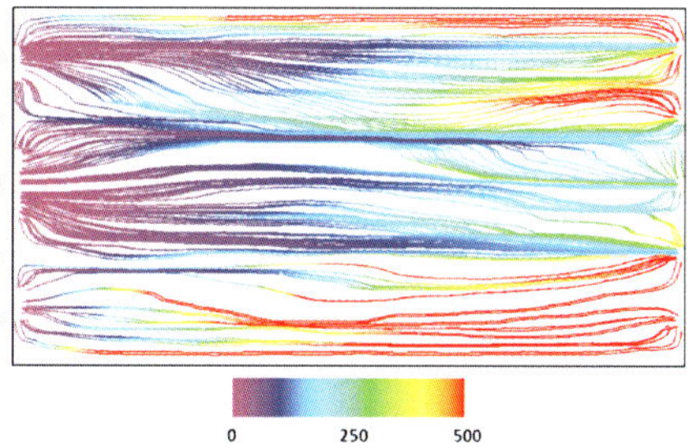

(b) Trajectories depicting the flow paths and time of flight.

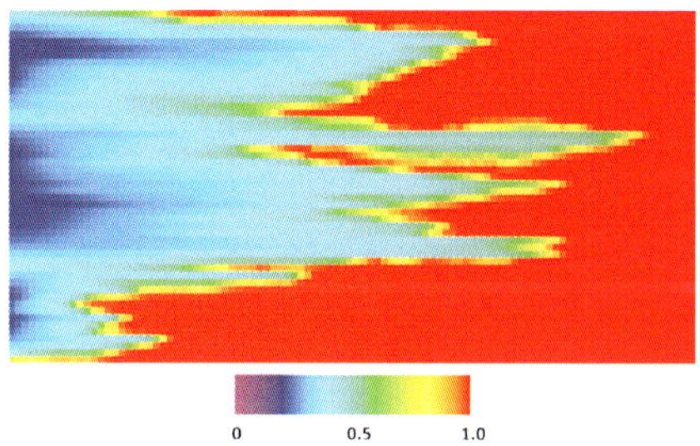

(c) The propagation of water saturation along the trajectories.

Figure 1.6 An illustration of multiphase transport calculations in heterogeneous media using the Buckeley–Leverett solution along one-dimensional characteristic curves [from Kam and Datta-Gupta (2015)]. Reprinted with permission. Copyright SPE.

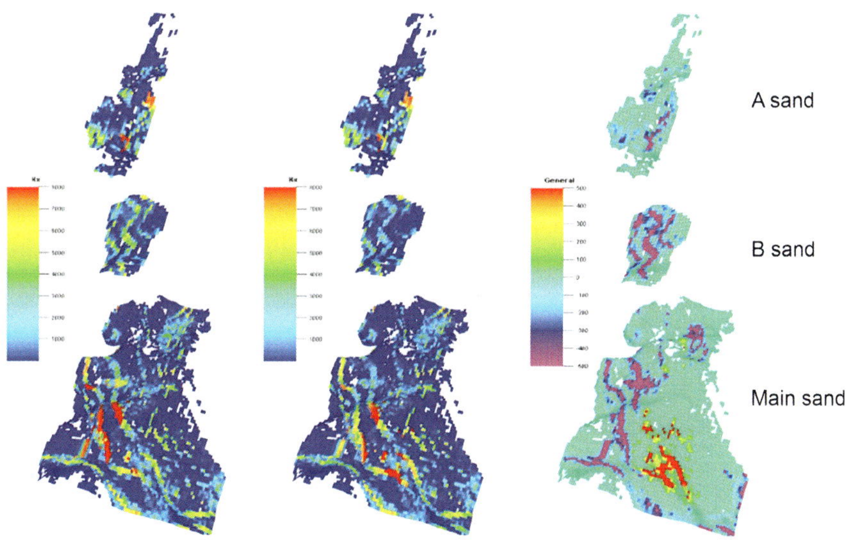

(a) (b) (c)

Figure 1.12 An illustration of geologic model calibration using multiphase flow data (a) Prior geologic model, (b) Updated model, and (c) Changes made during calibration [from Hohl et al., (2006)]. Reprinted with permission. Copyright SPE.

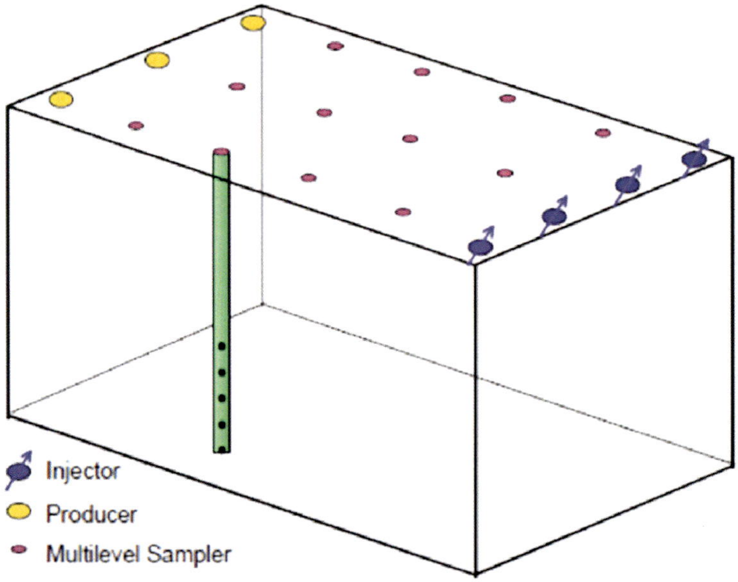

(a)

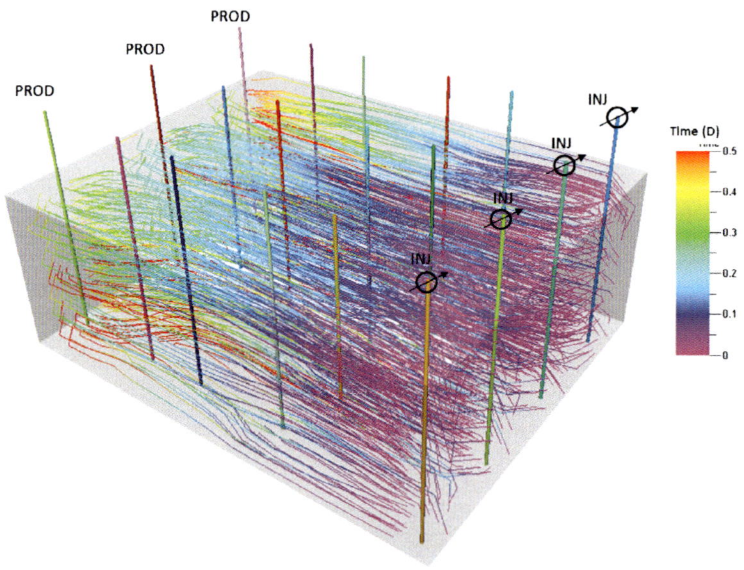

(b)

Figure 1.10 (a) Test cell diagram with multi-level samplers. (b) Tracer particle trajectories and travel times [from Yoon et al., (2001)]. Reprinted with permission. Copyright SPE.

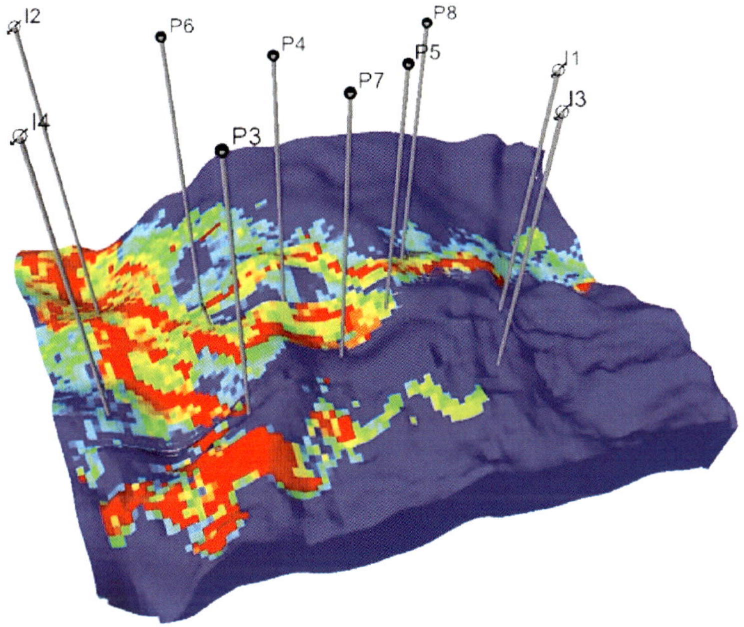

(a) Permeability distribution.

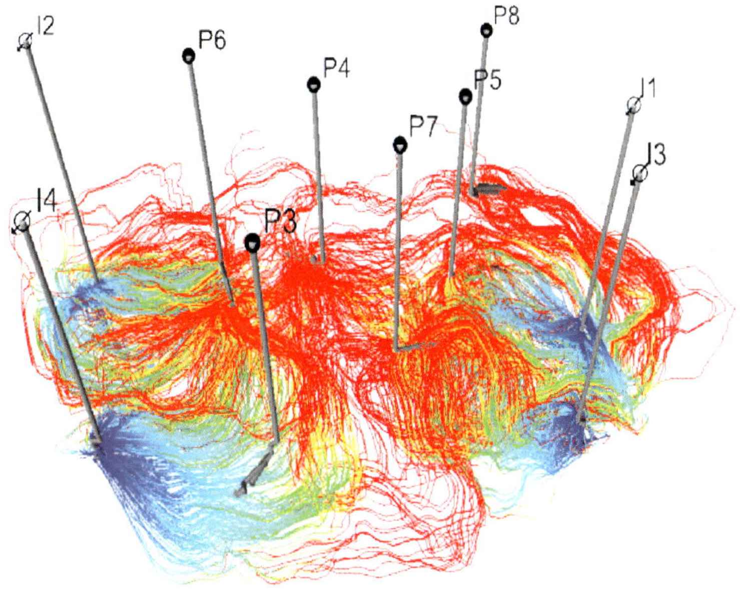

(b) Streamlines or flow paths.

Figure 1.11 Geologic model describing permeability distribution and a snapshot of streamlines for an offshore turbidite reservoir [from Hohl et al., (2006)]. Reprinted with permission. Copyright SPE.

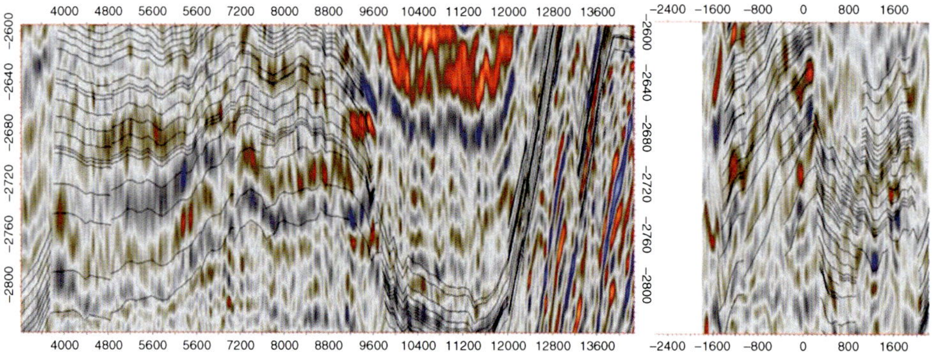

(a) Inline and crossline sections of seismic amplitudes from the Norne field.

(b) A three-dimensional perspective indicating the locations of the inline and crossline sections within the geologic model.

Figure 1.13 Seismic reflection observations from the Norne field (Watanabe et al., 2014). Reprinted with permission. Copyright SPE.

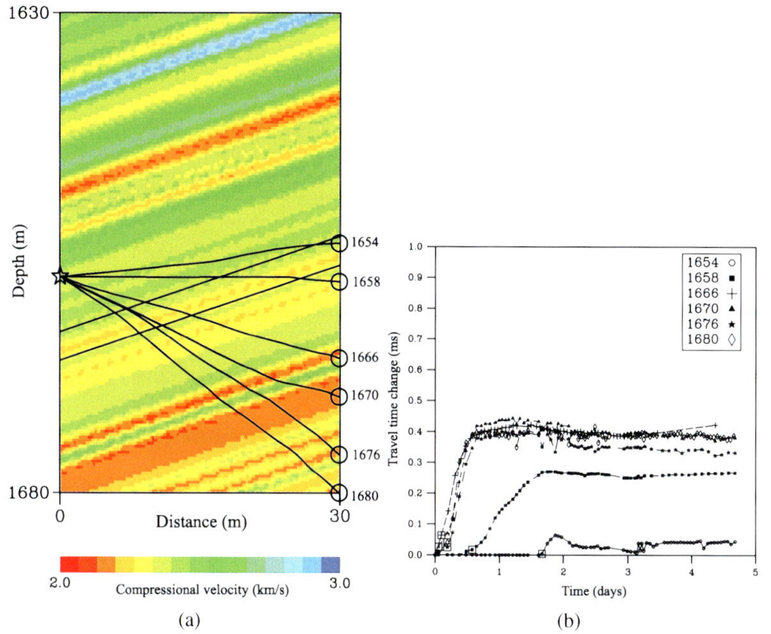

Figure 1.14 (a) Continuous active source seismic monitoring (CASSM) system with a single active source, indicated by the star on the left, and six receivers in the well on the right, denoted by open circles. The layer containing the injected carbon dioxide is denoted by two dipping parallel lines. The raypaths for the given velocity model are also plotted. (b) Seismic travel time changes observed at the receivers due to the injection of carbon dioxide.

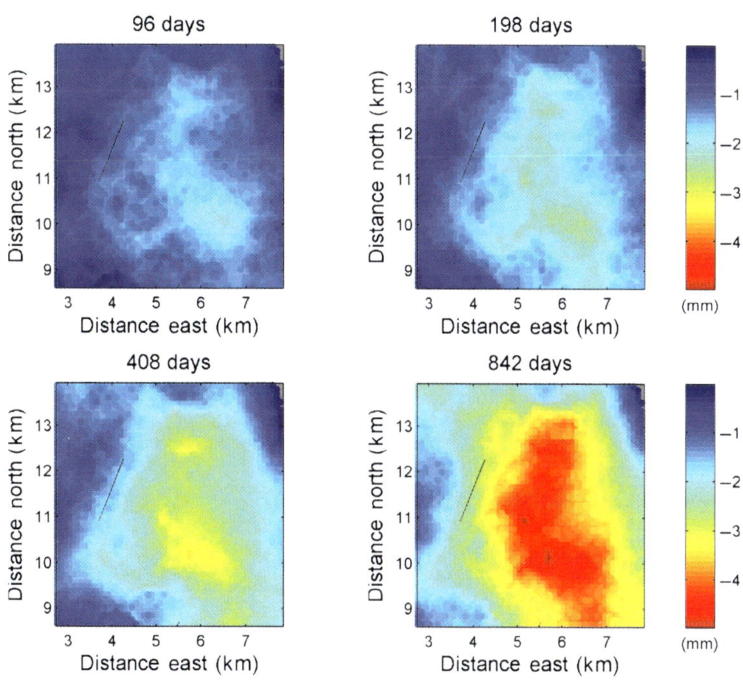

Figure 1.15 Range change in millimeters for four time intervals since the start of production at well KB-11. The segment of the production well within the reservoir is indicated by the black line.

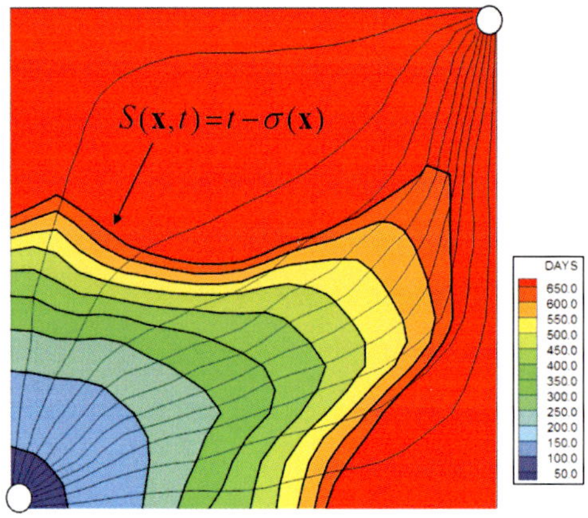

Figure 3.1 An illustration of a propagating tracer front for an injector (bottom left)-producer (top right) pair. Here, $\sigma(\mathbf{x})$represents the tracer travel time along trajectories (streamlines). The fronts are defined by the isochrones (equal travel time) and $S(\mathbf{x}, t)$ identifies the front location at a specific time.

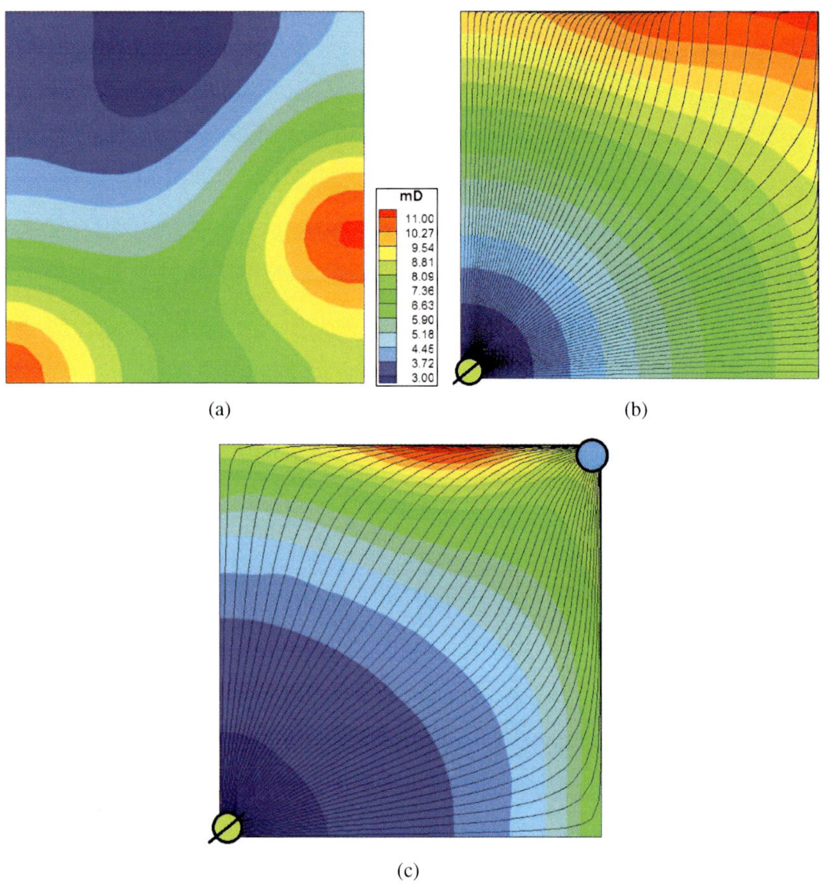

Figure 4.1 An illustration of the spatial variation of the phase function $\sigma(\mathbf{x})$ along trajectories for a heterogeneous permeable medium. (a) Permeability field (b) Trajectories and phase (c) Streamlines and convective time of flight for an injector (bottom left)-producer (top right) configuration.

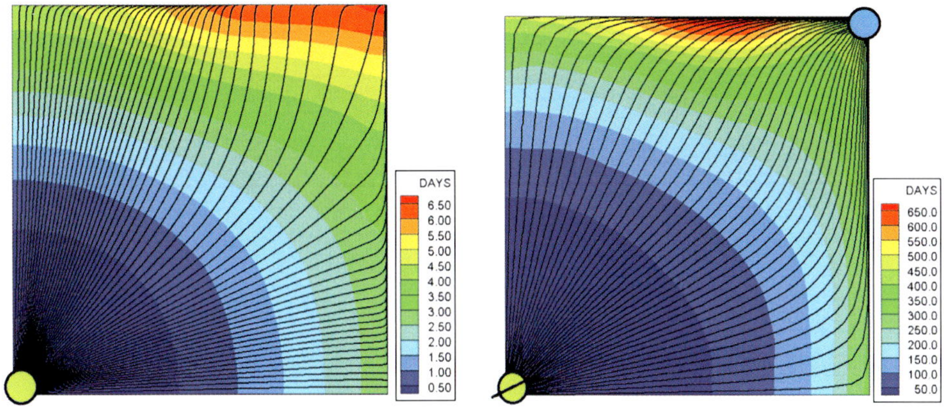

(a) (b)

Figure 4.2 The propagation of the diffusive pressure front (a) versus convective tracer front (b).

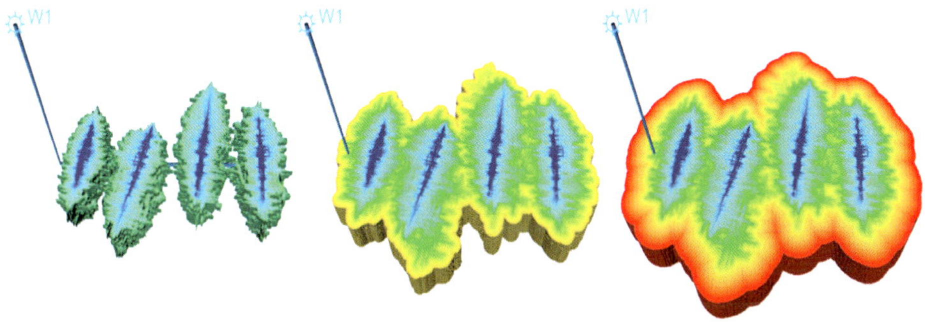

t = 50 days t = 1825 days t = 10950 days

Figure 4.5 Visualizing the evolution of the well drainage volume using the Fast Marching Method.

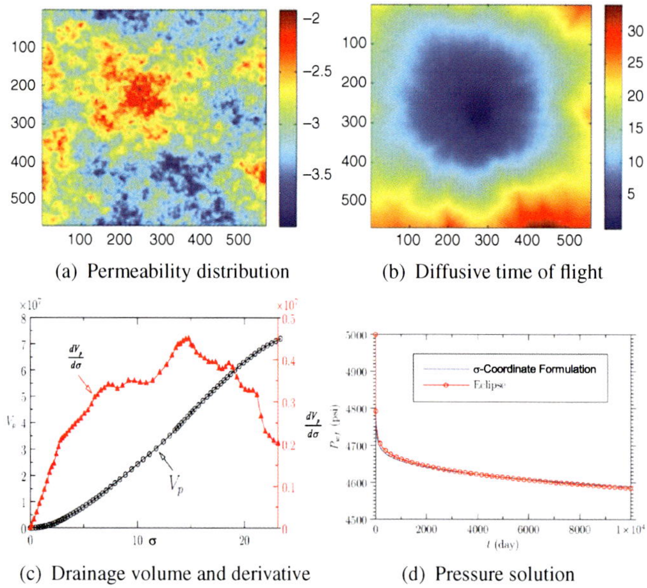

(a) Permeability distribution (b) Diffusive time of flight

(c) Drainage volume and derivative (d) Pressure solution

Figure 4.7 An example of calculated drainage pore volume using the Fast Marching Method and its derivative for a two-dimensional heterogeneous reservoir with a central well: (a) log permeability distribution; (b) calculated diffusive time of flight using the Fast Marching Method; (c) calculated drainage pore volume (V_p) and its derivative ($dV_p/d\sigma$); (d) comparison of wellbore pressure calculated using the 1-D solution in σ-coordinates and multidimensional numerical solution using a commercial simulator (*Eclipse*^{TM}) [from Zhang et al. (2014)]. Reprinted with permission. Copyright SPE.

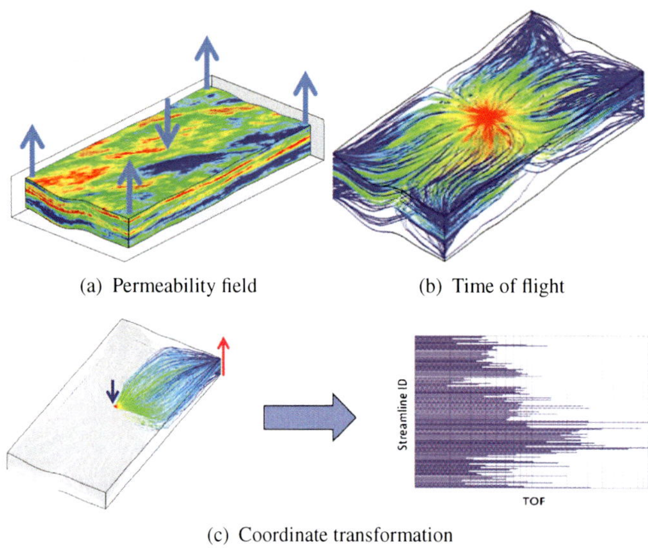

(a) Permeability field (b) Time of flight

(c) Coordinate transformation

Figure 5.5 An illustration of the streamline-based approach: (a) 3-D permeability fields and wells, (b) streamline trajectories colored by time of flight variation, and (c) transformation from 3-D space to the time of flight coordinates.

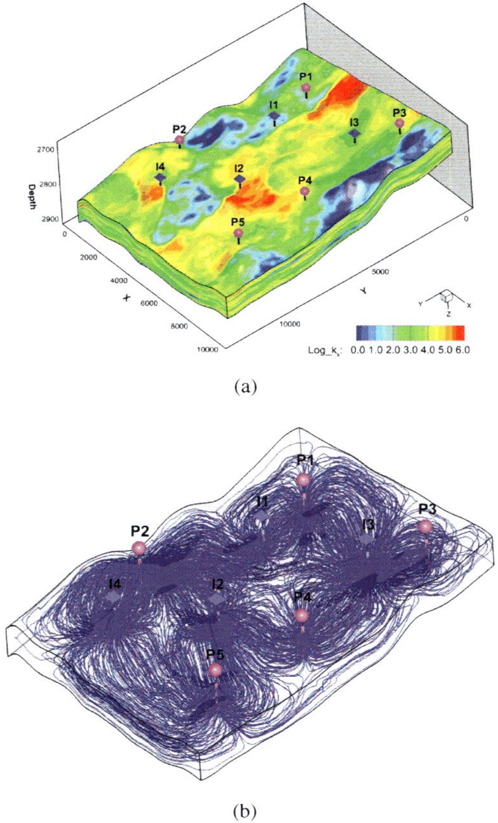

Figure 5.11 (a) Heterogeneous permeability field, displaying the well locations. (b) The streamlines associated with the flow field.

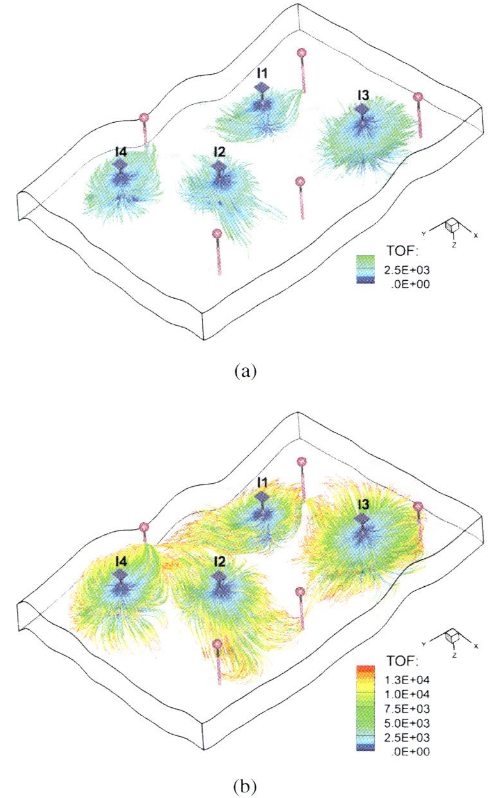

Figure 5.12 Streamline time of flight from the injector thresholded at two different times displaying the reservoir swept volume, the volume encountered by the injected fluid.

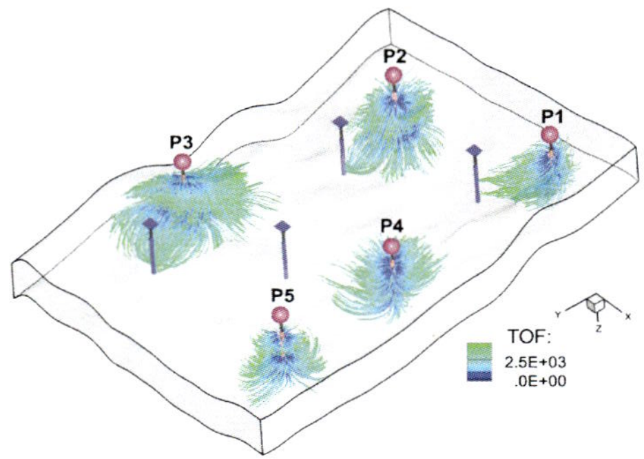

(a)

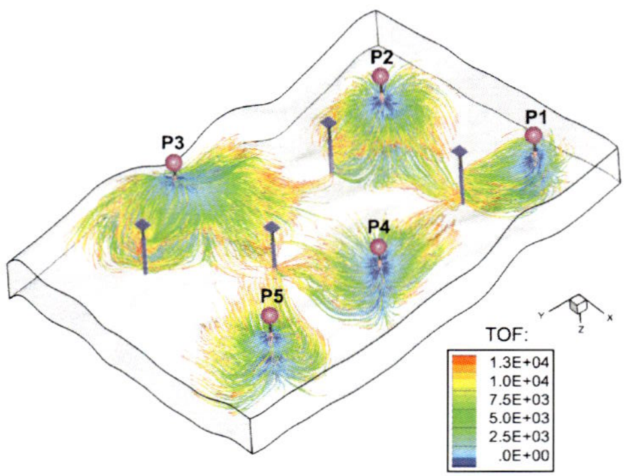

(b)

Figure 5.13 Streamline time of flight to the producer thresholded at two different times displaying the reservoir volume drained by the producing wells, the drainage volume.

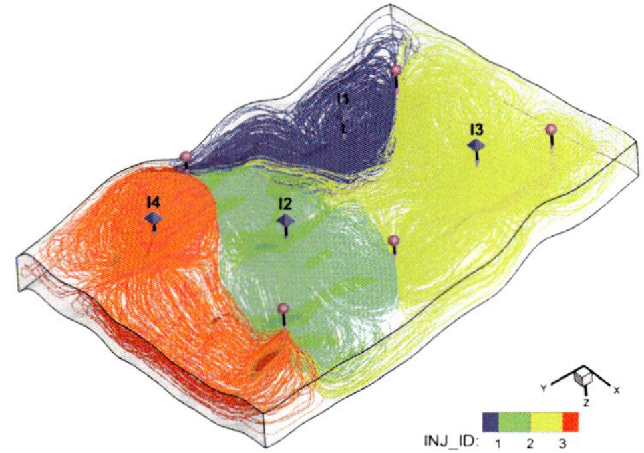

(a) Injector pore volume

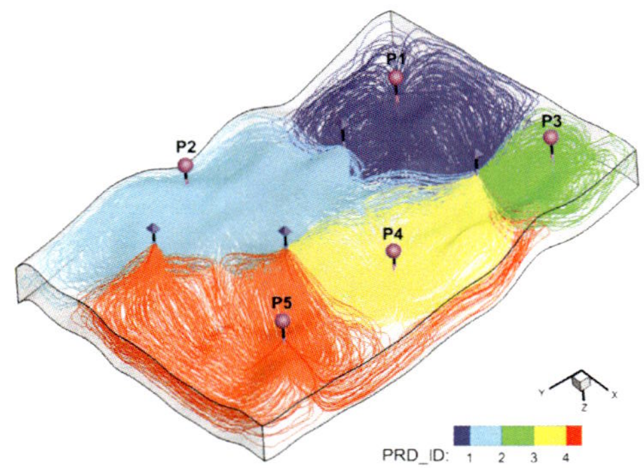

(b) Producer pore volume

Figure 5.14 Streamlines displaying the reservoir pore volumes associated with (a) injectors and (b) producers.

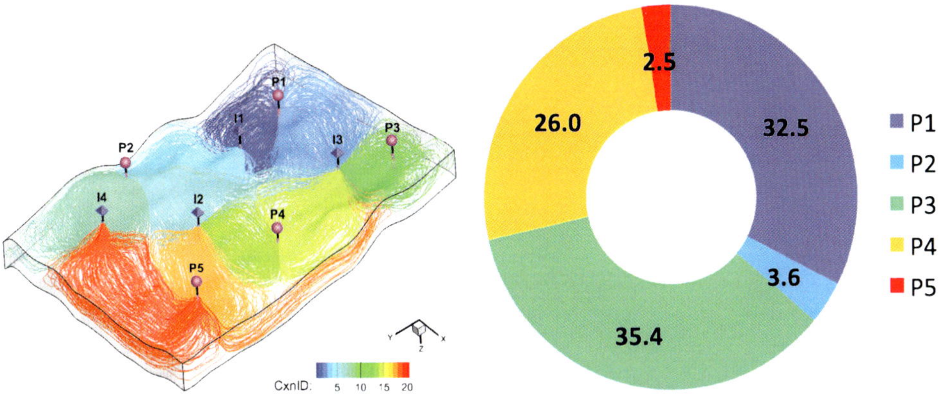

Figure 5.15 Streamlines displaying communication between the wells and the well allocation factor.

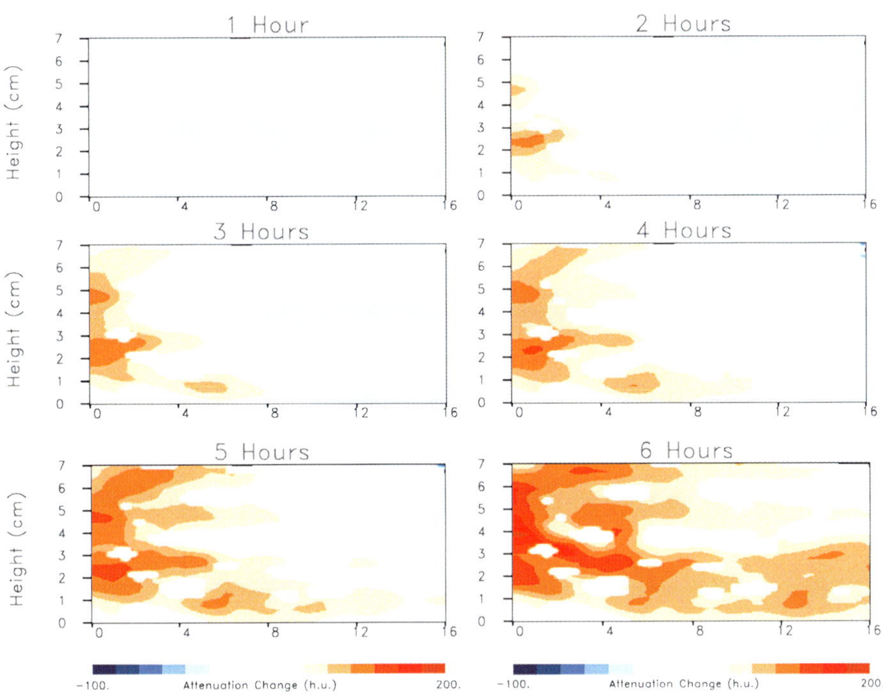

Figure 5.17 Tomographic x-ray attenuation data from the injection of a saline tracer into a sandstone core.

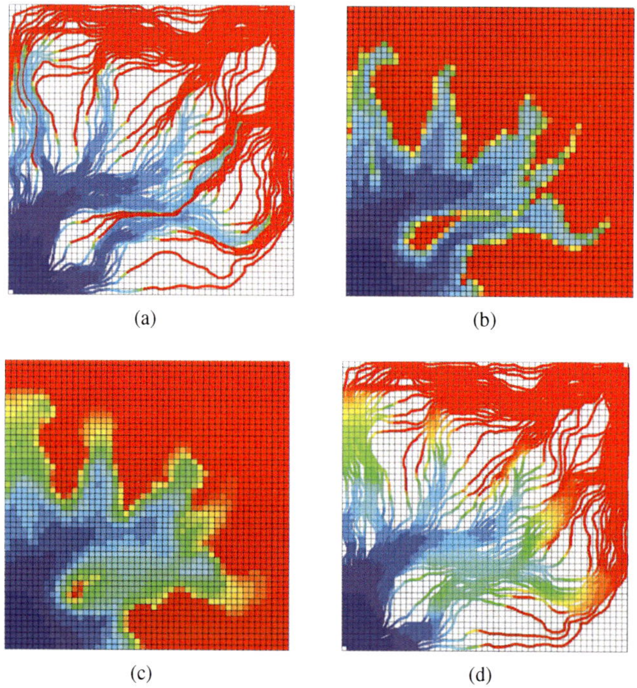

Figure 6.6 Stepwise illustration of the multiphase streamline approach in a quarter five spot with a single injector and a producer. (a) saturation transport along streamlines, (b) saturation mapping on to the grid (note that more streamlines were used for the mapping than are displayed here), (c) accounting for cross-streamline mechanisms (capillarity) on the grid, and (d) resampling of saturation on to the streamline trajectories for the next time interval calculations.

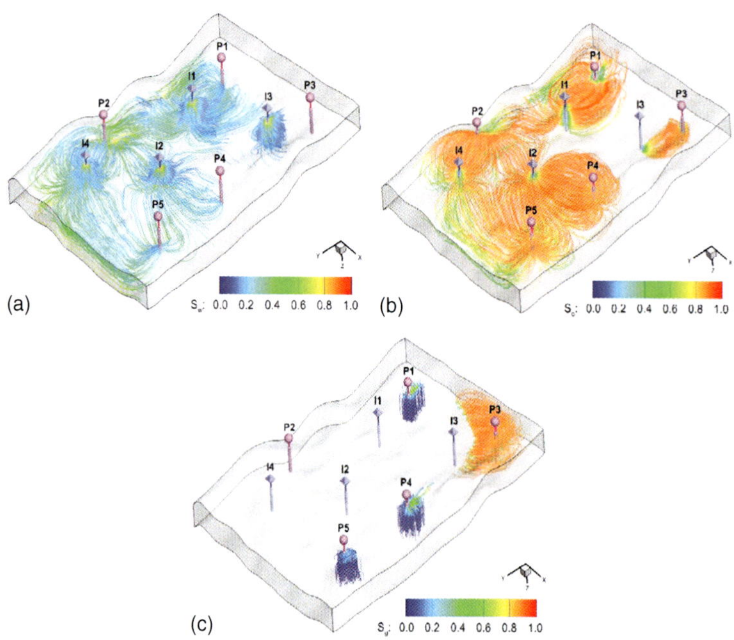

Figure 6.8 Visualization of phase streamlines. (a) Water phase streamlines showing movement of the injected water, (b) oil phase streamlines showing mobile oil being drained by producers, and (c) gas phase streamlines showing regions where depletion drive is active.

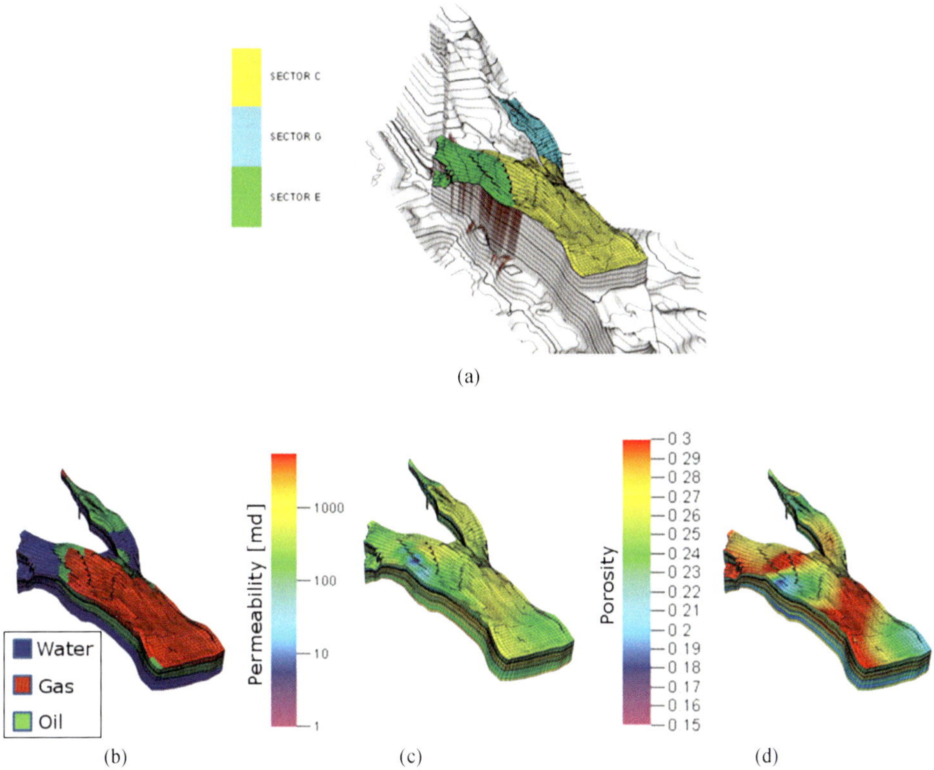

Figure 6.12 (a) Top surface of the reservoir indicating fault blocks, (b) initial fluid phase distribution, (c) initial horizontal permeability distribution, and (d) initial porosity distribution [from Rey et al. (2012)]. Reprinted with Permission. Copyright SEG.

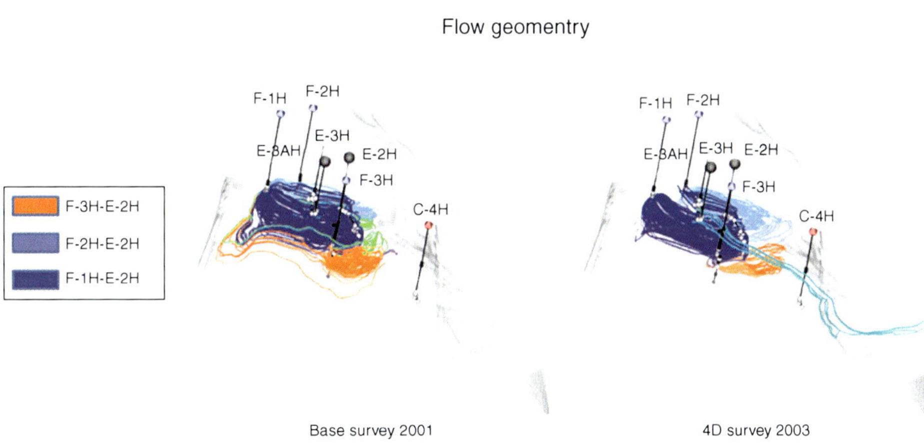

Figure 6.13 Well configuration and flow geometry during time lapse seismic surveys [from Rey et al. (2012)]. Reprinted with Permission. Copyright SEG.

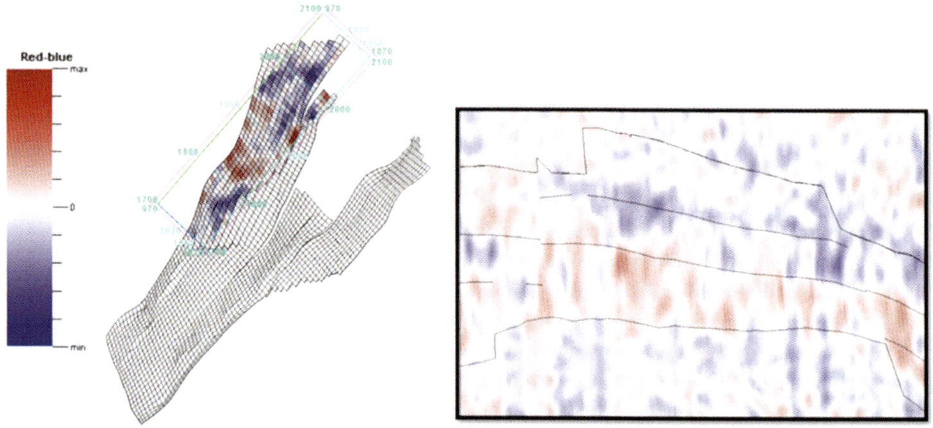

(a) (b)

Figure 6.14 (a) Time lapse seismic acoustic impedance changes displayed on the reservoir grid and (b) a transect along the section marked '1900' in (a) indicating hardening and softening of the rock because of reservoir production [from Rey et al. (2012)]. Reprinted with Permission. Copyright SEG.

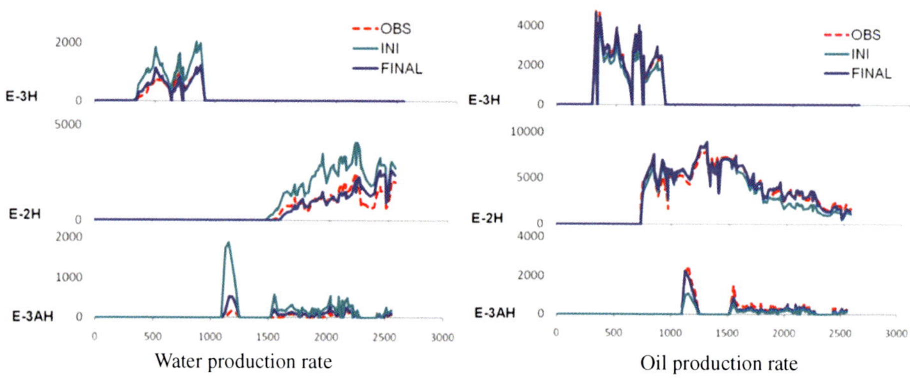

Figure 6.17 Comparison of the observed and calculated multiphase production response [from Rey et al. (2012)]. Reprinted with Permission. Copyright SEG.

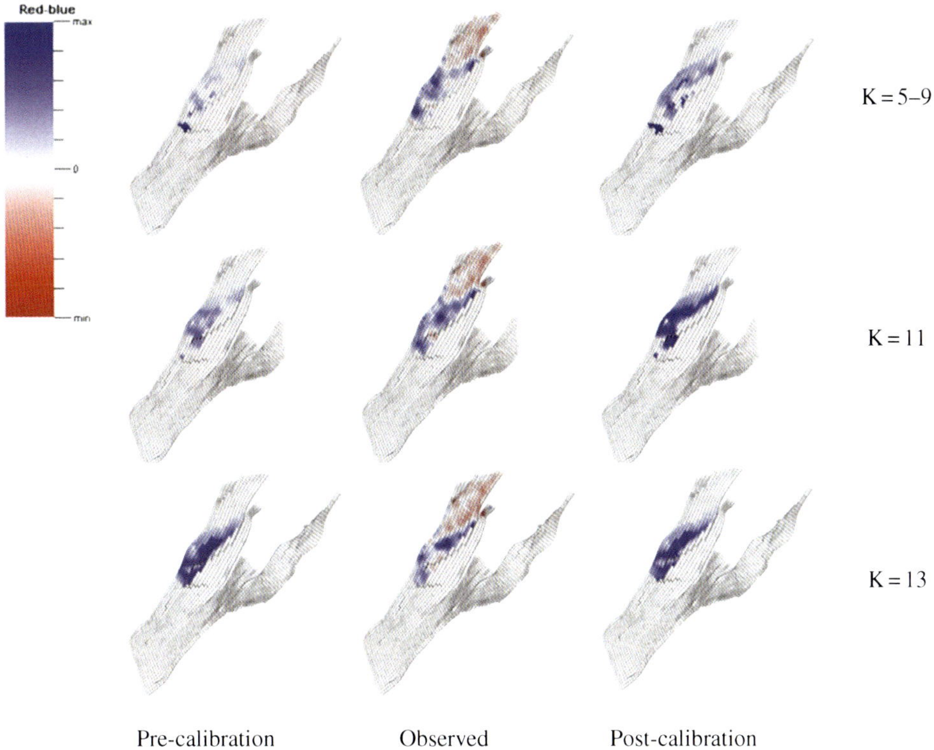

Figure 6.16 Acoustic impedence changes in selected layers before (pre-calibration) and after inversion (post-calibration) and their comparison with the observed values [from Rey et al. (2012)]. Reprinted with Permission. Copyright SEG.

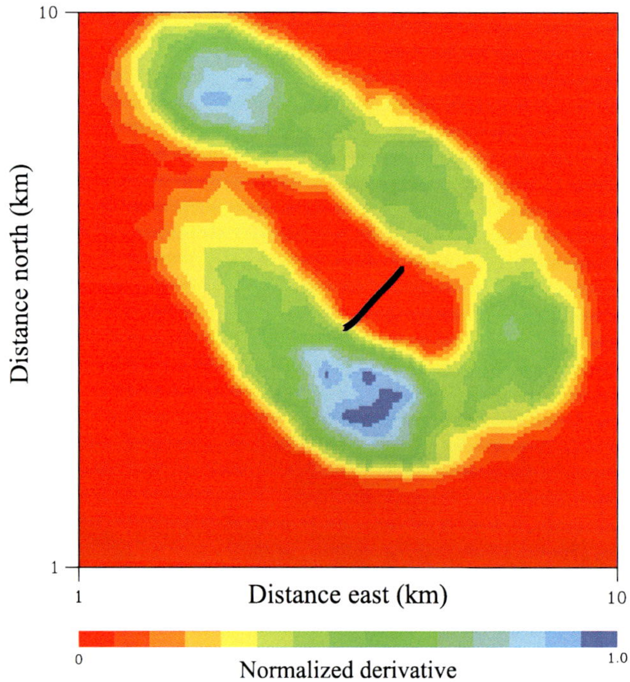

Figure 7.13 Normalized time derivative of reservoir volume change after approximately 300 days of injection. The intersection of the horizontal well and reservoir is indicated by the solid black line.

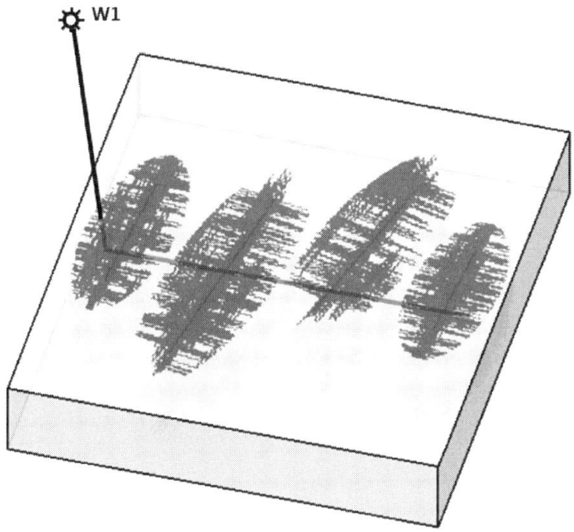

Figure 4.4 Permeability variations used in the well drainage volume calculation.

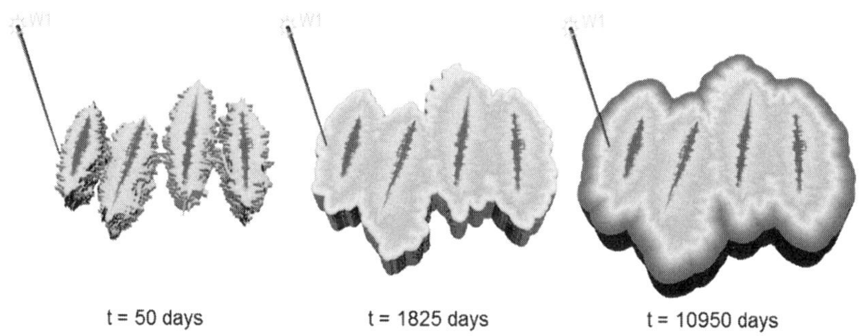

Figure 4.5 Visualizing the evolution of the well drainage volume using the Fast Marching Method. For the colour version, please refer to the plate section.

and the world energy market. The advent and growth of the development of these resources have been driven largely by the advances in technologies such as horizontal well drilling and multistage (multiple) hydraulic fracturing. However, reservoir engineers today still face significant challenges in optimizing production and predicting well performance in unconventional reservoirs.

Figure 4.4 shows a shale gas reservoir model with a matrix permeability of 100 nano-Darcy and a horizontal well with four hydraulic fractures. The natural fractures in the vicinity of the hydraulic fractures are also shown. The hydraulic fractures have different length, height, and orientation in the reservoir. We have shown the reservoir volume drained by the well at various times by computing the phase function using the Fast Marching Method and then thresholding it at the times of interest (Figure 4.5). This allows us to understand and visualize the interaction between the hydraulic fractures and natural fractures and most importantly, the onset of interference between different fracture stages. This can be crucial in optimizing the number of fracture stages in shale gas wells in order to maximize the well drainage volume.

Exercise 4.3. Use the software 'FRONTS3D' to solve the eikonal Equation and visualize the diffusive time of flight (phase) distribution for a homogeneous medium with a single hydraulic fracture in the middle (data file 'SFracture-Hom.fip'). Also, visualize the evolution of the pressure front in physical time by thresholding at various times. Repeat the exercise for the heterogeneous example(data file 'SFracture-Het.fip').

Exercise 4.4. Use the software 'FRONTS3D' to solve the eikonal Equation and visualize the diffusive time of flight (phase) distribution for a shale gas reservoir with multiple hydraulic fractures (data file 'MFracture-Hom.fip'). Also, visualize the evolution of the pressure front in physical time by thresholding at various times. Repeat the exercise for the heterogeneous example (data file 'MFracture-Het.fip').

4.3.2 Using the phase as a spatial coordinate

We have seen in the previous section that the impact of spatial heterogeneity is embedded in the phase function or the diffusive time of flight. There are computational and conceptual advantages if one uses the phase as a spatial coordinate. In fact, subject to our high-frequency assumption, rewriting the diffusion equation in trajectory-based coordinates reduces it to a simpler expression. Specifically, we can transform the pressure Equation (4.1) in three spatial variables, $\mathbf{x}$, and one temporal variable, t, to an equation in one spatial variable σ and time t. This leads to substantial savings in computation time in the numerical solution of the diffusion equation (Zhang et al., 2014; Fujita et al., 2015). This approach is analogous to the streamline formulation of fluid transport, discussed in chapters 5 and 6.

As a starting point, consider the eikonal Equation (4.32), a partial differential equation for the phase, rewritten as

$$|\nabla\sigma| = \frac{1}{\sqrt{\kappa(\mathbf{x})}}. \tag{4.53}$$

As indicated in Equation (4.29), for sufficiently rapid variations (high frequencies), the phase gradient vector ($\nabla\sigma$) is aligned with the pressure gradient and we have the approximation

$$\nabla p \approx \frac{\partial p}{\partial \sigma}\nabla \sigma = \frac{\partial p}{\partial \sigma}\frac{1}{\sqrt{\kappa}}\hat{\mathbf{p}}, \tag{4.54}$$

where $\hat{\mathbf{p}}$ represents the phase gradient vector (4.35) normalized by its magnitude. Note that $\hat{\mathbf{p}}$ is a unit vector orthogonal to the contours of $\sigma(\mathbf{x})$. The approximation (4.54) is a reasonable one for pressure propagation in a medium with smoothly varying heterogeneity. Using (4.54) to substitute for ∇p in the governing equation for pressure, Equation (4.1), results in

$$\nabla \cdot \left(S\sqrt{\kappa}\frac{\partial p}{\partial \sigma}\hat{\mathbf{p}} \right) = S\frac{\partial p}{\partial t}, \tag{4.55}$$

where we have used Equation (4.33) to substitute κS for λ. We would like to make a transformation from an equation in physical coordinates, (x, y, z) to one in trajectory-based coordinates (σ, ψ, χ), where σ is the diffusive phase. The other two coordinates, ψ and χ are defined on the contour surfaces of σ and are orthogonal to each other and to σ.

Box 4.2 **Transformation of the pressure equation into σ-coordinates**

In this box we transform the pressure Equation (4.55)

$$\nabla \cdot \left(S\sqrt{\kappa}\frac{\partial p}{\partial \sigma}\hat{\mathbf{p}} \right) = S\frac{\partial p}{\partial t}.$$

to an equation in (σ, t)-coordinates. In order to derive this equation we assumed that the pressure gradient aligns with the σ gradient direction [see Equation (4.29)]. This is equivalent to the requirement that isosurfaces of equal pressure p correspond to isosurfaces of the phase σ. The divergence of a vector $\mathbf{F}$ in 3-D Cartesian coordinates can be expressed as

$$\nabla \cdot \mathbf{F} = \frac{\partial F_x}{\partial x} + \frac{\partial F_y}{\partial y} + \frac{\partial F_z}{\partial z}.$$

Our goal is to transform from Cartesian coordinates (x, y, z) to trajectory-based coordinates (σ, ψ, χ), where σ is the diffusive time of flight. The variables ψ and χ are defined on the contour surfaces of σ and are orthogonal to each other and to σ. Now, the divergence operator in the (σ, ψ, χ) coordinate system can be written as

$$\nabla \cdot \mathbf{F} = \frac{1}{h_\sigma h_\psi h_\chi}\left[\frac{\partial(h_\psi h_\chi F_\sigma)}{\partial \sigma} + \frac{\partial(h_\sigma h_\chi F_\psi)}{\partial \psi} + \frac{\partial(h_\sigma h_\psi F_\chi)}{\partial \chi} \right],$$

where h_σ,h_ψ, and h_χ are the length of the vectors along each coordinate axis. For example, h_σ is the length of the coordinate vector, $\mathbf{t}_\sigma$ given explicitly by the expression

$$h_\sigma = |\mathbf{t}_\sigma| = \left|\frac{\partial \mathbf{x}}{\partial \sigma}\right| = \sqrt{\left(\frac{\partial x}{\partial \sigma}\right)^2 + \left(\frac{\partial y}{\partial \sigma}\right)^2 + \left(\frac{\partial z}{\partial \sigma}\right)^2}$$

The other two coordinate vector magnitudes are defined in a similar fashion. Rewriting our starting equation in trajectory-based coordinates and assuming that pressure gradients are aligned with phase (σ) gradients, so that $\hat{\mathbf{p}}$ only has a non-zero component along the σ axis, we can use the expression given above for the divergence operator to write

$$\frac{1}{h_\sigma h_\psi h_\chi}\frac{\partial}{\partial \sigma}\left(h_\psi h_\chi S\sqrt{\kappa}\frac{\partial p}{\partial \sigma}\right) = S\frac{\partial p}{\partial t}.$$

From the eikonal equation, $h_\sigma = |\mathbf{t}_\sigma| = \sqrt{\kappa}$ so that this equation becomes

$$\frac{1}{h_\sigma h_\psi h_\chi}\frac{\partial}{\partial \sigma}\left(h_\sigma h_\psi h_\chi S\frac{\partial p}{\partial \sigma}\right) = S\frac{\partial p}{\partial t}.$$

Since the product $h_\sigma h_\psi h_\chi$ is simply the Jacobian of the coordinate transformation, J, representing the ratio of volumes in the physical space to (σ, ψ, χ) coordinates, this equation reduces to

$$\frac{1}{J}\frac{\partial}{\partial \sigma}\left(JS\frac{\partial p}{\partial \sigma}\right) = S\frac{\partial p}{\partial t}.$$

As shown in Box 4.2, the coordinate transformation results in an equation for the fluid pressure

$$\frac{\partial}{\partial \sigma}\left(JS\frac{\partial p}{\partial \sigma}\right) = JS\frac{\partial p}{\partial t}, \tag{4.56}$$

that only contains derivatives with respect to σ and t. Here, J is the Jacobian of the transformation, representing the ratio of a volume element in the transformed coordinates to the volume element in the original coordinates (Marsden and Tromba, 1976, p. 258).

In Equation (4.56) the Jacobian of the coordinate transformation J and the storage S are functions of (σ, ψ, χ) in general. To eliminate any remaining dependence on ψ and χ we integrate both sides over these two coordinates. Because the pressure is assumed to depend solely upon σ and t, any terms containing $p(\sigma, t)$ can be moved outside of the integral. Furthermore, if we assume constant system compressibility c_t, then Equation (4.56) reduces to

$$\frac{1}{w(\sigma)}\frac{\partial}{\partial \sigma}\left[w(\sigma)\frac{\partial p}{\partial \sigma}\right] = \frac{\partial p}{\partial t}, \tag{4.57}$$

where

$$w(\sigma) = \int\int \phi(\sigma, \psi, \chi) J(\sigma, \psi, \chi) d\psi d\chi.$$

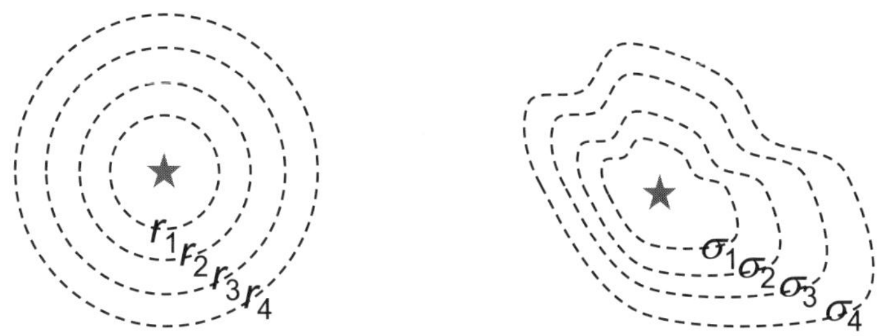

$$\frac{1}{r}\frac{\partial}{\partial r}\left(r\frac{\partial p}{\partial r}\right)=\frac{1}{\kappa}\frac{\partial p}{\partial t} \quad \xrightarrow{r\to\sigma} \quad \frac{1}{w(\sigma)}\frac{\partial}{\partial \sigma}\left(w(\sigma)\frac{\partial P}{\partial \sigma}\right)=\frac{\partial p}{\partial t}$$

Figure 4.6 Generalization of the radial pressure diffusivity equation to heterogeneous media using the diffusive time of flight as spatial coordinate.

The result is a single partial differential Equation (4.57), in two variables σ and t, that can be solved using standard numerical methods such as finite difference. Physically, we are using the fact that, for the high-frequency component, the changes in fluid pressure are in the direction of the phase gradient. Thus, along the isosurfaces of equal phase, the pressure amplitude is constant. The function $w(\sigma)$ can be related to the **drainage pore volume**, V_p, the volume of pore space in communication with a given well,

$$V_p(\sigma)=\int_{\sigma_0}^{\sigma}\left[\int\int \phi(\sigma,\psi,\chi)J(\sigma,\psi,\chi)d\psi d\chi\right]d\sigma=\int_{\sigma_0}^{\sigma}w(\sigma)d\sigma. \tag{4.58}$$

The coordinate transformation is illustrated schematically in Figure 4.6. Note that Equation (4.57) can be viewed as an extension of the radial pressure diffusivity equation for a homogeneous medium, given in the frequency-domain by Equation (4.6), to a medium that is heterogeneous. The two-dimensional case is illustrated in Figure 4.6. In place of the radial symmetry associated with a homogeneous medium, the heterogeneity is now captured by the deformation of the contours. Furthermore, the function $w(\sigma)$ is related to the area of an isosurface associated with a phase value of σ. Depending upon the symmetry of the medium and the source, the transformed pressure Equation (4.57) may be reduced to the standard linear, radial or spherical diffusion equation [see Exercise 4.6]. Analytical solutions are available for these symmetric cases (Lee, 1982).

Exercise 4.5. Starting with the definition of the Jacobian of the transformation,

$$dxdydz = Jd\sigma d\psi d\chi,$$

derive Equation (4.58).

Exercise 4.6. Show that for a homogeneous medium, Equation (4.57) reduces to the following equations:

(i) 1-D linear flow:

$$\frac{\partial^2 p}{\partial x^2} = \frac{1}{\kappa}\frac{\partial p}{\partial t}$$

(ii) 2-D radial flow:

$$\frac{1}{r}\frac{\partial}{\partial r}\left(r\frac{\partial p}{\partial r}\right) = \frac{1}{\kappa}\frac{\partial p}{\partial t}$$

(iii) 3-D spherical flow:

$$\frac{1}{r^2}\frac{\partial}{\partial r}\left(r^2\frac{\partial p}{\partial r}\right) = \frac{1}{\kappa}\frac{\partial p}{\partial t}$$

Hint: For 1-D linear flow,

$$\sigma = \frac{x}{\sqrt{\kappa}},$$

where $V_p(\sigma) = \varphi A x = \varphi A\sqrt{\kappa}\sigma$ and $w(\sigma) = dV_p/d\sigma = \varphi A\sqrt{\kappa}$.

Similarly, for 2-D radial flow,

$$\sigma = \frac{r}{\sqrt{\kappa}},$$

where $w(\sigma) = 2\pi\varphi\kappa h\sigma$, h is the thickness. For 3-D spherical flow we have

$$\sigma = \frac{r}{\sqrt{\kappa}},$$

and $w(\sigma) = 4\pi\varphi\kappa^{3/2}\sigma^2$.

Exercise 4.7. Generalize the transformation of the 3D pressure equation for an anisotropic hydraulic conductivity field,

$$\nabla \cdot \boldsymbol{\lambda}\nabla p = S\frac{\partial p}{\partial t},$$

where $\boldsymbol{\lambda} = \frac{\kappa(\mathbf{x})}{\mu}$ and $\kappa(\mathbf{x})$ is the permeability tensor, to diffusive time of flight coordinates. Show that the resulting equation has the same form as Equation (4.57) when the diffusive time of flight is computed for the anisotropic medium.

For more general cases involving spatially heterogeneous and anisotropic permeable medium, we need to solve Equation (4.57) numerically (Zhang et al., 2014; Fujita et al., 2015). For this, we require the drainage volume derivative $w(\sigma)$. As discussed before, the drainage pore volume $V_p(\sigma)$ can be calculated using the

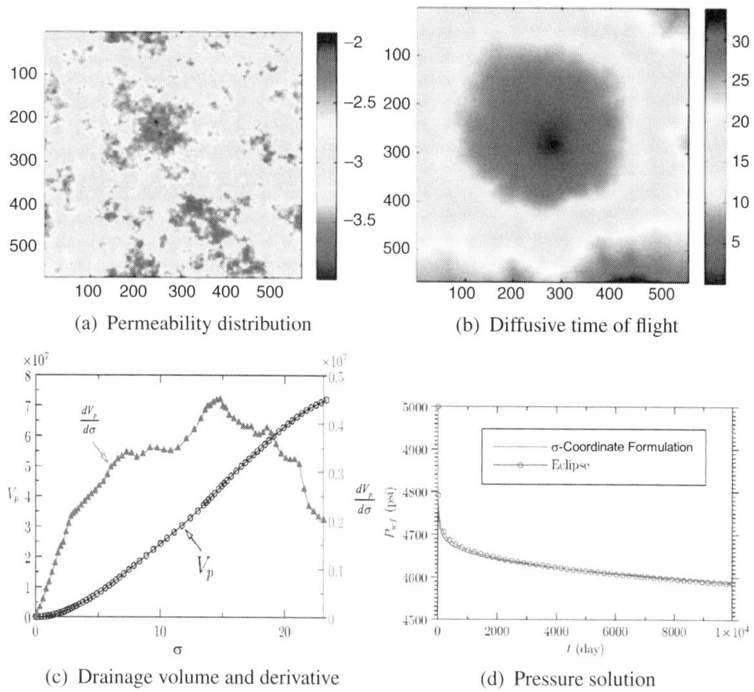

Figure 4.7 An example of calculated drainage pore volume using the Fast Marching Method and its derivative for a two-dimensional heterogeneous reservoir with a central well: (a) log permeability distribution; (b) calculated diffusive time of flight using the Fast Marching Method; (c) calculated drainage pore volume (V_p) and its derivative ($dV_p/d\sigma$); (d) comparison of wellbore pressure calculated using the 1-D solution in σ-coordinates and multidimensional numerical solution using a commercial simulator (*Eclipse*^{TM}) [from Zhang et al. (2014)]. Reprinted with permission. Copyright SPE. For the colour version, please refer to the plate section.

Fast Marching Method. Once the V_p versus σ data has been tabulated, we can calculate its derivative. Figure 4.7 shows an example of a drainage volume calculation. The drainage pore volume and its derivative for a two-dimensional heterogeneous reservoir are plotted as functions of the phase σ. Also shown is a comparison of results from the one-dimensional solution of Equation (4.57) and multidimensional solution of Equation (4.1). The close agreement validates the one-dimensional formulation. For field-scale applications involving high-resolution geologic models, the one-dimensional formulation can result in speed-ups of several orders of magnitude compared to the multidimensional solution (Zhang et al., 2014).

Exercise 4.8. Solve Equation (4.57) using finite differences with the stencil given in Figure 4.8 and the boundary conditions shown.

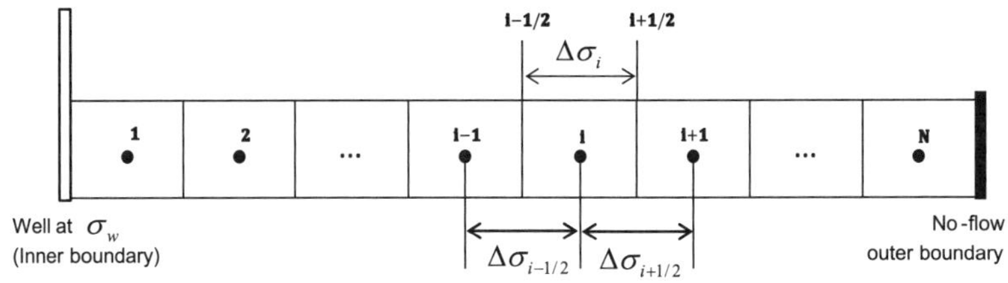

Figure 4.8 1-D finite difference solution.

Hint: The spatial derivative can be discretized as follows:

$$\frac{\partial}{\partial \sigma}\left(w(\sigma)\frac{\partial p}{\partial \sigma}\right) = \frac{T_{i-1/2}p_{i-1} - (T_{i-1/2} + T_{i+1/2})p_i + T_{i+1/2}p_{i+1}}{\Delta\sigma_i},$$

where $T_{i\pm1/2}$ represents the inter-cell transmissibility between i and $i \pm 1$.

$$T_{i\pm1/2} = \frac{w_{i\pm1/2}}{\Delta\sigma_{i\pm1/2}}; w_{i-1/2} = \left(\frac{dV_p}{d\sigma}\right)_{i-1/2} = \frac{V_{p,i} - V_{p,i-1}}{\sigma_i - \sigma_{i-1}}; w_{i+1/2} = \frac{V_{p,i+1} - V_{p,i}}{\sigma_{i+1} - \sigma_i}.$$

Discretizing the time derivative and rearranging will lead to a tri-diagonal system of equations.

Exercise 4.9. This is a continuation of Exercise 4.2. Using FRONTS3D, visualize the pressure distribution in the reservoir for the constant well production rate. Compare the contours of the diffusive time of flight and the pressure and observe the similarities. Examine the pressure response computed at the well bore using the phase as a spatial coordinate as discussed in the previous section. Repeat the exercise for a constant well bottomhole pressure and examine the rate response (data files 'Homogeneous_BHP.fip' and 'Heterogeneous_BHP.fip').

Exercise 4.10. This is a continuation of Exercise 4.4. Using FRONTS3D, visualize the pressure distribution in the shale gas reservoir at the given time. Observe the similarities between the contours of the diffusive time of flight and the pressure. Calculate the rate and the pressure response at the well bore using the phase as a spatial coordinate as discussed in the previous section.

4.3.3 Sensitivity computation

Model parameter sensitivities are a critical component of iterative imaging techniques (Menke, 1989; Aster et al., 2013). These sensitivities relate perturbations in model properties, such as $S(\mathbf{x})$ or $\lambda(\mathbf{x})$ in Equation (4.1), to deviations in the observed field data, such as $p(\mathbf{x},t)$ measured at a particular location. In order to compute model parameter sensitivities we take advantage of the fact that the perturbations are small. That is, because we are interested in an iterative construction of a model fitting the data, we will only perturb the model in small increments and by a small amount. Thus, we can assume that any perturbation in model properties, say $\delta\lambda(\mathbf{x})$, will be small, and will lead to a correspondingly small deviation in the field δp. Given the flexibility provided by the trajectory-based approach, partitioning the problem into a travel time calculation and an amplitude computation, it is fitting that we first consider perturbations in the travel times of a disturbance. After that we consider the more complicated relationship between model parameter perturbations and field amplitude deviations. However, we shall take advantage of the trajectory-based approach to gain both insight and efficiency in the computation of amplitude sensitivities.

Travel time sensitivities

The requisite governing equation is the expression (4.41) for $\sigma(\mathbf{x})$,

$$\sigma(\mathbf{x}_o) = \int_{\mathbf{x}_s}^{\mathbf{x}_o} \frac{1}{\sqrt{\kappa}} dx. \tag{4.59}$$

We revert to the expression for σ rather than the integral (4.51) for $\sqrt{T_{max}}$ in order to avoid the proliferation of factors of $1/\sqrt{6}$. However, it should be kept in mind that $\sigma = \sqrt{6T_{max}}$. The simple expression (4.59) hides the complications that arise from the fact that the integral is along the trajectory $\mathbf{x}(s)$, and is therefore a path integral. A perturbation in the reservoir model, say $\delta\kappa$ from a background model $\overline{\kappa}$, given by

$$\kappa = \overline{\kappa} + \delta\kappa, \tag{4.60}$$

will induce a perturbation in both the model $\overline{\kappa}(\mathbf{x})$ as well as a perturbation in the trajectory $\mathbf{x}(s)$

$$\mathbf{x}(s) = \overline{\mathbf{x}} + \delta\mathbf{x}. \tag{4.61}$$

Note that, while the ray path may vary between the end points $\mathbf{x}(0)$ and $\mathbf{x}(r_s)$, the end points themselves are considered to be fixed points, as the source and receiver locations are known. In the sensitivity calculation one perturbs κ in the eikonal Equation (4.32) and computes the corresponding deviations of the phase function $\sigma(\mathbf{x})$.

It has been shown in tomographic formulations that the effect of the perturbation in the path is second order in $\delta\kappa$ and may typically be neglected. Thus, to first order, we can integrate over the ray path $\overline{\mathbf{x}}$ in the background model $\overline{\kappa}$ and the effect of a perturbation in $\overline{\kappa}$ on σ will only involve a perturbation of the integrand in (4.59)

$$\delta\sigma = \int_{\mathbf{x}_s}^{\mathbf{x}_o} \delta\left[\frac{1}{\sqrt{\kappa}}\right] dx. \tag{4.62}$$

Carrying out the perturbation of the integrand gives

$$\delta\sigma = -\frac{1}{2}\int_{\mathbf{x}_s}^{\mathbf{x}_o} \frac{1}{\sqrt{\kappa}}\frac{\delta\kappa}{\kappa} dx. \tag{4.63}$$

Because κ is defined in terms of $S(\mathbf{x})$ and $\lambda(\mathbf{x})$, [see the definition (4.33)], we could write (4.63) in terms of perturbations in these component model parameters. It is evident from the definition (4.33) that there is a trade-off between variations in $\lambda(\mathbf{x})$ and $S(\mathbf{x})$. However, it might be possible to express $\lambda(\mathbf{x})$ in terms of $S(\mathbf{x})$ or to combine travel time observations with other types of data in order to reduce or eliminate the trade-off. Alternatively, the variations in $\lambda(\mathbf{x})$ might dominate any deviations in $S(\mathbf{x})$, or vice-versa, allowing us to neglect one of these parameters.

Diffusion tomography

With the relationship (4.63) we can devise an algorithm to estimate spatial variation in κ, given a collection of arrival times T_{max}. The method is simply a concrete realization of the intuitive tomographic approach described in Chapter 1. We shall outline the steps in general terms but refer to a specific crosswell geometry (Figure 4.9) that is suited to hydrological field tests.

In a crosswell configuration a finite segment of the well is isolated using inflatable packers, preventing flow outside the interval. The pressure within that segment is then varied. The exact time variation of the pressure depends upon the nature of the test. If the wells are close enough a pressure pulse may be transmitted from one to another. For example, one can conduct a crosswell extension of a **slug test** (Brauchler et al., 2010). In a slug test one monitors the recovery due to a prescribed and rapidly varying pressure transient (Karasaki et al., 1988; Butler et al., 2003). In a crosswell slug test one also monitors the pressure in nearby wells in order to infer flow properties further from the source. For distant wells it might be necessary to use a step function source, in which fluid is injected or withdrawn from the packed-off interval at a constant rate (Vasco et al., 2000). The step is due to the initiation of flow within the interval. For such tests one may consider the time derivative of the recorded pressure rather than the pressure itself, in order to determine a peak rate of pressure change and to define the arrival time, as noted in Chapter 1.

For the test represented in Figure 4.9 the wells were approximately 3.5 meters apart and the porous medium consisted of soft sediments. Thus, it was possible to transmit a pressure pulse to the observation well (Figure 4.10). From the recorded signals one extracts the time at which the first peak is observed, T_{max}, for each of the ten receivers. There were nine sources in all, situated in the well at the left-hand-side of the panel in Figure 4.9. The area between the wells is subdivided into

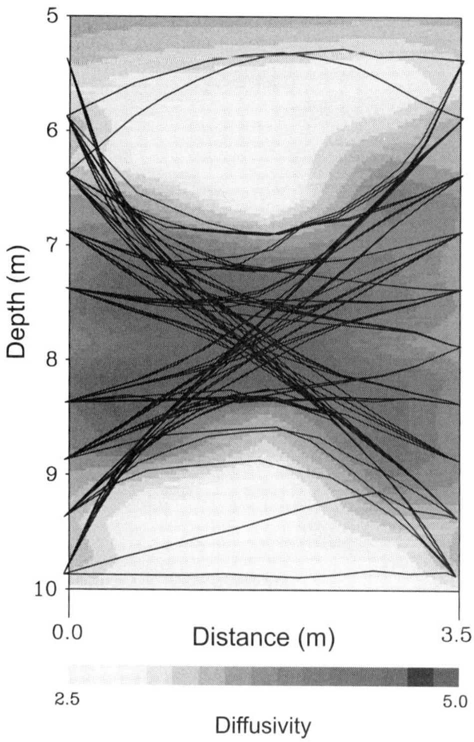

Figure 4.9 Crosswell pressure test geometry with the trajectories plotted as black curves and the final estimates of κ indicated by the variations in color.

a 5 (horizontal) by 20 (vertical) grid of cells for a total of 100 model parameters. Within each cell it is assumed that κ has a constant value. The value of κ within the i-th cell, denoted by κ_i, is constrained by the travel times of all trajectories intersecting the grid block. Stated another way, each travel time T_{max} constrains σ, and hence the values of κ, along the trajectory via the path integral (4.59). Similarly, given deviations in the travel time from a value calculated using some initial model, Equation (4.63) constrains perturbations of the model. For a piece-wise constant distribution of κ, the trajectory will be composed of straight-line segments within each grid block. Thus, the integral (4.63) may be written as a sum over all grid blocks intersected by the k-th trajectory:

$$\delta\sigma_k = -\frac{1}{2}\sum_{i\in X_k}\frac{\Delta x_{ik}}{\sqrt{\overline{\kappa_i}}}\frac{\delta\kappa_i}{\overline{\kappa_i}}. \tag{4.64}$$

where $\overline{\kappa}$ is the background value of κ for the i-th block, Δx_{ik} is the trajectory length in grid block i, and X_k is the collection of grid blocks traversed by the k-th trajectory.

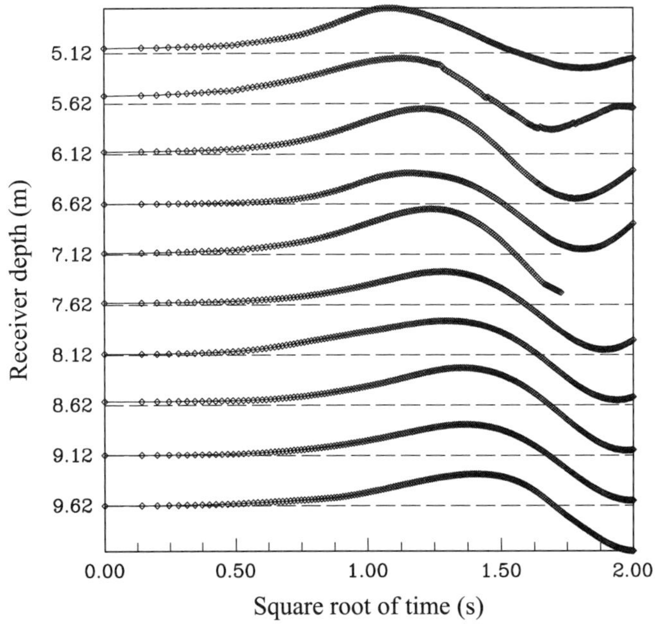

Figure 4.10 Pressure variations measured at ten receivers within isolated sections of the observation well.

The basic strategy is to repeatedly update an initial or starting model, at each iteration incorporating changes to improve the fit to the data. This approach has several desirable features. First, it plays to the strengths of the trajectory-based approach, providing a semi-analytic sensitivity that is easy to calculate. Second, the relationship between the observed data and the model parameters is quasi-linear so that the algorithm generally converges, even when the starting model is far from the true model (Cheng et al., 2005). Third, the approach provides some measure of data reduction, an entire transient pressure curve is reduced to a single quantity, the peak arrival time. These attributes indicate that the inversion of travel time might be a useful first step in imaging the internal structure of an object. Indeed, such travel time-based inversion is often the most cost effective form of tomographic imaging and is widely used in fields such as geophysics and medicine.

We begin with a starting model $\overline{\kappa}$. Next, trajectories are calculated for all the source and receivers, based upon the starting model. There are a number of ways to accomplish this. For example, one could solve the characteristic, or ray, Equations (4.36) and (4.37) using a numerical approach to integrate the ordinary differential equations (Press et al., 1992, p. 701). Alternatively, one could solve the eikonal Equation (4.32) directly using an efficient finite difference algorithm such as Fast Marching or a related approach (Sethian, 1999; Osher and Fedkiw, 2003).

Then one can use a technique for marching down the gradient of the travel time field, such as a Runge-Kutta algorithm (Press et al., 1992). That is the approach taken to generate the trajectories shown in Figure 4.9, using a smoothed and interpolated version of the spatial variation in κ between the wells. Finally, as noted in Section 4.2.4, one could use a reservoir simulator directly to compute the pressure history and calculate the arrival time field and phase function from the output of the simulation. Then one can trace a path down the gradient of the travel time field. This is not the most efficient approach but requires the least amount of programming, given an existing reservoir simulator. Regardless of the approach, one obtains the trajectories $\mathbf{x}(s)$ from the sources to the receivers and computes the sensitivities using the expression given in the sum (4.64).

A collection of source-receiver pairs, obtained from the crosswell pressure test, leads to a set of linear equations, each of the form (4.64). We can write this system of equations in matrix-vector form

$$\delta\boldsymbol{\sigma} = \mathbf{M}\delta\boldsymbol{\kappa} \tag{4.65}$$

where $\delta\boldsymbol{\sigma}$ is a vector of deviations of σ_i from values predicted using the background model. The length of this vector is equal to the number of source-receiver pairs. The vector $\delta\boldsymbol{\kappa}$ represents the changes in values of κ for each of the grid blocks in our model of the permeable medium between the wells. One could try and solve Equation (4.65) directly. However, simple accounting, with 90 observations and 100 model parameters, suggests that the system of equations is under-determined. Solutions of such under-determined systems are typically unstable with respect to errors in the observations and numerical noise in the computations.

Alternatively, one could solve Equation (4.65) using a method such as least squares (Lawson and Hanson, 1974; Dorny, 1983). That is, we could minimize the sum of the squares of the misfit to each observation:

$$\mu^2 = (\delta\boldsymbol{\sigma} - \mathbf{M}\delta\boldsymbol{\kappa})^t \cdot (\delta\boldsymbol{\sigma} - \mathbf{M}\delta\boldsymbol{\kappa}) \,. \tag{4.66}$$

The conditions for an extremum of this system are the vanishing of the derivatives of μ^2 with respect to each of the model parameters $\delta\boldsymbol{\kappa}$, that is

$$\nabla_\kappa \mu^2 = -\mathbf{M}^t\delta\boldsymbol{\sigma} + \mathbf{M}^t\mathbf{M}\delta\boldsymbol{\kappa} = 0 \tag{4.67}$$

or

$$\mathbf{M}^t\delta\boldsymbol{\sigma} = \mathbf{M}^t\mathbf{M}\delta\boldsymbol{\kappa}. \tag{4.68}$$

Unfortunately, even these **normal equations** for the solution of the least squares problem are prone to instability when the original system of equations is under-determined. That is, our estimate of $\delta\boldsymbol{\kappa}$, obtained by solving (4.68), can vary dramatically in the face of data noise and numerical errors.

The most common strategy for stabilizing the solution of the inverse problem, the solution of the system (4.68), involves introducing additional constraints or attributes that we seek in our solution. One attribute is suggested by the fact that we have a sparse data set and cannot expect to resolve fine details of the spatial variation of $\delta\boldsymbol{\kappa}$. Therefore, we should probably try and find the smoothly varying component of the spatial distribution of $\delta\boldsymbol{\kappa}$. One way to do this in the context of finding a least squares solution, is to introduce a measure of the smoothness, or conversely the roughness, as a penalty term in the minimization problem (Menke, 1989; Aster et al., 2013). For example, we can introduce a matrix $\mathbf{L}$ that is a difference operator on the vector of model parameters $\delta\boldsymbol{\kappa}$. In particular, $\mathbf{y} = \mathbf{L}\delta\boldsymbol{\kappa}$ is a vector of differences between values of $\delta\kappa_i$ for adjacent grid blocks, a discrete approximation to the gradient $\nabla\delta\kappa$. Adding $\beta\mathbf{y}^t\mathbf{y}$ to the misfit term μ^2 and minimizing results in a modification of the normal Equations (4.68):

$$\mathbf{M}^t\delta\sigma = \left(\mathbf{M}^t\mathbf{M} + \beta\mathbf{L}^t\mathbf{L}\right)\delta\boldsymbol{\kappa}, \tag{4.69}$$

this is the **regularized inverse problem**. The scalar weighting factor β controls the importance of a smooth solution with relative to matching the observations. The presence of the roughness penalty term stabilizes the solution of Equation (4.69) (Aki and Richards, 1980a; Menke, 1989). Note that other penalty terms are possible, such as a norm penalty or a term weighting the changes by the distance from a given point. Because the trajectories typically intersect a small number of grid blocks in the model, the matrix $\mathbf{M}$ is sparse, primarily consisting of zeros, and algorithms for solving the system (4.69) can take advantage of that structure (Paige and Saunders, 1982; Saad, 2003).

After solving Equation (4.69) for $\delta\boldsymbol{\kappa}$, we then complete the update by adding it to the background model. Because the problem is non-linear, the trajectories determining the coefficients in $\mathbf{M}$ depend upon $\kappa(\mathbf{x})$, we must iterate, taking additional linearized steps in order to solve the full problem. At each step we recalculate the trajectories and recompute the residuals. Nineteen iterations were required to construct the final model plotted in Figure 4.9. The initial and final fits to the 90 travel times are plotted in Figure 4.11. We discuss further such model updating via inversion including model assessment in Chapters 5 and 6 for single and multiphase transport problems. Note that this iterative approach is the least sophisticated technique for solving the inverse problem with little control on its convergence characteristics. There are other algorithms, such as the conjugate gradient routine, specifically designed for solving non-linear problems (Luenberger, 1973; Gill et al., 1982).

Waveform sensitivities and amplitude inversion

Let us move on to the consideration of sensitivities for the time-varying field itself, that is for the waveform $p(\mathbf{x}, t)$. In the discussion that follows we shall work in the

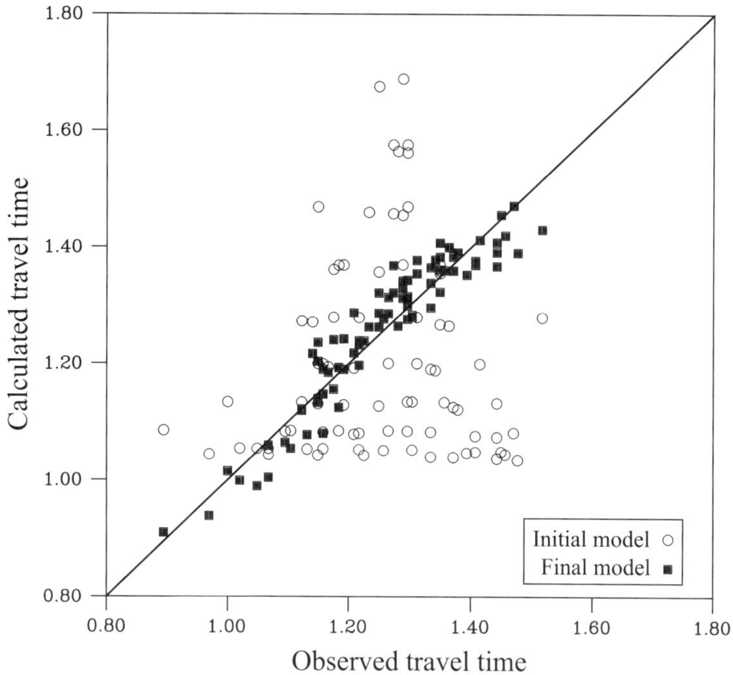

Figure 4.11 Initial and final fits to the observed travel time data. Travel times were calculated using the initial background model (Initial) and the final model that resulted after 19 iterations (Final). The calculated values are plotted against the observed values.

frequency domain and focus on $P(\mathbf{x}, \omega)$, as this leads to a cleaner derivation that is easier to follow. We can transform back if it is necessary to obtain time-domain expressions. In deriving waveform sensitivities there are at least two approaches to consider. First, we could treat the asymptotic expression (4.46) directly, perturbing a parameter such as permeability and following through to determine the resulting perturbations in phase and amplitude. This leads to the question of perturbations of the ray paths that was touched upon in the previous section, and to the topic of ray perturbation theory. Such derivations, though they lead to more efficient algorithms, can be rather elaborate and we shall not consider this approach further. As an alternative approach, we can consider a general perturbation technique beginning with the governing equation itself. We then introduce an asymptotic approximation at the end of our derivation, providing some of the efficiency associated with a trajectory-based approach.

The general perturbation approach is outlined in Box 4.3, given below. Our starting point is the diffusion equation in the frequency domain, Equation (4.3) where we have multiplied through by an assumed constant fluid viscosity μ,

$$\nabla \cdot (k(\mathbf{x})\nabla P) + i\omega\mu S(\mathbf{x})P = 0. \tag{4.70}$$

Because model parameter sensitivities relate small variations in model parameters to small changes in observed field quantities, we can use a perturbation approach for their calculation. In this spirit, consider the calculation of conductivity sensitivities. That is, let us calculate the changes in $P(\mathbf{x}, \omega)$ that arise when $k(\mathbf{x})$ is perturbed from a background value $\bar{k}(\mathbf{x})$, by a small amount $\delta k(\mathbf{x})$:

$$k(\mathbf{x}) = \bar{k}(\mathbf{x}) + \delta k(\mathbf{x}). \tag{4.71}$$

We will leave the simpler calculation of $S(\mathbf{x})$ sensitivities as an exercise.

Box 4.3 **A perturbation solution**

In this box we present an overview of a perturbation technique for computing a solution of a linear equation of the form

$$Lf = 0$$

where L is a linear operator and $f(\mathbf{x}, \omega)$ is the solution. For our particular application L is given by the differential operator with spatially varying coefficients: $L = \nabla \cdot k(\mathbf{x}) \nabla + i\omega S(\mathbf{x})$. We hope that by keeping the presentation general, the reader can see what is going on without the many distracting details necessary for our particular application below.

We will assume that L is the perturbation of a linear operator $\bar{L}$,

$$L = \bar{L} + \delta L,$$

where $\bar{L}$ defines an equation

$$\overline{Lf} = 0$$

that we can solve. Furthermore, we write the solution to the original equation, $Lf = 0$, as a perturbation of $\bar{f}$

$$f = \bar{f} + \delta f.$$

Hence, we can write the original equation as

$$\left(\bar{L} + \delta L\right)\left(\bar{f} + \delta f\right) = 0.$$

Carrying out the multiplication gives

$$\overline{Lf} + \bar{L}\delta f + \delta L\bar{f} + \delta L\delta f = 0$$

Because $\bar{f}$ satisfies the equation $\overline{Lf} = 0$, the first term vanishes. In addition, because the perturbations δL and δf are assumed to be small, we can neglect the product term $\delta L\delta f$, giving the equation

$$\bar{L}\delta f = -\delta L\bar{f}.$$

This is the background operator $\overline{L}$ applied to δf with a non-zero right-hand-side. The source term on the right consists of the operator δL applied to the solution of the background equation, $\overline{f}$. We may construct a formal solution to this equation using a Green's function approach (Roach, 1970; Stakgold, 1979), where the Green's function, $G(\mathbf{x}, \omega; \mathbf{y})$, is the solution of the background equation with a Dirac delta function source term:

$$\overline{L}G(\mathbf{x}, \omega; \mathbf{y}) = \delta(\mathbf{y}).$$

The formal solution for the perturbation, δf, is then

$$\delta f(\mathbf{x}, \omega) = -\int G(\mathbf{x}, \omega; \mathbf{y}) \delta L \overline{f}(\mathbf{y}, \omega) d\mathbf{y}.$$

The derivation proceeds in two stages. First, we derive a general expression for $\delta P(\mathbf{x}, \omega)$ in terms of $\delta k(\mathbf{x})$. Second, we use our asymptotic expression for $P(\mathbf{x}, \omega)$ to derive an efficient method for computing the sensitivities. The introduction of the perturbation in the model, $\delta k(\mathbf{x})$, will result in a perturbation of the field, which we write as the sum

$$P(\mathbf{x}, \omega) = \overline{P}(\mathbf{x}, \omega) + \delta P(\mathbf{x}, \omega) \tag{4.72}$$

where $\overline{P}$ is the field associated with the background model, and the perturbation, δP, is assumed to be small. Substituting the expressions (4.71) and (4.72) into the governing differential Equation (4.70) gives

$$\nabla \cdot \left[\left(\overline{k} + \delta k \right) \nabla \left(\overline{P} + \delta P \right) \right] + i\omega\mu\overline{S} \left(\overline{P} + \delta P \right) = 0 \tag{4.73}$$

or, expanding the products and neglecting terms of second order in the perturbations

$$\nabla \cdot \left(\overline{k} \nabla \overline{P} \right) + i\omega\mu\overline{S}\overline{P} + \nabla \cdot \left(\overline{k} \nabla \delta P \right) + \nabla \cdot \left(\delta k \nabla \overline{P} \right) + i\omega\mu\overline{S}\delta P = 0. \tag{4.74}$$

Because $\overline{P}$ satisfies the diffusion equation for the background model, the first two terms vanish and we are left with the equation

$$\nabla \cdot \left(\overline{k} \nabla \delta P \right) + i\omega\mu\overline{S}\delta P = -\nabla \cdot \left(\delta k \nabla \overline{P} \right). \tag{4.75}$$

This is a linear partial differential equation for the perturbation δP, that is identical to the governing Equation (4.70) but with a non-zero source term on the right-hand-side.

As noted in Box 4.3, we may formally solve the inhomogeneous partial differential Equation (4.75) using a Green's function, $G(\mathbf{x}, \omega; \mathbf{y})$, which is the solution of

$$\nabla \cdot \left(\overline{k}(\mathbf{x})\nabla G\right) + i\omega\mu\overline{S}(\mathbf{x})G = \delta(\mathbf{y}). \tag{4.76}$$

where $\delta(\mathbf{y})$ is a Dirac delta function situated at $\mathbf{y}$. The solution of Equation (4.75), written in terms of the Green's function is

$$\delta P(\mathbf{x},\omega) = -\int_V G(\mathbf{x},\omega;\mathbf{y})\nabla \cdot \left(\delta k \nabla \overline{P}\right) d\mathbf{y}, \tag{4.77}$$

where the integration is over the volume V (Stakgold, 1979). Typically, the volume V is the entire modeling domain and will extend a great distance from the point $\mathbf{x}$. Note that Equation (4.77) is in the form of the final equation in box 4.3.

At this point we are close to our ultimate goal, however, due to the presence of the divergence operator, we do not yet have a direct linear relationship between the perturbation in P and the perturbation in k. In fact, we are just an exercise in vector analysis away from our final expression. In Exercise 4.11 we ask the reader to derive the final representation of $\delta P(\mathbf{x},\omega)$:

$$\delta P(\mathbf{x},\omega) = \int_V \left[\nabla G(\mathbf{x},\omega;\mathbf{y}) \cdot \nabla \overline{P}(\mathbf{y},\omega)\right] \delta k(\mathbf{y}) dV, \tag{4.78}$$

a formula relating a perturbation in k at point $\mathbf{y}$ to a perturbation in P at the observation point $\mathbf{x}$. The sensitivity, the quantity contained within the square brackets, takes the particularly simple form of the scaler product of the gradient of the Green's function with the gradient of the background pressure field.

Exercise 4.11. Fill in the steps necessary to go from Equation (4.77) to Equation (4.78), in other words show that:

$$\int_V G\nabla \cdot \left(\delta k \nabla \overline{P}\right) dV = -\int_V \left(\nabla G \cdot \nabla \overline{P}\right) \delta k dV.$$

Hint: Start with the Gauss' divergence theorem

$$\int_V \nabla \cdot \mathbf{F} dV = \int_\Sigma \mathbf{F} \cdot \mathbf{n} d\Sigma$$

taking $\mathbf{F} = G\mathbf{g}$. Adopt the particular form $\mathbf{g} = \delta k \nabla \overline{P}$ for the vector $\mathbf{g}$. Note that it is typically assumed that the perturbation δk vanishes on the boundary, either because k is known on the boundary or because the boundary is so far away that a perturbation there will not influence our results at point $\mathbf{x}$.

Up to this point we have not made use of the asymptotic expression for $P(\mathbf{x},\omega)$ or $p(\mathbf{x},t)$, given by Equations (4.46) and (4.47), respectively. Any method could be used to find the functions $G(\mathbf{x},\omega;\mathbf{y})$ and $\overline{P}(\mathbf{y},\omega)$, that contribute to the integral (4.78). In fact, the form of the integrand provides an entry point into adjoint-state techniques for calculating model parameter sensitivities (Sun, 1994). In order to see this, consider the component terms within the square brackets of the integrand. The Green's function represents propagation from the point $\mathbf{y}$ to the observation point $\mathbf{x}$. The function $\overline{P}(\mathbf{y},\omega)$ denotes the field at the point $\mathbf{y}$ due to the source. Using reciprocity, one can calculate the Green's function using a source at the observation point $\mathbf{x}$ and observation points at all points $\mathbf{y}$ within model. Thus, one may compute the sensitivities for a given data point, observed at $\mathbf{x}$, and a given source location $(\mathbf{x}_s)$, using two forward simulations, with sources at $\mathbf{x}$ and $\mathbf{x}_s$, respectively.

Let us take advantage of the semi-analytic zeroth-order asymptotic solution (4.46) with a slight variation on the notation

$$\overline{P}(\mathbf{y},\omega) = A\left(\mathbf{x}_s,\mathbf{y}\right) e^{-\sqrt{-i\omega}\sigma(\mathbf{x}_s,\mathbf{y})}. \tag{4.79}$$

where $(\mathbf{x}_s,\mathbf{y})$ in the amplitude and phase terms signifies that the source is at $\mathbf{x}_s$ and the observation is at $\mathbf{y}$. Similarly, for the Green's function we have

$$G(\mathbf{x},\omega;\mathbf{y}) = A\left(\mathbf{y},\mathbf{x}\right) e^{-\sqrt{-i\omega}\sigma(\mathbf{y},\mathbf{x})}, \tag{4.80}$$

for a source at $\mathbf{y}$ and an observation at $\mathbf{x}$. These solutions are defined on trajectories extending from the source to the receiver. The trajectories will curve in response to spatial variations in $k(\mathbf{x})$. In order to compute the sensitivities using the expression in square brackets in (4.78) we need the gradients of $\overline{P}$ and G. In Exercise 4.12 we ask the reader to derive these gradients and to substitute them into Equation (4.78) to derive the asymptotic approximation of the pressure sensitivity

$$\delta P(\mathbf{x},\omega) = -i\omega \int_V \nabla\sigma\left(\mathbf{x}_s,\mathbf{y}\right)\cdot\nabla\sigma\left(\mathbf{y},\mathbf{x}\right) A\left(\mathbf{x}_s,\mathbf{x}\right) e^{-\sqrt{-i\omega}\sigma(\mathbf{x}_s,\mathbf{x})}\delta k(\mathbf{y})dV. \tag{4.81}$$

In this equation $A\left(\mathbf{x}_s,\mathbf{x}\right)$ (given by Equation 4.45) and $\sigma\left(\mathbf{x}_s,\mathbf{x}\right)$ are composite terms due to the propagation from the source, $\mathbf{x}_s$, to the point at which the sensitivity is to be calculated, $\mathbf{y}$, and from $\mathbf{y}$ to the observation point $\mathbf{x}$, given by

$$A\left(\mathbf{x}_s,\mathbf{x}\right) = A\left(\mathbf{x}_s,\mathbf{y}\right) A\left(\mathbf{y},\mathbf{x}\right) \tag{4.82}$$

and

$$\sigma\left(\mathbf{x}_s,\mathbf{x}\right) = \sigma\left(\mathbf{x}_s,\mathbf{y}\right) + \sigma\left(\mathbf{y},\mathbf{x}\right). \tag{4.83}$$

Exercise 4.12. Apply the gradient operator to the asymptotic expressions for $\overline{P}(\mathbf{y},\omega)$ and $G(\mathbf{x},\omega;\mathbf{y})$, Equations (4.79) and (4.80), respectively. Substitute the resulting expressions into Equation (4.78)

$$\delta P(\mathbf{x},\omega) = \int_V \left[\nabla G(\mathbf{x},\omega;\mathbf{y})\cdot\nabla\overline{P}(\mathbf{y},\omega)\right]\delta k(\mathbf{y})dV,$$

to derive the high-frequency asymptotic expression (4.81) relating a perturbation in the model to a perturbation in the observable pressure.

Hint: For the high-frequency asymptotic solution only terms of the highest order in $\sqrt{-i\omega}$ are important.

To recapitulate, in order to compute a model parameter sensitivity we require two trajectory-based solutions. The first is from the source to the location of the point of interest. The second is from the point of interest to the observation point. The total perturbation δP is a volume integration over all of the model parameter perturbations δk.

The semi-analytic form (4.81) provides insight into the nature of the sensitivity kernel, relating a perturbation in $k(\mathbf{y})$ to a perturbation in the observed pressure $P(\mathbf{x}, \omega)$. An example of a sensitivity kernel is plotted in Figure 4.12. The sensitivity

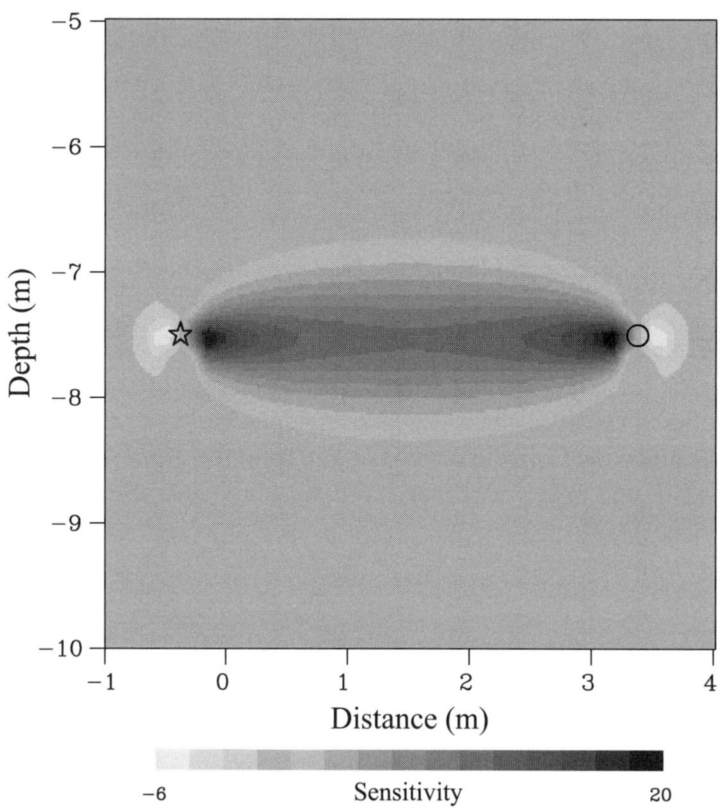

Figure 4.12 Permeability sensitivity for an observation gathered 0.16s after the start of injection. The sensitivity was computed using the perturbation approach described in this section and an asymptotic solution. The source and receiver are denoted by an unfilled star and an open circle, respectively.

kernel is given by the integrand of (4.81) along with the outer factor $-i\omega$. From the structure of the integrand one notices that the sensitivity kernel involves a propagation from the source at $\mathbf{x}_s$, to the point of interest $\mathbf{y}$ and then onto the point $\mathbf{x}$ at which the pressure is measured. This quantity is weighted by the scalar product of the gradients of the phase for each path, $\nabla\sigma(\mathbf{x}_x, \mathbf{y}) \cdot \nabla\sigma(\mathbf{y}, \mathbf{x})$. The scalar product is a strong function of distance from the source and observation points and also changes sign when the angle between the two paths at $\mathbf{y}$ is between 90 and 270 degrees. These features are evident in the Figure 4.12 where we see negative sensitivities behind the source and observation points. Intuitively, this corresponds to the fact that lowering the permeability behind the source and/or the observation point will increase the observed pressure response. The $-i\omega$ term in the Equation (4.81) transforms into a temporal derivative in time domain. It reflects the fact that the slope of the pressure curve is important for the high-frequency sensitivity. The slope of the pressure curve provides a relationship between a time shift of the pulse, because an increase or decrease in $k(\mathbf{y})$ results in a change in σ and hence a time shift, leading to a change in the amplitude of the curve.

4.4 Summary

Though diffusion and wave propagation are dramatically different physical processes, they can be studied and understood using common approaches. In particular, asymptotic series and a trajectory-based formulation can be applied to both. This is helpful conceptually, allowing for the partitioning of the problem into a travel time calculation and an amplitude computation. Furthermore, the common framework is welcome because many physical problems do not simply fall into either category, and may change character depending on the range of their parameters.

5

Equations governing advection and transport

5.1 Introduction

An important characteristic of a fluid in motion is its ability to transport dissolved or microscopic material. This property forms the basis for numerous important physical processes in porous media. Furthermore, by monitoring the movement of advected objects one can image the structure of the host medium and infer its flow properties. Intuitively, one might postulate that the concentration of advected material will depend in a fundamental way upon the trajectories associated with the fluid flow. In this chapter we develop a trajectory-based solution for the concentration of an advected material as a function of time and spatial position. Though our derivation in the next section is along the lines of the asymptotic approach developed previously, solutions of this type have a long history. We connect the asymptotic results to current streamline-based modeling techniques. We conclude the chapter with applications illustrating the use of such techniques in visualization, in sensitivity computation, and in imaging spatial variations in flow properties.

5.2 The governing equation

The equation governing the concentration of an advected chemical component, $c(\mathbf{x}, t)$, was given in Chapter 2. It can be written as the partial differential Equation [see Equation (2.104)]

$$\phi \frac{\partial c}{\partial t} = \nabla \cdot (\mathbf{D} \nabla c - \mathbf{q} c), \tag{5.1}$$

where ϕ is the kinematic porosity, $\mathbf{D}$ is the dispersion tensor, and $\mathbf{q}$ is the Darcy flow velocity. The dispersion tensor was introduced in Chapter 2 and some of its characteristics are noted in Box 5.1. Equation (5.1), the advection-dispersion equation, can be of mixed character. That is, the equation may have solutions that display both diffusive and wave-like behavior, depending upon the relative magnitude of the coefficients. As noted in Chapter 2, this is the governing equation for the simplest case of a non-interacting tracer, where the advected material does not significantly influence the flow field. In that case we refer to (5.1) as the equation governing **tracer transport**. However, there are a host of more complicated processes, such as reactive transport and multiphase flow, that may be modeled by equations of the form of (5.1). Some of these phenomena render the governing equations nonlinear. However, we will tackle the simpler linear case first, leaving the nonlinear multiphase case for the Chapter 6.

In order to complete the specification, we need to relate the flow velocity vector $\mathbf{q}$ to the properties of the medium and the pressure field. For that task we invoke Darcy's law, Equation (2.90), relating the flow velocity to the gradient of the pressure field p:

$$\mathbf{q} = -\frac{\mathbf{k}}{\mu} \cdot \nabla p$$

where $\mathbf{k}(\mathbf{x})$ is the permeability tensor. From this equation and Equation (5.1) one can see that the concentration of advective fluid depends upon the kinematic porosity, the dispersion matrix, the hydraulic conductivity tensor, and the gradient of the fluid pressure. The presence of the pressure field, implicit in Equation (5.1), means that we must also solve its governing equation, the diffusion equation treated in Chapter 4, [see Equation (4.1)] or its steady-state version

$$\nabla \cdot \frac{\mathbf{k}}{\mu} \cdot \nabla p = Q(\mathbf{x}) \tag{5.2}$$

where $Q(\mathbf{x})$ is a source term. This must be done either first, if the concentration does not change the flow field, or simultaneously, when the presence of the advected material influences the flow field. The dependence on the fluid pressure field is a hallmark of advective processes that must always be kept in mind. It is particularly meaningful for the inverse problem, when using tracer for imaging properties of the porous medium. In that case, changes in the flow properties can produce variations in both the pressure field and the concentration distribution.

Box 5.1 **Hydrodynamic dispersion**

The hydrodynamic dispersion of a solute describes the spreading of the material due to the local heterogeneity of both the porous material as well as the flow field. Early studies emphasized both diffusion and the mixing due to variable velocity for flow in small tubes (Taylor, 1953). Scheidegger (1954) conducted an early examination of the small scale mixing caused by flow variation in a microscopically heterogeneous porous medium. After initial work, it was realized that dispersion is inherently anisotropic even if the medium is isotropic and the principle directions are determined by the flow direction (de Josselin de Jong, 1958; Nikolaevskii, 1959; Saffman, 1959; Bear, 1961; Scheidegger, 1961). The contributions of stochastic methods to hydrology (Warren and Skiba, 1964; Schwartz, 1977; Gelhar et al., 1979; Dagan, 1982; Gelhar and Axness, 1983) have advanced our understanding of transport in heterogeneous permeable media. As a result of these and other efforts, there are now several models of dispersion in a porous medium, and work in this area continues (Liu and Si, 2008; Bruining et al., 2012). Here we adopt an established approach that accounts for both molecular diffusion and hydrodynamic dispersion. Experimental evidence, as cited by de Marsily (1986, p. 237), indicates that in most flow regimes and for an isotropic medium, the dispersion along (D_l) and transverse (D_t) to the flow direction, are well approximated by

$$D_l = \varphi_k d_m + \alpha_l q$$

and

$$D_t = \varphi_k d_m + \alpha_t q$$

where d_m is the molecular diffusion coefficient, q is the magnitude of the Darcy velocity vector. The parameters α_l and α_t are the **intrinsic dispersion coefficients**, or **dispersivities**, in the longitudinal (parallel to the flow) and transverse (perpendicular to the flow) directions. The approach described here is appropriate for a general anisotropic porous medium in which the dispersion in the principle directions of the dispersion tensor are given by three scalars, say D_1, D_2 and D_3. Hence, our development will be in terms of a general tensor $\mathbf{D}$.

The dispersion tensor is commonly assumed to be symmetric. In fact, the asymmetry of the dispersion tensor was demonstrated by Koch and Brady (1987), and recently verified by Auriault et al. (2010) in the case of low Pectlet numbers. The general asymmetry of $\mathbf{D}$ has not been emphasized in hydrology. This may be due to the fact that the symmetry holds in a homogeneous media, but not in one that is heterogeneous (Carbonell and Whitaker, 1983). To see this, consider equation 5.1 when the dispersion tensor does not depend upon $\mathbf{x}$. The symmetry of the array of second-order spatial derivatives of C, and the invariance of the other terms in the equation, implies the symmetry of the dispersion tensor.

5.3 An asymptotic solution

There are several ways to derive a trajectory-based solution to a governing equation, such as (5.1). In Chapter 4 we adopted an approach based upon an expansion in powers of $1/\omega$, a very common avenue to such solutions. This approach is typically implemented in the limit of high frequency. There are situations in which the frequency might be expected to vary over a wider range of values and ω may not always be large. The multiple-scale asymptotic method for a smoothly-varying medium, introduced in Chapter 3, will allow us to derive a trajectory-based solution that is slightly more general. The approach may be used in the frequency domain, as is done here, or in the time domain, as in Chapter 6.

The general idea is that the leading edge of the concentration front is sharp, relative to the controlling features of the problem, in this case the heterogeneity of the medium. Note that this condition is only required in regions between interfaces. That is, as in the manner of geometrical optics and ray methods (Chapman, 2004), we may include rapid variations in medium properties as boundary conditions. Away from such boundaries, it is assumed that the properties vary over a length scale that we shall characterize by the parameter L. In essence, we are stipulating that all heterogeneity or background variations are over length scales that are greater or equal to L. The propagation of the advected material is defined by a front, over which the concentration varies from an initial background value to a value characteristic of the injected material. The length scale over which this transition is achieved is denoted by l. The requirement that the front is sharp, or that the heterogeneity is smoothly-varying, is formulated in terms of the inequality:

$$l \ll L. \tag{5.3}$$

Note that this condition may be applied to either spatial or temporal length scales or to both. From the inequality (5.3) we can define a scale parameter, based upon the ratio of length scales:

$$\varepsilon = \frac{l}{L} \ll 1. \tag{5.4}$$

To facilitate the scale separation we introduce coordinates that are designed to represent the slowly-varying background properties. These coordinates are defined in relation to the physical coordinates $\mathbf{x}$ and t, in terms of the scale parameter:

$$\mathbf{X} = \varepsilon \mathbf{x} \tag{5.5}$$

and/or

$$T = \varepsilon t. \tag{5.6}$$

So, as the physical coordinates x, t vary by a unit amount, the coordinates $\mathbf{X}, T$ vary by a correspondingly small multiple ε. The asymptotic solution is a power series representation of the concentration variation $c(\mathbf{x}, t)$, in our case an expansion in powers of ε,

$$c(\mathbf{x}, t) = c(\mathbf{X}, T, \varepsilon) = \sum_{n=0}^{\infty} \varepsilon^n c_n(\mathbf{X}, T, \theta). \tag{5.7}$$

For a given value of ε, the series (5.7) may ultimately diverge as n grows large. However, in the limit as ε approaches zero, the first few terms in the series provide an increasingly accurate in representation of $c(\mathbf{x}, t)$ (Dingle, 1973).

In our analysis of concentration variations we shall work in the frequency domain. In doing so, we shall maintain some continuity with the results of Chapter 4, and provide a transition to the applications in Chapter 6 and 7. Furthermore, it is easier to motivate aspects of a propagating front, such as the front phase, in the frequency domain. As in Chapter 4, we apply the Fourier transform to the governing equation, in this case Equation (5.1)

$$\nabla \cdot (\mathbf{D}\nabla C - \mathbf{q}C) + i\omega\phi C = 0 \tag{5.8}$$

where $C(\mathbf{x}, \omega)$ denotes the Fourier transform of $c(\mathbf{x}, t)$

$$C(\mathbf{x}, \omega) = \frac{1}{\sqrt{2\pi}} \int_{-\infty}^{\infty} c(\mathbf{x}, t) e^{i\omega t} dt \tag{5.9}$$

(Bracewell, 2000; Debnath, 2005, p. 35) which is a complex quantity. The asymptotic representation (5.7) may be partitioned into phase and amplitude components:

$$C(\mathbf{x}, \omega, \varepsilon) = e^{\theta(\mathbf{x},\omega)} \sum_{n=0}^{\infty} \varepsilon^n C_n(\mathbf{X}, \omega), \tag{5.10}$$

where $\theta(\mathbf{x}, \omega)$ is the phase and $C_n(\mathbf{X}, \omega)$ are successive amplitude terms. In the frequency domain the phase shift associated with a propagating transient disturbance or wave takes the form of an exponential factor. Thus, one can consider the representation (5.10) as a generalized wavefront expansion of a propagating disturbance (Aki and Richards, 1980a). Note that the phase, representing the jump across the propagating front, varies rapidly in space, and therefore depends upon the physical coordinates $\mathbf{x}$. The amplitude terms $C_n(\mathbf{X}, \omega)$ represent the much more gradual decay of the concentration with distance, which varies over the length scale of the heterogeneity and is thus a function of the $\mathbf{X}$ coordinates. The partitioned form provides additional flexibility in both the forward modeling and in solving the inverse problem, allowing for a form of travel time tomography, as illustrated at the end of this chapter.

In substituting $C(\mathbf{x}, \omega, \varepsilon)$ into the governing equation (5.8) some care must be exercised when computing the spatial derivatives. In particular, the relationship between $\mathbf{x}$ and $\mathbf{X}$ in Equation (5.5) needs to be accounted for, as does the implicit dependence upon $\mathbf{x}$, through the presence of the phase function $\theta(\mathbf{x}, \omega)$. Thus, the partial derivative with respect to x_i is given by

$$\frac{\partial C}{\partial x_i} = \frac{\partial X_i}{\partial x_i}\frac{\partial C}{\partial X_i} + \frac{\partial \theta}{\partial x_i}\frac{\partial C}{\partial \theta}$$
$$= \varepsilon\frac{\partial C}{\partial X_i} + \frac{\partial \theta}{\partial x_i}C, \tag{5.11}$$

where we have used the fact that the solution is taken to have the form (5.10) and the partial derivative with respect to θ returns the original function:

$$\frac{\partial C}{\partial \theta} = C.$$

Writing Equation (5.8) in terms of $C(\mathbf{X}, \omega, \varepsilon)$ and its derivatives with respect to x_i, results in the expression

$$\begin{aligned}&\varepsilon^2\bar{\nabla}\cdot(\mathbf{D}\cdot\bar{\nabla}C)\\&+\varepsilon\left[\bar{\nabla}\cdot(\mathbf{D}\cdot\nabla\theta C)+\nabla\theta\cdot(\mathbf{D}\cdot\bar{\nabla}C)-\mathbf{q}\cdot\bar{\nabla}C-C\bar{\nabla}\cdot\mathbf{q}\right]\\&+(\nabla\theta\cdot\mathbf{D}\cdot\nabla\theta)\,C-\nabla\theta\cdot\mathbf{q}C+i\omega\phi C=0,\end{aligned} \tag{5.12}$$

where the overbar signifies that the gradient is with respect to the $\mathbf{X}$ coordinate. Because we are constructing a solution in a medium with smoothly varying heterogeneity, the condition $\varepsilon \ll 1$ holds and terms of lowest order in ε are the most significant. We consider these terms in some detail in the sub-section that follows.

Exercise 5.1. Derive the asymptotic expression

$$\begin{aligned}&\varepsilon^2\bar{\nabla}\cdot(\mathbf{D}\cdot\bar{\nabla}C)\\&+\varepsilon\left[\bar{\nabla}\cdot(\mathbf{D}\cdot\nabla\theta C)+\nabla\theta\cdot(\mathbf{D}\cdot\bar{\nabla}C)-\mathbf{q}\cdot\bar{\nabla}C-C\bar{\nabla}\cdot\mathbf{q}\right]\\&+(\nabla\theta\cdot\mathbf{D}\cdot\nabla\theta)\,C-\nabla\theta\cdot\mathbf{q}C+i\omega\phi C=0,\end{aligned}$$

starting from the governing equation (5.8) and using expression (5.11) for the spatial derivatives of C.

5.3.1 Zeroth-order terms: expressions for the trajectories and the phase

Note that each term in (5.12) is linear in C. Therefore, after we substitute the series (5.10) for C, the terms of lowest order in ε, those of zeroth-order, are found in the third line of Equation (5.12):

$$[\nabla\theta\cdot\mathbf{D}\nabla\theta-\nabla\theta\cdot\mathbf{q}+i\omega\phi]\,C_0=0. \tag{5.13}$$

The terms of order-zero are obtained when we consider the first term of the asymptotic series (5.10). From the above equation it is clear that the quantity in square brackets must vanish in order to have a non-trivial solution:

$$\nabla\theta\cdot\mathbf{D}\nabla\theta-\nabla\theta\cdot\mathbf{q}+i\omega\phi=0, \tag{5.14}$$

producing a first-order, non-linear, scalar partial differential equation for $\theta(\mathbf{x}, \omega)$.

At first sight, it may seem like a drawback to go from a linear, scalar partial differential equation (5.1) to our current equation (5.14). However, there is a compelling reason to make this transition. The original governing equation (5.8) is second-order partial differential equation while the current equation is only first-order. As indicated in Chapter 3, there is a rich literature (Courant and Hilbert, 1962; Sneddon, 2006, p. 44) on solving such first-order equations. These techniques provide important physical insight and will allow us to define solutions along one-dimensional trajectories. Introducing the phase gradient vector $\mathbf{p} = \nabla\theta$, we can write Equation (5.14) as

$$F(\mathbf{x}, \mathbf{p}) = \mathbf{p} \cdot \frac{\mathbf{D}}{\phi}\mathbf{p} - \mathbf{p} \cdot \frac{\mathbf{q}}{\phi} + i\omega = 0. \tag{5.15}$$

Equation (5.15) may be treated using the method of characteristics, introduced in Section 3.5 of Chapter 3, where the independent variables $\mathbf{x}$ and $\mathbf{p}$ are considered to be functions of a scalar quantity s. The scalar s is a parameter signifying position along a trajectory extending from the source to the observation point. For F given by Equation (5.15) the characteristic equations are

$$\begin{aligned} \frac{dx_l}{ds} &= \frac{\partial F}{\partial p_l} = 2\frac{D_{lj}}{\phi}p_j - \frac{q_l}{\phi}, \\ \frac{dp_l}{ds} &= -\frac{\partial F}{\partial x_l} = -p_i\frac{\partial}{\partial x_l}\left(\frac{D_{ij}}{\phi}\right)p_j + \frac{\partial}{\partial x_l}\left(\frac{q_i}{\phi}\right)p_i, \end{aligned} \tag{5.16}$$

[see the ray equations in the box following Section 3.5]. This form was derived in an early analysis of dispersion governed by the advection-diffusion equation containing a constant scalar diffusion coefficient in place of our tensor $\mathbf{D}$ (Smith, 1981). We can also write these equations in matrix-vector form:

$$\begin{aligned} \frac{d\mathbf{x}}{ds} &= 2\frac{\mathbf{D}}{\phi}\mathbf{p} - \frac{\mathbf{q}}{\phi} \\ \frac{d\mathbf{p}}{ds} &= -\mathbf{p}^T \cdot \nabla\left(\frac{\mathbf{D}}{\phi}\right)\mathbf{p} + \nabla\left(\frac{\mathbf{q}}{\phi}\right) \cdot \mathbf{p} \end{aligned} \tag{5.17}$$

(Vasco et al., 2016). Given the coefficients $\mathbf{D}$, $\mathbf{q}$, and ϕ and the appropriate initial conditions such as the source point, $\mathbf{x}_s$ and the initial vector $\mathbf{p}_s$ or, equivalently, the take-off angle of the trajectory, the ray is uniquely determined by the coupled set of ordinary differential equations (5.17). Thus, taking the ray-based approach, one simply solves the system of equations (5.17) numerically using software for the integration of ordinary differential equations (Press et al., 1992, p. 701). We should not lose sight of our main task in this section, to determine the phase $\theta(\mathbf{x}, \omega)$, the solution of Equation (5.14). For this, one may use the third ray equation [see the box following Section 3.5]:

$$\frac{d\theta}{ds} = \mathbf{p} \cdot \nabla_p F = 2\mathbf{p} \cdot \frac{\mathbf{D}}{\phi}\mathbf{p} - \mathbf{p} \cdot \frac{\mathbf{q}}{\phi}, \tag{5.18}$$

which may be integrated numerically, in conjunction with the system of Equations (5.17).

There is another intriguing approach that leads us on an entirely new path. Specifically, we can adopt the trajectory defined by a simplification of the first set of ray equations (5.17)

$$\frac{d\mathbf{x}}{ds} = -\frac{\mathbf{q}}{\phi} \tag{5.19}$$

as our path, rather than the exact ray path. This equation follows from the ray equations if the dispersion vanishes. Given the flow field $\mathbf{q}$, say one obtained from a numerical solution of the pressure equation (5.2), the computation of the trajectory $\mathbf{x}(s)$ based upon Equation (5.19) is an elementary undertaking. The trajectory has a useful physical interpretation as a streamline of the flow field. There is a vast literature on streamline computation and streamline-based fluid flow modeling (Bear, 1972; Crane and Blunt, 1999; Datta-Gupta and King, 2007). We will have more to say about streamline modeling later in this chapter. However, the reader should keep in mind that streamlines are completely determined by the flow field and are not necessarily paths taken by the tracer itself. This is clear in the dipole flow field portrayed in Figure 5.1, where we plot both the trajectories $\mathbf{x}(s)$ defined by the Equation (5.17) on the left and corresponding streamlines on the right. In this

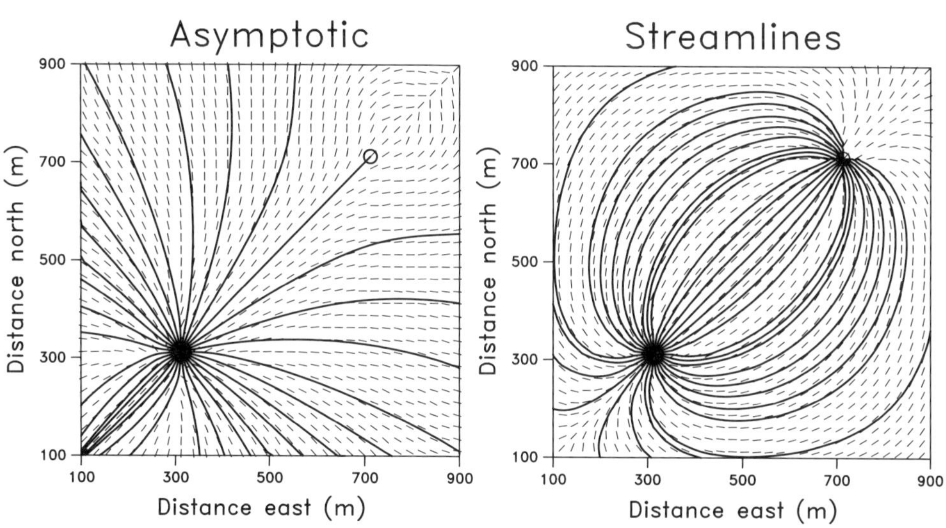

Figure 5.1 A comparison of the ray paths(left) and streamlines(right) for a dipole flow field.

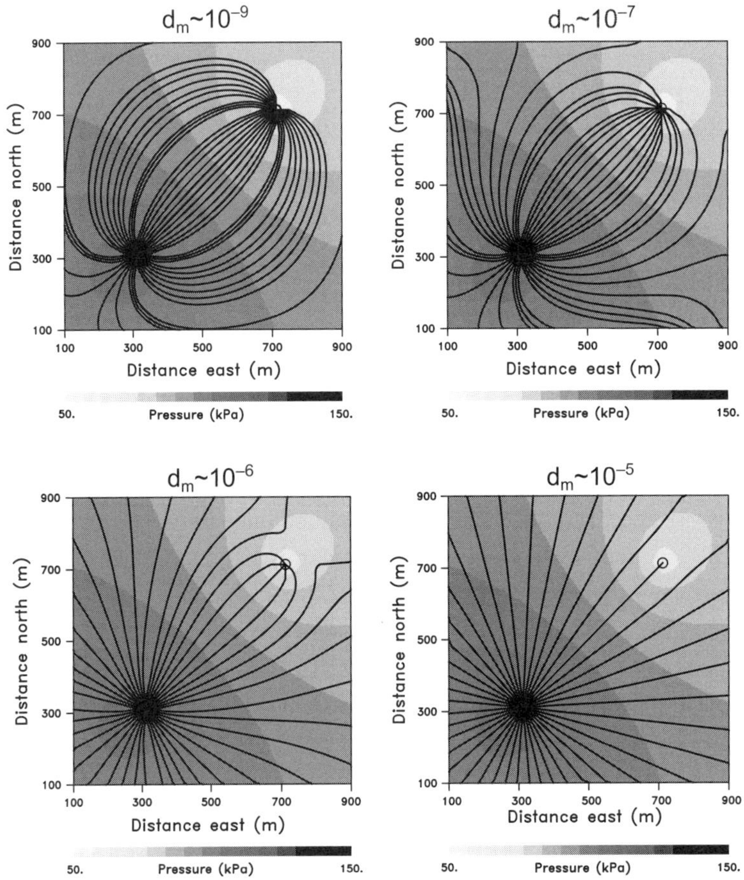

Figure 5.2 A comparison of the trajectories for varying values of molecular diffusion, d_m.

case the longitudinal and transverse dispersion are taken to be equal to exaggerate the effects. As indicated by the first equation (5.17), the tangent to the asymptotic trajectory $\mathbf{x}(s)$ is determined by the phase gradient vector, $\mathbf{p} = \nabla\theta$, as well as by the flow velocity vector $\mathbf{q}$. In certain situations the vector field defined by the right-hand-side of the first equation (5.17) can differ significantly from $\mathbf{q}$. In the left panel of Figure 5.1 we plot the normalized tangent vectors as short line segments. They are notably different from the normalized flow directions $\hat{\mathbf{q}}$ plotted in the right panel. Admittedly, this is an extreme case, typically the longitudinal dispersion is much larger than the transverse dispersion. But this example illustrates that tracer trajectories can deviate from streamlines, and the first equation of (5.17) quantifies this deviation. In Figure 5.2 we plot the asymptotic trajectories for varying values of molecular diffusion, d_m. As the molecular diffusion increases the trajectories increasingly deviate from conventional streamlines. Later in this chapter we discuss a technique for including transverse dispersion in streamline calculations.

Exercise 5.2. Starting with Equation (5.15), derive the trajectory equations given by Equation (5.17) and (5.18). **Hint**: Refer to the ray equations in Box 3.1.

Exercise 5.3. Use the software 'TRACE3D' to visualize the streamlines for a homogeneous medium with a five spot well configuration (a central injector and four surrounding producers). Use the data files in the folder 'Examples/Simulation/5-Spot-Hom' by creating a project in the same folder and running the software. Repeat the exercise for the heterogeneous example (in 'Examples/Simulation/5-Spot-Het'). View the heterogeneous permeability field and the streamlines side by side to examine how the trajectories are impacted by the heterogeneity.

5.3.2 Terms of order ε: an expression for the amplitude

The next level of approximation involves consideration of terms of first order in Equation (5.12). The only terms that could possibly be of order ε, after substituting in the series (5.10), are contained in the last two lines. Recalling Equation (5.14) from the previous section, we find that the expression in brackets in the bottom line of Equation (5.12) vanishes. We are left with everything within the brackets in the second line of (5.12). If we insert the asymptotic series (5.10) into the resulting equation then the only terms of order ε will be those corresponding to $n = 0$. Factoring out e^{θ} under the assumption that it does not vanish, and grouping terms that contain C_0 and ∇C_0 results in

$$\Upsilon \cdot \nabla C_0 = \upsilon C_0. \tag{5.20}$$

where

$$\Upsilon = \mathbf{Dp} + \mathbf{p} \cdot \mathbf{D} - \mathbf{q} \tag{5.21}$$

$$\upsilon = \nabla \cdot (\mathbf{Dp}) + 2\,(\mathbf{Dp})\,\mathbf{p} - \mathbf{q} \cdot \mathbf{p} - \nabla \cdot \mathbf{q}. \tag{5.22}$$

Equation (5.20) is a linear, first-order partial differential equation for the zeroth-order amplitude $C_0(\mathbf{X}, \omega)$, commonly known as the transport equation. As there are numerous texts on the numerical solution of linear partial differential equations in general and on the solution of the transport equation in particular, we shall not discuss the solution of (5.20) any further. We end by pointing out that one may derive a semi-analytical solution of the transport equation in ray coordinates (Kravtsov and Orlov, 1990).

5.3.3 Construction and interpretation of a zeroth-order expression for the concentration

The expressions for the phase $\theta(\mathbf{x}, \omega)$ and the zeroth-order amplitude $C_0(\mathbf{X}, \omega)$ provide the ingredients that we need to construct the lowest-order approximation of the concentration as a function of position and frequency:

$$C(\mathbf{x}, \omega) = e^{\theta(\mathbf{x}, \omega)} C_0(\mathbf{x}, \omega), \tag{5.23}$$

the first term of the asymptotic series (5.10). A time-domain solution is obtained by applying the inverse Fourier transform to (5.23) (Bracewell, 2000). Because the inverse Fourier transform of a product of two functions is equivalent to the convolution of their inverse Fourier transforms, the time-domain solution will have the form

$$c(\mathbf{x}, t) = \mathcal{F}^{-1}\left[e^{\theta(\mathbf{x}, \omega)}\right] * c_0(\mathbf{x}, t), \tag{5.24}$$

where $\mathcal{F}^{-1}[F(\omega)]$ denotes the inverse Fourier transform of the function $F(\omega)$ and $*$ denotes a convolution. If the phase function $\theta(\mathbf{x}, \omega)$ does not have an analytic form, for example it is defined by the differential equation (5.18), then it is not possible to compute a closed form inverse Fourier transform. That is not an issue, because it is trivial to compute Fourier transforms numerically, using the fast Fourier transform algorithm (Press et al., 1992, p. 498). In certain situations, such as a rapid variation in concentration or a non-dispersive porous medium, one can perform an explicit evaluation of the inverse transform. The resulting semi-analytic solutions provide some insight into the contributions to the concentration variation at a particular observation point. We discuss these two cases in the sub-sections that follow.

The differential equation (5.18) is helpful in obtaining a numerical estimate but does not offer much insight into the structure of the phase. To make further progress in finding a useful expression for the tracer concentration we will need an alternative form for the phase. It is fruitful to return to the defining equation (5.15). Because the velocity vector $\mathbf{q}$, the porosity ϕ, and the dispersion tensor $\mathbf{D}$ are known parameters, the real value of Equation (5.15) is that it provides a semi-analytic relationship between the phase $\theta(\mathbf{x})$ and the properties of the medium. To see this, we first use the singular value decomposition of the dispersion tensor (Noble and Daniel, 1977, p. 306)

$$\mathbf{D} = \mathbf{U}\Lambda\mathbf{V}^T,$$

where $\mathbf{U}$ and $\mathbf{V}$ are orthonormal matrices (their columns are vectors of unit magnitude) and Λ is a diagonal matrix, to write Equation (5.15) as a scalar equation for $p = |\mathbf{p}|$, the magnitude of the vector $\mathbf{p}$:

$$\beta p^2 - \gamma p + i\omega\varphi_k = 0. \tag{5.25}$$

The coefficient β of this quadratic equation is given by

$$\beta = \sum_{i=1}^{3} D_i \cos \mu_i \cos \nu_i$$

where D_i is the *i-th* element on the diagonal of Λ, $\cos \mu_i = \mathbf{u}_i \cdot \hat{\mathbf{p}}$ is the cosine of the angle between $\hat{\mathbf{p}}$, the unit vector in the direction of $\mathbf{p}$, and the *i-th* column of $\mathbf{U}$, Similarly, $\cos \nu_i = \mathbf{v}_i \cdot \hat{\mathbf{p}}$ denotes the cosine of the angle, ν_i, between $\mathbf{v}_i$ and $\mathbf{p}$. Note that diagonal elements D_i are discussed in Box 5.1, where they are related to the molecular diffusion and the intrinsic dispersion coefficients. The coefficient γ is sensitive to the fluid velocity vector $\mathbf{q}$,

$$\gamma = q \cos \eta, \tag{5.26}$$

where $q = |\mathbf{q}|$ and η is the angle between the vectors $\mathbf{p}$ and $\mathbf{q}$. In the exercises we ask the reader to derive Equation (5.25). It is straight-forward to solve the quadratic equation and find the two solutions for p:

$$p = \frac{\gamma}{2\beta} \pm \frac{\sqrt{\gamma^2 - 4\beta\varphi_k i\omega}}{2\beta}. \tag{5.27}$$

In the characteristic coordinate system, determined by the trajectories $\mathbf{x}(s)$, Equation (5.27) may be integrated with respect to s, the distance along the trajectory. Thus, we can write the phase $\theta(\mathbf{x}, \omega)$ succinctly as

$$\theta(\mathbf{x}, \omega) = \xi(\mathbf{x}) \pm \chi(\mathbf{x}, \omega), \tag{5.28}$$

where we have defined a function that only depends upon spatially varying quantities,

$$\xi(\mathbf{x}) = \int_{\mathbf{x}} \frac{\gamma}{2\beta} ds \tag{5.29}$$

and a function that describes the frequency-dependence,

$$\chi(\mathbf{x}, \omega) = \int_{\mathbf{x}} \frac{\sqrt{\gamma^2 - 4\beta\varphi_k i\omega}}{2\beta} ds, \tag{5.30}$$

as well as the spatial variation contained in γ, β, and φ_k. The time-domain solution is obtained by substituting this expression for $\theta(\mathbf{x}, \omega)$ into Equation (5.24) and performing the inverse Fourier transform followed by the convolution.

As an illustration, we consider the simple case of an impulsive injection of tracer into a uniform, two-dimensional flow field. Cleary and Ungs (1978) provided an analytic solution for this case, as presented in Javandel et al. (1984). The solution corresponds to a pulse-like injection along a linear segment oriented perpendicular to a uniform flow field. The dispersion is characterized by transverse (D_t) and longitudinal (D_l) coefficients, which are 1.0×10^{-6} m^2/s and 2.0×10^{-3} m^2/s, respectively. In Figure 5.3 we plot the temporal variation in concentration observed

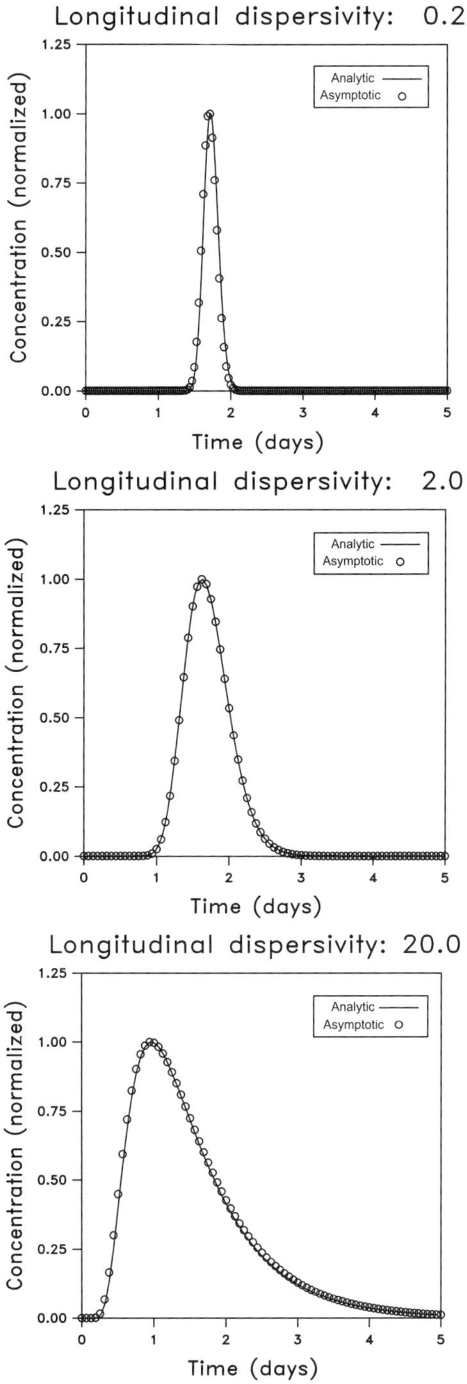

Figure 5.3 A comparison of the semi-analytic solutions for concentration with exact analytic solutions.

at a point 21 m directly downstream of the source. For comparison, we also plot the asymptotic solution, given by the expression (5.24) where $\theta(\mathbf{x}, \omega)$ is provided by Equation (5.28), which are very similar.

Exercise 5.4. Using the singular value decomposition (Noble and Daniel, 1977, p. 306).

$$\mathbf{D} = \mathbf{U}\Lambda\mathbf{V}^T$$

reduce the equation for the phase $\theta(\mathbf{x}, \omega)$,

$$\mathbf{p} \cdot \mathbf{D}\mathbf{p} - \mathbf{q} \cdot \mathbf{p} + i\omega\phi = 0,$$

to a quadratic equation for p. [**Hint**: The columns of **U** and **V** are orthornormal vectors.]

A rapidly varying concentration

For a general solution of the form (5.24) we can derive an expression for the concentration. However, due to the complexity of the frequency-dependence of the phase [see Equations (5.28), (5.29), and (5.30)], we cannot derive a time-domain semi-analytic expression for the concentration. If we consider a concentration that is varying rapidly in time, relative to the background or natural variation, then a semi-analytic expression is possible. In such a case, one can assume that the frequency of the variation, ω, approaches a large value. Therefore, we can consider the function $\chi(\mathbf{x}, \omega)$, defined by the expression (5.30), in the limit of large frequencies. First, Equation (5.30) is rearranged by factoring out $\sqrt{-i\omega}$ from the square root

$$\chi(\mathbf{x}, \omega) = \sqrt{-i\omega} \int_{\mathbf{x}} \sqrt{1 + \omega^{-1}\delta} \sqrt{\frac{\phi}{\beta}} ds \tag{5.31}$$

where

$$\delta = \frac{i\gamma^2}{4\beta\phi}. \tag{5.32}$$

Then the limit as ω takes on large values is

$$\lim_{\omega \to \infty} \chi(\mathbf{x}, \omega) = \lim_{\omega \to \infty} \sqrt{-i\omega} \int_{\mathbf{x}} \sqrt{1 + \omega^{-1}\delta} \sqrt{\frac{\phi}{\beta}} ds$$

$$= \sqrt{-i\omega} \int_{\mathbf{x}} \sqrt{\frac{\phi}{\beta}} ds. \tag{5.33}$$

Thus, we can write the limit in a partitioned form

$$\lim_{\omega \to \infty} \chi(\mathbf{x}, \omega) = \sqrt{-i\omega}\chi(\mathbf{x}) \tag{5.34}$$

where we have denoted the spatially dependent component of $\lim_{\omega\to\infty} \chi(\mathbf{x}, \omega)$ as

$$\chi(\mathbf{x}) = \int_{\mathbf{x}} \sqrt{\frac{\phi}{\beta}} ds. \tag{5.35}$$

The inverse Fourier transform of e^{θ} may now be evaluated analytically (Virieux et al., 1994),

$$\mathcal{F}^{-1}\left[e^{-\sqrt{-i\omega}\chi(\mathbf{x})}\right] = \frac{\chi(\mathbf{x})}{2\sqrt{\pi t^3}} \exp\left[\frac{-\chi^2(\mathbf{x})}{4t}\right] H(t) \tag{5.36}$$

where $H(t)$ is the Heaviside (step) function that is 0 for negative values and 1 for non-negative values. Using Equation (5.28), the complete expression (5.24) for the time-varying concentration is given by

$$c(\mathbf{x}, t) = \frac{\chi(\mathbf{x})}{2\sqrt{\pi t^3}} \exp\left[\xi(\mathbf{x}) - \frac{\chi^2(\mathbf{x})}{4t}\right] H(t) * c_0(\mathbf{x}, t) \tag{5.37}$$

for an impulsive source-time function. Where the negative sign was chosen to ensure that the concentration is positive. For a general time-varying source, we simply convolve expression (5.37) by the appropriate source-time function. In Figure 5.4 we plot the approximation (5.37) against the full asymptotic solution. The dispersion and flow field are identical to those used for the calculations associated with the results plotted in Figure 5.3. The approximation (5.36) generally agrees with the full asymptotic result.

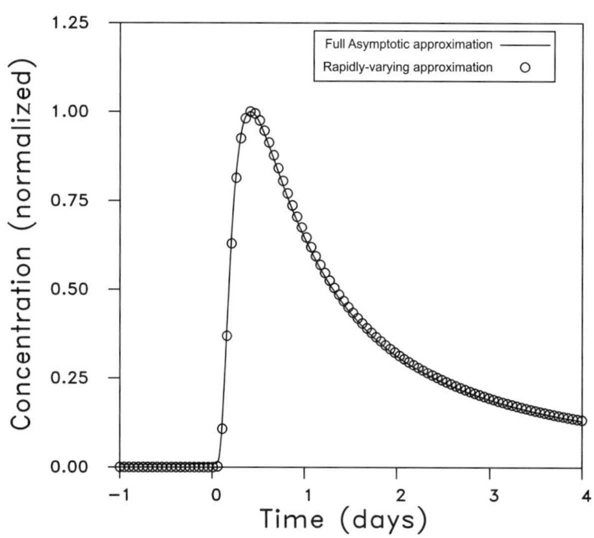

Figure 5.4 A comparison of the semi-analytic solution for concentration with the rapidly varying approximation (5.36) for an impulsive source function.

Exercise 5.5. We can gain some physical insight into the meaning of the phase function, at least for a rapidly varying concentration, if we consider the solution (5.37) for the case of an impulsive source:

$$c(\mathbf{x}, t) = e^{\xi(\mathbf{x})} \frac{\chi(\mathbf{x})}{2\sqrt{\pi t^3}} e^{-\chi^2(\mathbf{x})/4t}$$

at non-negative times. Following a procedure similar to that taken in Chapter 4, use this relationship and the explicit expression for the phase function, to relate T_{max} to the properties along the path from the injection point to the observation point. In particular, show that

$$\sqrt{T_{max}} = \frac{1}{\sqrt{6}} \int_0^{r_s} \sqrt{\frac{\phi}{\beta}} dr$$

in the case of non-zero dispersion β. The general dependence upon β and the decrease in arrival time with increased dispersion is evident in Figure 5.3 and discussed in more detail in Vasco et al. (2016).

A non-dispersive porous medium

As an aside, note that if the dispersion can be neglected then the matrix $\mathbf{D}$ in Equation (5.14) vanishes, or equivalently, β in Equation (5.25) is zero. Therefore Equation (5.25) reduces to

$$p = i\omega \frac{\phi}{\gamma}, \tag{5.38}$$

where γ is given by (5.26). We orient our coordinate system along the trajectory, use expression (5.38) to integrate p along the path, and derive an equation for the phase

$$\theta(\mathbf{x}, \omega) = i\omega\tau(\mathbf{x}) \tag{5.39}$$

where we have defined the function

$$\tau(\mathbf{x}) = \int_{\mathbf{x}} \frac{\phi}{\gamma} ds. \tag{5.40}$$

The zeroth-order expression for the concentration may now be written as

$$C(\mathbf{x}, \omega) = e^{i\omega\tau(\mathbf{x})} C_0(\mathbf{x}, \omega). \tag{5.41}$$

Taking the inverse Fourier transform gives

$$c(\mathbf{x}, t) = c_0(\mathbf{x}, t - \tau(\mathbf{x})), \tag{5.42}$$

indicating that the concentration is a shifted version of the Fourier transform of the zeroth-order amplitude function. Thus, $\tau(\mathbf{x})$ as the non-dispersive travel time, is also known as the convective 'time of flight'. Note that there will generally be changes in the amplitude with propagation due to attenuation introduced by the transport equation (5.20), hence the dependence of the amplitude on the position along the trajectory $\mathbf{x}(s)$.

Exercise 5.6. This is a continuation of Exercise 5.3. Use 'TRACE3D' to visualize the 'time of flight' (phase field) from the injector for the homogeneous five spot pattern. Use the data files in the folder 'Examples/Simulation/5-Spot-Hom' by creating a project in the same folder. Notice that the time of flight displays the convective tracer front propagation in the medium. Also, visualize the evolution of the front by thresholding at various times. Repeat the exercise for the heterogeneous example (in 'Examples/Simulation/5-Spot-Het') to visualize how the tracer front propagation is impacted by the permeability heterogeneity.

5.4 The streamline approach for transport modeling

As alluded to earlier, one may interpret streamlines as trajectories derived by integrating the ray equation (5.17) in the absence of dispersion. The streamline approach to transport modeling approximates three-dimensional fluid flow and advection calculations using a sum of one-dimensional solutions along these trajectories. The choice of streamline directions for the one dimensional calculations makes the approach especially effective for modeling advection dominated transport. Such is the case when heterogeneity primarily governs flow behavior. A key concept in the streamline approach is to divorce the effects of geologic heterogeneity from the underlying physics of flow calculations. Mathematically, this is accomplished with the help of a coordinate transformation in which the streamline **time of flight** or the travel time of a tracer particle is used as a spatial coordinate variable. The exact definition of the time of flight, the travel time of a tracer in a non-dispersive porous medium, follows directly from equations (5.40) and (5.26)

$$\tau = \int \frac{\phi}{|\mathbf{q}|} ds \tag{5.43}$$

because in the absence of dispersion $\gamma = q = |\mathbf{q}|$. With τ as the spatial variable along the trajectory, the effect of any heterogeneity is contained in the time of flight as well as in the geometry of the streamline. The task of modeling the physical processes associated with the advection and dispersion of a transported component is reduced to finding solutions along trajectories. The details of the coordinate transformation will be discussed later. Because the streamlines are generally distributed in space with higher resolution than the underlying spatial grid, one can obtain excellent transverse resolution in regions of faster flow. Transport calculations decouple from any underlying three dimensional spatial grid and may be carried out with little or no intrinsic numerical time-step and stability limitations

(Datta-Gupta and King, 2007). To illustrate these concepts, the basic steps involved in the trajectory-based approach are outlined below with an example.

- Given a heterogeneous spatial variation of properties and specified boundary conditions, we can use numerical methods to trace the trajectories for a given flow velocity field. The velocity field is obtained by solving the governing equation (5.2) for fluid pressure, as noted at the beginning of this Chapter. Figure 5.5 shows a heterogeneous permeability field and the streamlines corresponding to a five-spot well pattern with a central injection well and four surrounding extraction wells. Notice that the streamlines tend to cluster along the high permeability streaks, providing higher resolution along preferential flow paths.
- Along streamlines, we compute the travel time of a neutral particle – the streamline time of flight. Figure 5.5 shows the color-coded time of flight field defined on the trajectories. The time of flight contours or isochrones correspond to fluid fronts of transported material, allowing for a quantitative form of flow visualization. The time of flight also serves as a measure of distance, and will be used as a spatial coordinate during fluid transport calculations.
- Finally, we solve the transport equations for saturation and/or concentration along the streamlines. These calculations are performed in the time of flight coordinate

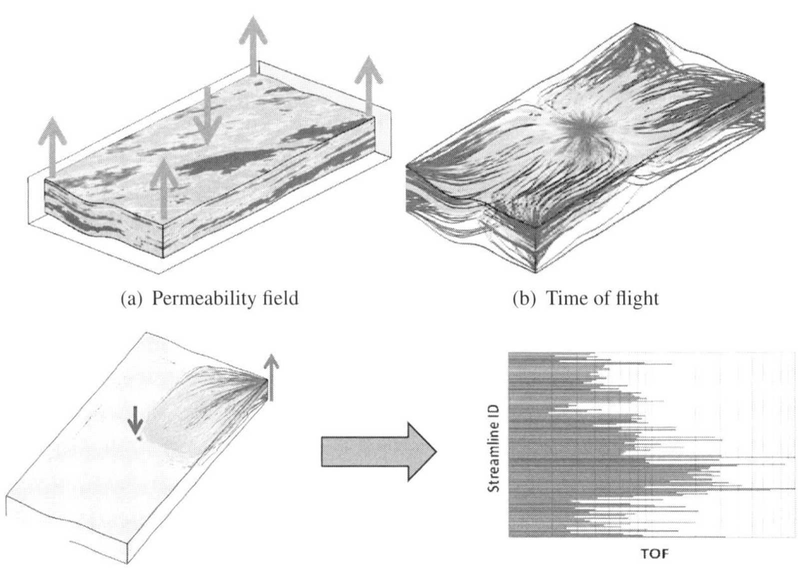

(a) Permeability field (b) Time of flight

(c) Coordinate transformation

Figure 5.5 An illustration of the streamline-based approach: (a) 3-D permeability fields and wells, (b) streamline trajectories colored by time of flight variation, and (c) transformation from 3-D space to the time of flight coordinates. For the colour version, please refer to the plate section.

system, effectively decoupling heterogeneity effects and significantly simplifying the calculations. As shown in Figure 5.5, in the transformed coordinates the streamlines are straight lines and the heterogeneity is reflected in the variable domain lengths. That is, the heterogeneity leads to variations in the time of flight along the different trajectories and the time of flight acts as the domain length for the 1-D calculation.

Exercise 5.7. This exercise illustrates the steps involved in streamline simulation. Run 'TRACE3D' using the data files in the folder 'Examples/Simulation/5-Spot-Hom' by creating a project in the same folder. Visualize the pressure distribution, streamline trajectories, time of flight from the injector and producers, water saturation distribution, oil-water production history and oil recovery history for water injection into an oil reservoir. The details of the two-phase displacement calculations will be discussed in Chapter 6. Repeat the exercise for the heterogeneous example (in 'Examples/Simulation/5-Spot-Het') to examine the impact of heterogeneity in the two-phase displacement calculations.

5.4.1 Time of flight as a spatial coordinate

A key feature of trajectory-based modeling is that it isolates the effects of spatial heterogeneity from the transport calculations themselves. We have already noted in general terms how this is done: by transforming to a coordinate system in which the time of flight is the independent variable parameterizing position along the trajectory. Effectively, given the trajectories, we solve problems in two dimensions with time and time of flight as coordinates, rather than in four dimensions with time and three spatial variables $\mathbf{x}$ as coordinates. Let us look in some detail at the mechanics of this process. To start with, we rewrite Equation (5.43) in a differential form as follows

$$\mathbf{q} \cdot \nabla\tau = \phi. \tag{5.44}$$

We can relate this equation back to our asymptotic derivation in the previous section. In particular, if we consider Equation (5.14) for the phase θ in the absence of dispersion, then Equation (5.44) follows if

$$\theta(\mathbf{x}, \omega) = i\omega\tau(\mathbf{x})$$

which is just the expression for the phase in a non-dispersive medium [see Equation (5.39)] and τ is simply the spatial variation of the phase. Notice that the equivalence of Equations (5.43) and (5.44) is apparent from the following,

$$\frac{\phi}{|\mathbf{q}|} = \frac{d\tau}{ds} = \frac{\mathbf{q}}{|\mathbf{q}|} \cdot \nabla\tau. \tag{5.45}$$

Because the time of flight gradients are defined along the streamlines, we can invoke an orthogonal coordinate system that is defined in terms of the time of flight and two orthogonal bases ψ and χ called the bi-stream functions (Bear, 1972; Datta-Gupta and King, 2007). Formally, a streamline in 3D is defined by the intersection of two stream function surfaces, with each given by a constant value for ψ and a constant value for χ, respectively. The velocity field for a general three-dimensional medium is related to the bi-stream functions in the following manner:

$$\mathbf{q} = \nabla\psi \times \nabla\chi \tag{5.46}$$

(Bear, 1972). Note that this definition automatically satisfies the continuity Equation (2.60) for a steady-state flow field and an incompressible fluid, via a well-known vector identity for arbitrary scalar functions ψ and χ:

$$\nabla \cdot \mathbf{q} = \nabla \cdot (\nabla\psi \times \nabla\chi) = 0 \tag{5.47}$$

(Marsden and Tromba, 1976, p. 166).

A coordinate system determined by t, τ, ψ, and χ is rooted in the physics of flow within the porous medium (Bear, 1972). A number of useful properties follow from this fact (Datta-Gupta and King, 2007) and we will outline a few of them here. For example, the spatial gradient along the streamline assumes a very simple form in the time of flight coordinates. That is, using the (τ, ψ, χ) spatial coordinate system, the gradient operator can be expressed as

$$\nabla = (\nabla\tau)\frac{\partial}{\partial\tau} + (\nabla\psi)\frac{\partial}{\partial\psi} + (\nabla\chi)\frac{\partial}{\partial\chi}. \tag{5.48}$$

Because $\mathbf{q}$ is orthogonal to both $\nabla\psi$ and $\nabla\chi$, a consequence of (5.46), we have the following operator identity

$$\mathbf{q} \cdot \nabla = (\mathbf{q} \cdot \nabla\tau)\frac{\partial}{\partial\tau} = \phi\frac{\partial}{\partial\tau}, \tag{5.49}$$

where we have invoked Equation (5.44). The operator identity in Equation (5.48) forms the basis for coordinate transformation from physical space to the streamline and time of flight space as we will see shortly. Because of Equation (5.44), the Jacobian of the transformation, relating the volume elements in the two coordinate systems, assumes a very simple form

$$\left\| \frac{\partial(\tau, \psi, \chi)}{\partial(x, y, z)} \right\| = \nabla\tau \cdot (\nabla\psi \times \nabla\chi) = \nabla\tau \cdot \mathbf{q} = \phi. \tag{5.50}$$

Another advantage of the τ coordinate system becomes evident when we consider the transport equation (5.1) in the absence of dispersion.

$$\phi\frac{\partial c}{\partial t} + \nabla \cdot (\mathbf{q}c) = 0. \tag{5.51}$$

This equation can be expanded and transformed with τ as the spatial coordinate using Equation (5.49), resulting in the following equation

$$\frac{\partial c}{\partial t} + \frac{\partial c}{\partial \tau} = 0, \tag{5.52}$$

describing transport along the streamline. In deriving Equation (5.52) we have assumed that the fluid is incompressible and made use of the Equation (5.47) to remove the term containing $\nabla \cdot \mathbf{q}$. With this coordinate transformation, we have decomposed Equation (5.51) in three spatial variables into a series of Equations (5.52) in one spatial variable τ defined along the streamlines. Individual elements of the family of trajectories are determined by particular values of the two bi-stream parameters ψ and χ. This is similar to the asymptotic solution described in the previous section, characterized by the initial conditions used to specify the trajectory. The initial conditions are typically the take-off angles, the direction at which the trajectory leaves the source. Equation (5.52) is equally valid in one, two and three spatial dimensions, and in both homogeneous and heterogeneous media. The τ transformation includes all of these effects. All that is required for the implementation is the velocity field, obtained by solving the pressure equation (5.2), and the evaluation of the line integral in Equation (5.43). This integration is relatively straight-forward, given an efficient algorithm for tracing streamlines, as described in the next section.

5.4.2 Tracing trajectories and computing the time of flight

We now describe the computational aspects of evaluating the integral (5.43). A critical component of this calculation is the construction of a trajectory within a grid block. We presume that the associated pressure field has been calculated, typically using numerical methods, for example a finite difference solution of Equation (5.2). From the pressure solution, and Darcy's law, we know the volumetric flux across the faces of each cell of the model (Figure 5.6). To trace the trajectory across a cell, we use the simplest possible representation of the velocity field that allows us to smoothly interpolate the fluxes across the faces. As will be shown, these equations are exactly integrable in time, describing the trajectory of a particle within a cell. The most commonly used trajectory calculations follow an approach known as Pollock's algorithm (Pollock, 1988), based upon a linear velocity model within each grid block. In particular, each velocity component varies linearly between the values on appropriate pairs of cell faces:

$$\begin{aligned} q_x &= q_{x_1} + c_x(x - x_1) \\ q_y &= q_{y_1} + c_y(y - y_1) \\ q_z &= q_{z_1} + c_z(z - z_1), \end{aligned} \tag{5.53}$$

(see Figure 5.6). The velocity gradients in each direction are computed from the difference of Darcy velocities on the grid block faces, for example in the x direction we have: $c_x = (q_{x_2} - q_{x_1})/\Delta x$. Note that the velocity interpolation in Equation (5.53)

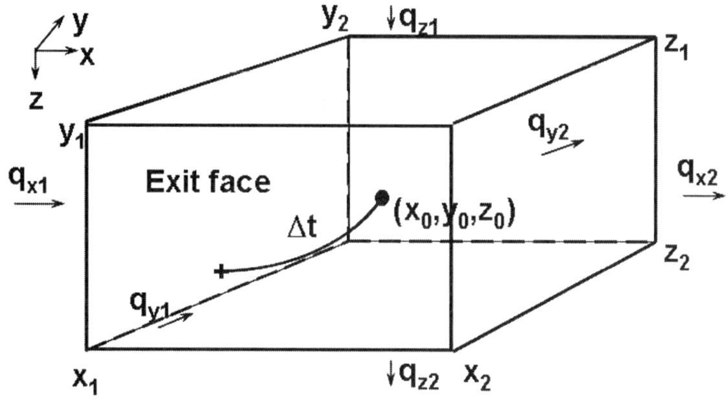

Figure 5.6 Illustration of grid block time of flight calculations.

is quite general and applies to compressible as well as incompressible flow. For incompressible flow, and away from any sources or sinks, the divergence-free nature of the velocity field requires that

$$c_x + c_y + c_z = 0. \tag{5.54}$$

For compressible flow the sum (5.54) will not vanish because fluid and rock expansion and compression act effectively as source and sink terms. For the velocity models in Equation (5.53), the time of flight can be integrated explicitly, and independently, in each direction. For example, the integral solution in the x-direction starting from location x_0 within the cell (Figure 5.6) will be given by,

$$\frac{\Delta\tau_{x_i}}{\phi} = \int_{x_0}^{x_i} \frac{dx}{q_{x0} + c_x(x - x_0)} \tag{5.55}$$

$$= \frac{1}{c_x} ln \left[\frac{q_{x_i}}{q_{x0}}\right]. \tag{5.56}$$

The index $i = 1, 2$ indicates the grid block faces in the x-direction. Identical constructions will arise when integrating in the y- and z-direction. Thus, the actual cell time of flight for the particle $\Delta\tau$, will be given by the minimum positive arrival time over allowable edges (Pollock, 1988) that is the minimum of $\Delta\tau_{x_1}$, $\Delta\tau_{x_2}$, $\Delta\tau_{y_1}$, $\Delta\tau_{y_2}$, $\Delta\tau_{z_1}$, and $\Delta\tau_{z_2}$, where only positive values are considered. After computing the particle time of flight, its exit coordinates can be obtained by simply rearranging Equation (5.56) as follows.

$$x = x_0 + q_{x0}\left(\frac{e^{c_x\Delta\tau/\phi} - 1}{c_x}\right). \tag{5.57}$$

Notice that the term within parentheses in Equation (5.57) increases with time and can be thought of as a pseudo-time,

$$\eta = \frac{e^{c_x \Delta\tau/\phi} - 1}{c_x}.$$

In terms of η, the Equation (5.57) has now a simple linear form,

$$x = x_0 + q_{x0}\eta. \tag{5.58}$$

When the cell velocity is uniform (constant) in a given direction, for example in the x-direction, then $c_x = 0$. In this limit,

$$\frac{1}{c_x} ln \left[\frac{q_x}{q_{x0}} \right] \rightarrow \frac{x - x_0}{q_{x0}} \tag{5.59}$$

$$\eta_x = \frac{e^{c_x \Delta\tau/\phi} - 1}{c_x} \rightarrow \frac{\Delta\tau}{\phi}$$

$$x = x_0 + q_{x0} \frac{\Delta\tau}{\phi}. \tag{5.60}$$

As expected, the position varies linearly with time for a constant velocity. The trajectory and time of flight calculations discussed above can be extended to complex and unstructured grids that are often used to represent reservoir geometry, stratigraphy, and faults (Datta-Gupta and King, 2007).

5.4.3 Transport equation and amplitudes

Let us now return to the transport equation (5.51) and its one-dimensional form (5.52) in streamline and time of flight coordinates. If the injection rate varies as a function of time then Equation (5.52) constitutes a time-dependent boundary value problem. One may recall that we have seen expression (5.52) before, as Equation (1.5) in our discussion of propagating disturbances in Chapter 1. It was an example of the simplest possible wave equation, describing propagation through a uniform medium. Equation (5.52) only appears to describe propagation through a uniform medium because we have embedded all of the heterogeneity into the time of flight variable τ. As shown in the previous section on non-dispersive transport [see Equation (5.42)], the tracer concentration at the observation point at time t will simply be the injected concentration delayed by the time it takes to travel along the streamline:

$$c(t) = c_0(t - \tau) = c_0 \left(t - \int \frac{\phi}{|\mathbf{q}|} ds \right), \tag{5.61}$$

where we have substituted our definition (5.43) for τ. For the special case of a step in the rate of injection, jumping from zero to a fixed injection rate at $t = 0$, the solution will be particularly simple:

$$c(t) = \begin{cases} 0, & t < \tau \\ c_0, & t \geq \tau, \end{cases} \tag{5.62}$$

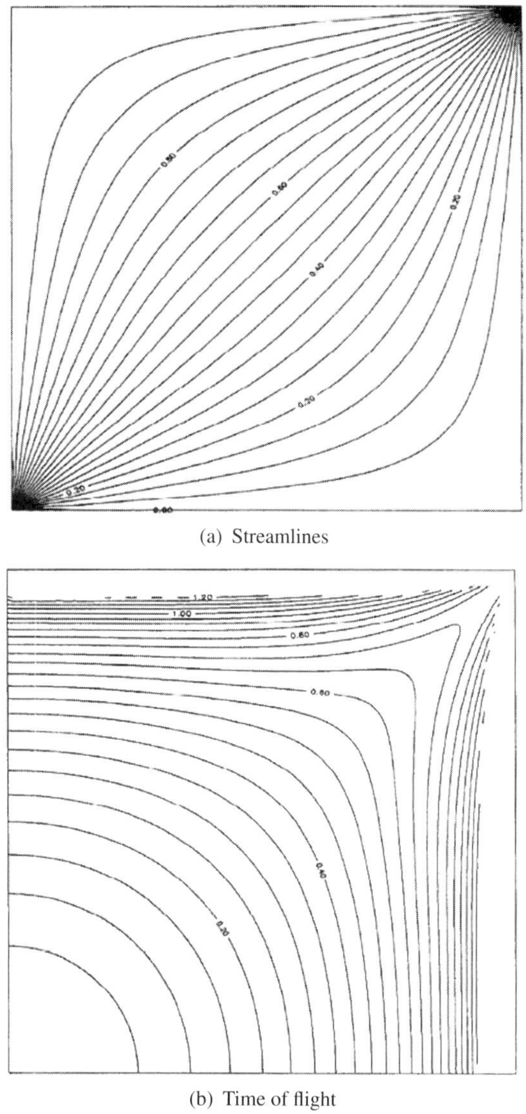

Figure 5.7 (a) Streamlines and (b) time of flight in a quarter five-spot pattern [from Datta-Gupta and King (1995)]. Reprinted with permission. Copyright Elsevier Limited.

a step function at $t = \tau$. The solution for a common pulse test, with injection from time t_0 to t_1, can be obtained by simply replacing the source function (5.62) by

$$c(t) = c_0 H(t - t_0) H(t_1 - t), \tag{5.63}$$

where $H(t)$ is the Heaviside step function, which is zero for $t < 0$ and one otherwise.

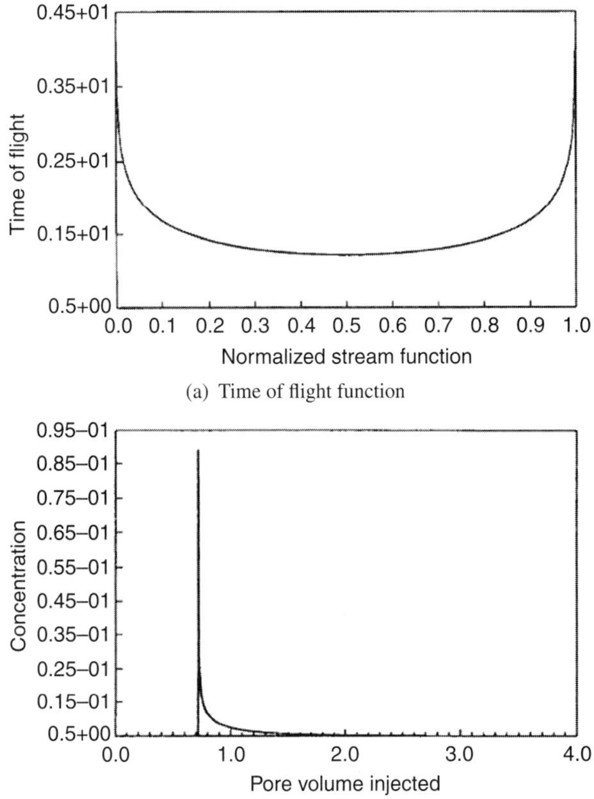

(a) Time of flight function

(b) Calculated tracer response

Figure 5.8 Tracer response calculations in a homogeneous porous medium. (a) The time of flight function. (b) Tracer response at the observation point for a pulse injection [From Datta-Gupta and King (1995)]. Reprinted with permission. Copyright Elsevier Limited.

The solution of the transport equation, given by the expression (5.61), corresponds to the contribution from a single streamline. In certain situations that may be the only significant component at a particular observation point. However, in most cases the observation point will be at an extraction well and the trajectories will converge at that location. Thus, there will be a family of streamlines connecting the source and observation points, as shown in Figure 5.7. The overall tracer response at an observation point will be given by the sum of contributions from all the streamlines in that family,

$$\begin{aligned} c(t) &= \int c_0\left(t - \tau(\Psi)\right) d\Psi \Big/ \int d\Psi \\ &= \frac{1}{Q} \int c_0\left(t - \tau(\Psi)\right) d\Psi, \end{aligned} \tag{5.64}$$

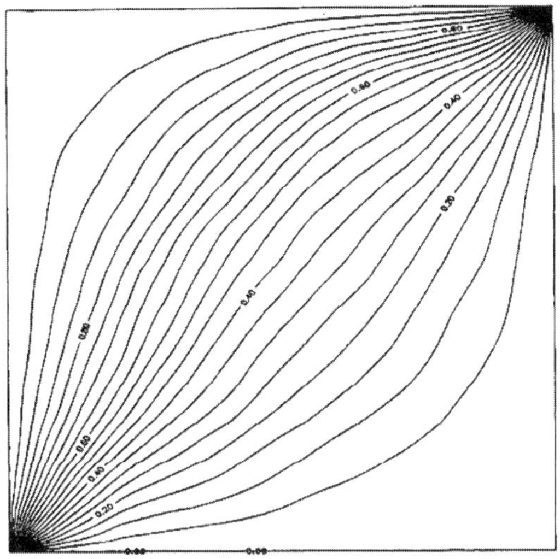

(a) Streamlines

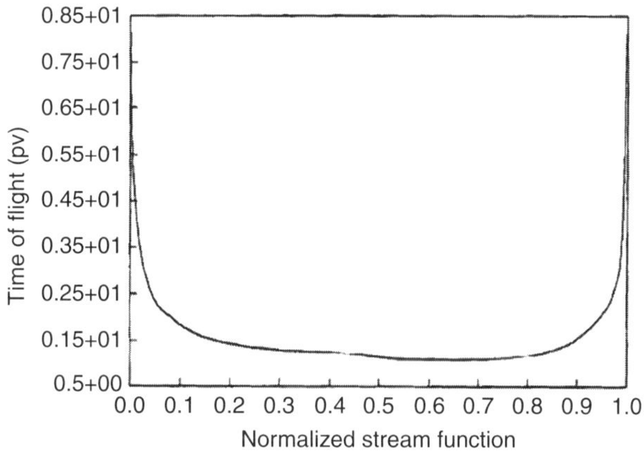

(b) Time of flight function

Figure 5.9 Tracer response calculations in a heterogeneous quarter five-spot pattern (a) streamlines (b) the time of flight function, $\tau(\Psi)$ at the producer [From Datta-Gupta and King (1995)]. Reprinted with permission. Copyright Elsevier Limited.

where Ψ identifies members of the family of streamlines. In general, the integration will be over the two-dimensional surface that parameterizes the family of streamlines. Thus, $d\Psi$ will signify a surface area element of the form $d\psi d\chi$ specifying the volumetric rate associated with the trajectory and ψ and χ are the bi-stream coordinates (Datta-Gupta and King, 2007). The denominator Q in the expression is the total fluid production at the well. Hence, Equation (5.64) will incorporate concentration dilution effects.

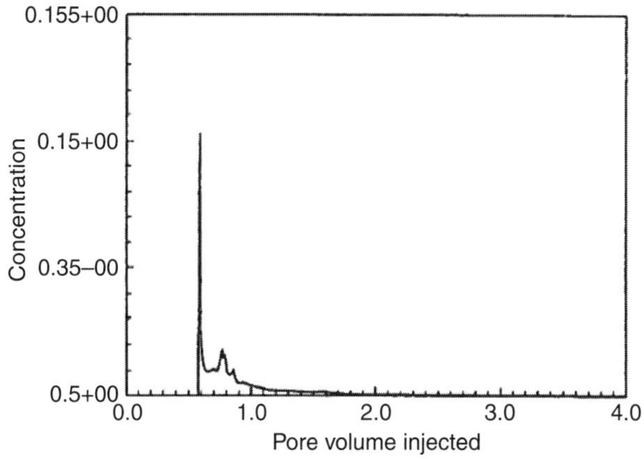

Figure 5.10 Tracer response for pulse injection in a heterogeneous quarter five-spot pattern [From Datta-Gupta and King (1995)]. Reprinted with permission. Copyright Elsevier Limited.

As an illustration, consider the injection of a passive tracer into a homogeneous porous medium, with an injection well on the bottom left and an extraction well at the top right (Figure 5.7). No flow is allowed across the edges of the grid, a 'no-flow' boundary condition. This pattern constitutes the well known 'quarter five spot' configuration that is often used in applications due to its simplicity. The streamlines and the time of flight are shown in the top and bottom panels, respectively. The time of flight distribution function, $\tau(\Psi)$, associated with this quarter five spot example, is plotted in Figure 5.8(a). For this homogeneous case, the central streamline, which is the shortest path, arrives first and the travel time, or the time of flight, increases monotonically towards the boundary streamlines that are progressively longer. Furthermore, the flow velocity slows near the boundaries, leading to a further increase in the travel time associated with the outermost streamlines. Notice the 'flatness' of the $\tau(\Psi)$ function with a single minimum in the middle region in Figure 5.8(a). The near absence of curvature and lack of dip signifies the arrival of a large number of streamlines in a relatively short time, resulting in a sharp peak in concentration [Figure 5.8(b)] for a narrow injection pulse given by Equation (5.63). Heterogeneity can be easily introduced into the tracer response calculations through the $\tau(\Psi)$ function which simply represents the arrival time distribution for various streamlines. In general, and particularly in the presence of high permeability contrasts, flow channels or barriers, the $\tau(\Psi)$ function will be non-monotonic, potentially with a number of peaks and valleys, leading to a tracer response containing multiple arrivals for each local minimum (see Figures 5.9 and 5.10). In Box 5.2 we reach back into the section on the asymptotic approach to explain why each minimum (valley) gives rise to a peak in the tracer concentration.

This argument, based upon the method of stationary phase, is just one illustration of the interplay between the asymptotic approach and streamline methods that provides insight into the nature of tracer transport.

Box 5.2 The method of stationary phase and multiple arrivals

For simplicity, consider a tracer front propagating in two-dimensions, thus each trajectory associated with the front is characterized by a single number Ψ. For example, Ψ may represent the 'take-off' angle, the angle (direction) at which a trajectory leaves the injection point, the initial angle at which the streamlines leave the source in Figures 5.7(a) and 5.8(a). As indicated by Equation (5.64) an arbitrary variation in concentration is synthesized by an integration over the trajectories. Our starting point will be this integral in the frequency domain, using the frequency-domain representation (5.41) of the asymptotic solution for the tracer concentration in a non-dispersive porous medium:

$$C(\mathbf{x}, \omega) = \int C_0(\Psi, \omega) e^{i\omega\tau(\Psi)} d\Psi.$$

To simplify the presentation we are setting the normalizing constant $1/Q$, which has no effect on our argument, to unity. In order to obtain a solution in the time-domain we take the inverse Fourier transform of $C(\mathbf{x}, \omega)$,

$$\begin{aligned} c(x, t) &= \int_{-\infty}^{\infty} \int C_0(\Psi, \omega) \exp\left[i\omega\tau(\Psi) - i\omega t\right] d\Psi d\omega \\ &= \int_{-\infty}^{\infty} \int C_0(\Psi, \omega) \exp\left[-i\omega\chi(\Psi)\right] d\Psi d\omega \end{aligned}$$

where

$$\chi(\Psi) = t - \tau(\Psi).$$

Using a physical argument made by Lord Kelvin (Thomson, 1887), known as the method of stationary phase, one can approximate the integral as a sum of arrivals from a discrete set of trajectories. The idea is that for relatively large values of ω the exponential term in the integrand is highly oscillatory. The superposition of such waves in any given interval $\Delta\Psi$ will destructively interfere except where there is a concentration of waves with nearly the same phase. Such a concentration occurs when $\chi(\Psi)$ is stationary with respect to changes in Ψ, that is where:

$$\chi' = \frac{d\chi}{d\Psi} = \frac{-d\tau}{-d\Psi} = 0,$$

where the prime denotes a derivative with respect to Ψ. We can denote such a stationary point by Ψ_s. Thus, the main contributions to the integral over Ψ will be from all stationary points of τ, from all local minima of the time of flight curve.

This approach, along with alternative techniques collectively known as saddle-point methods (Dingle, 1973, p. 134), are helpful in evaluating integrals containing a rapidly-varying exponential factor multiplied by a function. The basic idea is similar, the exponential contribution decays rapidly away from critical points (extrema) in its argument, localizing the contribution to the integral.

Exercise 5.8. This exercise illustrates continuous tracer injection in a quarter five-spot pattern with one injector and one producer. The fluid properties are chosen to emulate a unit mobility ratio, piston-like displacement along streamlines. Run 'TRACE3D' using the data files in the folder 'Examples/Simulation/Q5-Spot-Hom' by creating a project in the same folder. Visualize the pressure distribution, streamline trajectories, time of flight from the injector, and the oil-water production history. The time of flight displays the tracer front propagation and the fractional water production (water-cut) emulates the tracer concentration history for a continuous injection. Repeat the exercise for the heterogeneous example (in 'Examples/Simulation/Q5-Spot-Het') to examine the impact of heterogeneity on the tracer concentration history.

Including dispersion in the streamline solution

So far we have neglected physical dispersion in our streamline calculations. Longitudinal dispersion can be accommodated easily along streamlines (Abbaszadeh and Brigham, 1984). However, transverse dispersion across streamlines will require special treatment. Typically, operator splitting methods have been used to account for transverse mechanisms whereby different physical mechanisms are incorporated in sequence, some along the trajectories and some on the grid (Datta-Gupta and King, 2007). That is one advantage of the asymptotic solution given in the previous section, one can incorporate a general dispersion tensor directly into the trajectory calculations themselves, resulting in the ray equations (5.17). Such an approach offers additional insight into the factors controlling dispersion and the physical path of the dispersing material.

5.4.4 Partitioning tracer

Partitioning chemical species, which partially dissolve into a hydrocarbon or non-aqueous phase liquid (NAPL) form an important class of tracers for numerous applications. This is a form of reactive tracer in which the injected chemical species interacts with an in-situ fluid phase, a topic discussed in Section 2.5.2 of Chapter 2. For example, in the presence of an organic phase a tracer travel time may increase due to dissolution or partitioning into the phase. The resulting tracer delay, referenced to a non-partitioning tracer, can be used to estimate the in-situ saturation distribution of the phase. The tracer equations presented above can be easily generalized to allow for partitioning, by redefining the tracer transit time as

$$\tau = \int \frac{\phi(S_a + K_n S_n)}{|\mathbf{q}|} ds \tag{5.65}$$

[see the related Equations (2.113) to (2.115) in Chapter 2]. The quantities S_a and S_n in Equation (5.65) are aqueous and non-aqueous saturations, respectively. The coefficient K_n is the partitioning or distribution coefficient, defined as the ratio of the tracer concentration in the non-aqueous phase to that in the aqueous phase. For a non-reactive tracer, K_n is unity and the equation reverts back to the time of flight definition in Equation (5.43).

5.5 Applications

5.5.1 Flow visualization: trajectories, time of flight and tracer fronts

One attractive feature of trajectory-based methods is their intuitive appeal. This makes such approaches particularly well-suited for visualizing the interaction between subsurface heterogeneity and the flow field imposed by the injection/extraction wells. The flow diagnostics generated from such quantitative visualization can have a variety of applications in subsurface characterization, reservoir management, process optimization, and field development strategies. In particular for oil and gas applications, geologic models routinely consist of millions of cells, making detailed flow simulation a computationally expensive task. The simplified diagnostic plots discussed here can be generated relatively fast and can provide valuable insights for comparing numerous geologic scenarios or multiple reservoir development strategies. It is worth pointing out that during subsurface flow modeling, we encounter many situations and decision points that do not call for the full complexity of the underlying process physics. Under such conditions, flow diagnostic plots can provide valuable insight, particularly when the transport process is dominated by subsurface heterogeneity.

The trajectories provide a visual representation of the velocity field. These flow paths are analogous to highways along which the fluids travel. The streamlines tend to cluster in regions of high velocity (flux) and spread apart in regions of low velocities, thus, automatically giving higher resolution in regions of faster flow (Figure 5.11). The time of flight or the tracer travel time along streamlines provides a quantitative aspect to the flow visualization, allowing us to ask questions like "Where is the injected fluid going from a specific injector?" This is simply a depiction of the time of flight starting from the injector (Figure 5.12)

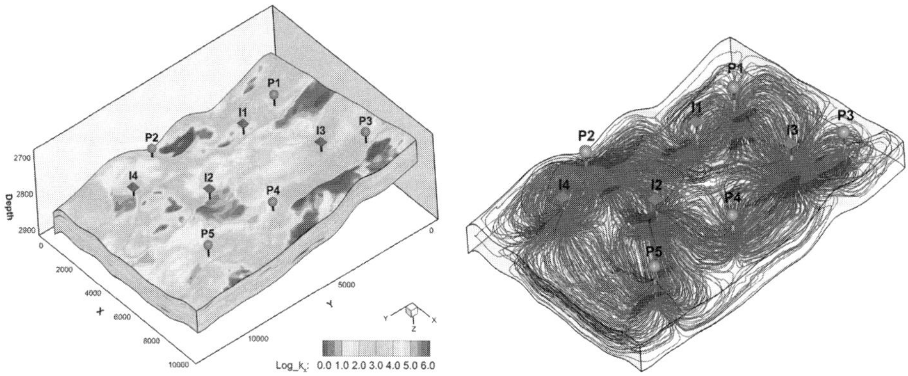

(a) (b)

Figure 5.11 (a) Heterogeneous permeability field, displaying the well locations. (b) The streamlines associated with the flow field. For the colour version, please refer to the plate section.

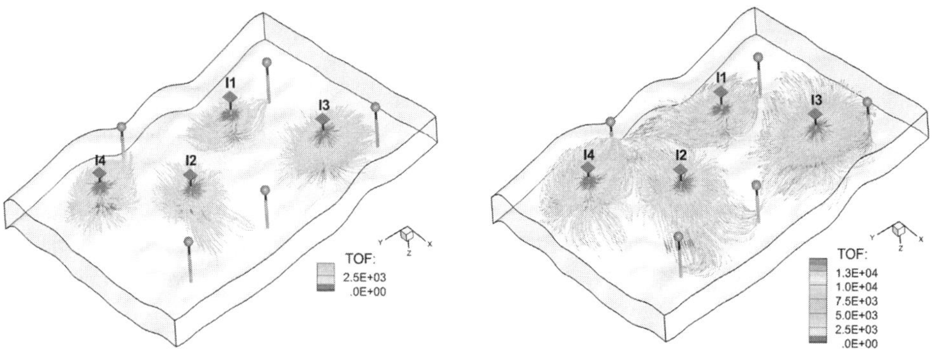

(a) (b)

Figure 5.12 Streamline time of flight from the injector thresholded at two different times displaying the reservoir swept volume, the volume encountered by the injected fluid. For the colour version, please refer to the plate section.

and is commonly referred to as the ‘swept volume’. Swept volume calculations are particularly useful for optimizing fluid injection strategies and the placement of injection wells. Similarly, we can display the region contributing to the fluid production for a given extraction well, otherwise known as the ‘drainage volume’. This will be the time of flight computed by reversing the flow field and starting from the producer (Figure 5.13). Visualization of the drainage volume can be useful in optimizing fluid extraction strategies by the proper placement of the production wells. By thresholding the time of flight, we can visualize the evolution of the flood front. Understanding this evolution can have a variety of uses in dynamic reservoir characterization. For example, the streamlines can also be used to partition

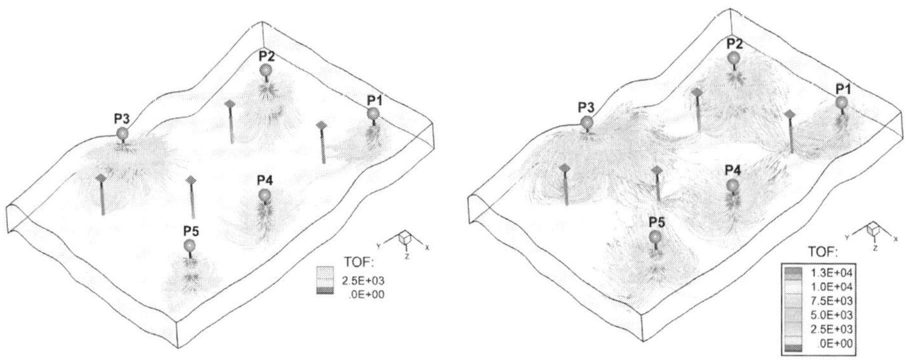

(a) (b)

Figure 5.13 Streamline time of flight to the producer thresholded at two different times displaying the reservoir volume drained by the producing wells, the drainage volume. For the colour version, please refer to the plate section.

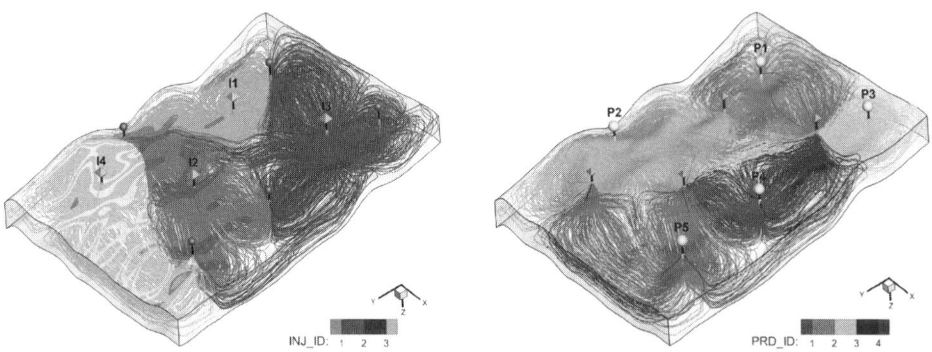

(a) Injector pore volume (b) Producer pore volume

Figure 5.14 Streamlines displaying the reservoir pore volumes associated with (a) injectors and (b) producers. For the colour version, please refer to the plate section.

the domain and identify the pore volumes associated with individual injectors and producers (Figure 5.14). An important piece of information that naturally comes out of streamlines is the communication between the wells (Figure 5.15). Such information is not readily accessible from standard finite difference and finite element simulation models. Because each streamline carries a known volume of fluid, the knowledge of well communications also allows computing the well allocation factors, that is, how the injected fluid is distributed among various producers. These allocation factors are extremely useful for balancing fluid injection and production rates in order to maintain reservoir pressure and improve the performance of fluid injection/extraction strategies.

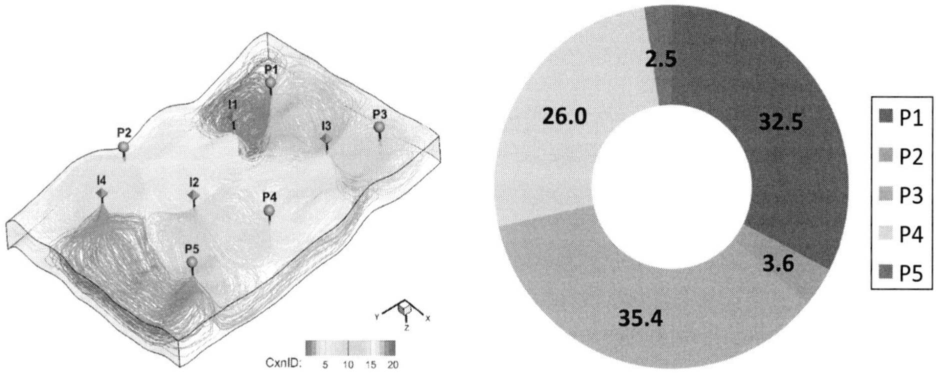

Figure 5.15 Streamlines displaying communication between the wells and the well allocation factor. For the colour version, please refer to the plate section.

Exercise 5.9. This exercise illustrates the practical application of the streamline-based modeling using a 3D example with multiple injection and extraction wells in a highly heterogeneous reservoir. Run 'TRACE3D' using the data files in the folder 'Examples/Simulation/3D-Multiwell-Het' by creating a project in the same folder. Visualize the pressure distribution, streamline trajectories, time of flight from the injector (injector swept volume) and producer (producer drainage volume), pore volumes associated with individual injectors and producers, well communications and allocation factors.

5.5.2 Sensitivity computation

An added advantage of a trajectory-based approach is the efficient computation of parameter sensitivities used for reservoir characterization and inverse modeling (Vasco and Datta-Gupta, 1999; Datta-Gupta et al., 2002; Vasco and Finsterle, 2004). The goal of inverse modeling is to estimate subsurface properties such as permeability, porosity, and fluid saturation distribution in the subsurface. Model parameter sensitivities constitute an integral part of most inverse modeling techniques that we discuss later. Sensitivities are partial derivatives that comprise the coefficients in a linear relationship between a perturbation in a model parameter and a perturbation in an observation. Several methods for computing sensitivities are described in the literature. Among these techniques numerical perturbation methods, sensitivity equation methods, and adjoint methods (Sun and Yeh, 1990; Yeh, 1990) are the most prominent. The methods are limited by their computational costs and the degree of algorithm complexity required for their implementation. The streamline approach provides an extremely efficient means for computing parameter sensitivities using a single flow simulation. Sensitivities can be computed semi-analytically, requiring the evaluation of one-dimensional integrals along streamlines. Thus, trajectory-based methods scale favorably with increasing problem size, roughly as the growth in the number of cells intersected by a line through the model. The semi-analytic expressions that appear in the sensitivities are relatively simple in form and easy to implement in a computer code. This makes the streamline approach, and trajectory-based methods in general, particularly effective for high-resolution inverse modeling.

The trajectory-based formulation provides some additional flexibility in formulating the inverse problem. This follows from the partitioning of the modeling into two sub-problems: a travel time or time of flight calculation and an amplitude computation. These two problems are particularly evident in the asymptotic approach. The governing equation for the travel time is associated with the zeroth-order terms of the asymptotic expansion, while the transport equation (5.20) for the amplitude results when first-order terms are considered. Sensitivities for the arrival time are typically much simpler to compute than are amplitude sensitivities. For example, the first arrival time is determined by the travel time associated with the path that has the smallest travel time, while the amplitude is typically controlled by the family of trajectories with similar arrival times, sometimes leading to non-local calculations.

In deriving the trajectory-based sensitivities, and to exploit an analogy with seismic rays, we will rewrite the travel time (5.43) in terms of slowness, the reciprocal of speed [see Equation (5.45)], as is commonly done in seismology (Nolet, 1987). For tracer, the definition of the slowness is

$$s(\mathbf{x}) = \frac{\phi(\mathbf{x})}{|q(\mathbf{x})|} = \frac{\phi(\mathbf{x})\mu}{k(\mathbf{x})|\nabla p(\mathbf{x})|}, \tag{5.66}$$

with ϕ denoting the porosity, μ the fluid viscosity, k the permeability , and p the pressure. In Equation (5.66), we have used Darcy's law, given at the beginning of the chapter. The travel time along the trajectory, as in Equation (5.43), can be written as

$$\tau(\Psi) = \int_{\Psi} s(\mathbf{x}) d\mathbf{x}, \tag{5.67}$$

where Ψ identifies the trajectory and may represent the two bi-stream coordinates (ψ, χ) or the two take-off angles. The integral is along the trajectory or streamline $\mathbf{x}(r)$ where r is the position along the trajectory. The concentration at an observation point is given by Equation (5.64), an integration over all streamlines connecting this point with the injection:

$$c(t) = \frac{1}{Q} \int c_0 \left[t - \tau(\Psi)\right] d\Psi. \tag{5.68}$$

Consider a small perturbation in a subsurface property, for example, permeability or porosity. The perturbation will induce changes in the observed tracer concentration

$$c(t) = c^o(t) + \delta c(t), \tag{5.69}$$

where c^o refers to the concentration associated with an initial model. Due to the linearity of the sensitivities and the linearity of the integration over the streamlines, we can compute the effect of a perturbation on a single streamline and then integrate over all the streamlines to get the total perturbation. Thus, for a single streamline indexed by Ψ, we simply have to consider the effect of a perturbation in reservoir properties on the integrand of (5.68). For a fixed trajectory the perturbation follows from an application of a Taylor's series expansion and the chain rule,

$$\delta c_\Psi(t) = -\dot{c_0}\left[t - \tau^o(\Psi)\right]\delta\tau(\Psi), \tag{5.70}$$

where $\tau^o(\Psi)$ is the time of flight in the background model. In Equation (5.70), $\dot{c_0}$ is the time derivative of the injection history, acting as a weighting term. From Equation (5.70) we would expect that those portions of the injection history with the steepest slope would have the greatest influence. This makes physical sense because Equation (5.70) relates a perturbation in the travel time to a perturbation in the size of the concentration. A time shift will have the most significant effect on the amplitude precisely in those regions where the slope of the concentration curve is greatest.

The next order of business is to relate the perturbation in the travel time, $\delta\tau(\Psi)$ in Equation (5.70), to the perturbations in reservoir flow properties. In order to do this we need the integral relationship (5.67) where the integrand $s(\mathbf{x})$, the slowness, is given by Equation (5.66). However, before attacking this problem we must address what might seem like a subtle point. In perturbing the flow properties, such as porosity and permeability, we will change the flow field and therefore the trajectories. Thus, the path of integration Ψ in the expression (5.67) will also change and this perturbation should be accounted for. This issue has been examined in other fields, such as electromagnetic and elastic wave propagation, and the effect has been shown to be second order in the model perturbations (Nolet, 1987). We can also provide a plausibility argument, based upon the fact that the perturbation is small in magnitude and is over a single grid block. Thus, for a single small grid block, a perturbation will likely induce a small lateral shift in the trajectory. This lateral shift will change the length of the trajectory by a very small amount. The small change in length is unlikely to have much effect on the total travel time and we can use the path in the unperturbed medium. Having addressed this issue, we now perturb the integral relationship (5.67) to arrive at

$$\delta\tau(\Psi) = \int_{\Psi_o} \delta s(\mathbf{x})d\mathbf{x}, \tag{5.71}$$

where Ψ_o is the trajectory in the unperturbed model. Because the slowness $s(\mathbf{x})$ is given by the expression (5.66), its variation with respect to perturbations in the flow properties ϕ and k can be written as

$$\delta s(\mathbf{x}) = \frac{\partial s(\mathbf{x})}{\partial k}\delta k(\mathbf{x}) + \frac{\partial s(\mathbf{x})}{\partial \phi}\delta \phi(\mathbf{x}) \tag{5.72}$$

if we neglect the implicit dependence of $p(\mathbf{x})$ upon ϕ and k. Arguments for neglecting the implicit dependence of the steady-state pressure field $p(\mathbf{x})$ upon flow properties are similar to those for ignoring the perturbation of the ray path. Because the steady-state pressure is governed by an elliptic equation and is strongly controlled by the large-scale permeability distribution and boundary conditions, a small perturbation in permeability over a small grid block is unlikely to change the pressure field significantly. We can compute the partial derivatives using the analytic form (5.66) for $s(\mathbf{x})$

$$\begin{aligned} \frac{\partial s(\mathbf{x})}{\partial k} &= -\frac{\phi(\mathbf{x})\mu}{k^2(\mathbf{x})|\nabla p|} = -\frac{s(\mathbf{x})}{k(\mathbf{x})}, \\ \frac{\partial s(\mathbf{x})}{\partial \phi} &= \frac{\mu}{k(\mathbf{x})|\nabla p|} = \frac{s(\mathbf{x})}{\phi(\mathbf{x})}. \end{aligned} \tag{5.73}$$

These partial derivatives provide semi-analytic expressions for the model parameter sensitivities. Note that these sensitivities, and Equation (5.71) can also be used to relate perturbations in flow properties to deviations in tracer breakthrough times. The final task is to sum up the contributions from all the trajectories connecting the source and the observation point. Thus, we must evaluate the integral

$$\delta c(t) = -\frac{1}{Q}\int \dot{c}_0\left[t - \tau^o(\Psi)\right]\delta\tau(\Psi)d\Psi \tag{5.74}$$

where the range of integration spans all relevant trajectories.

It is important to note that the sensitivity computations involve quantities that are readily available along the trajectories. Furthermore, the expressions are given by line integrals along the trajectories, requiring little additional computation. Figure 5.16 shows the concentration sensitivity to perturbations in porosity and permeability, computed using the semi-analytic, trajectory-based approach. The well configuration consists of a source in the top left corner and an observation point in bottom right corner. For comparison, on the right we have plotted the results of a numerical perturbation approach, whereby each parameter is perturbed in succession and the calculated concentration is recomputed using a distinct forward simulation. Thus, for N parameters the numerical perturbation approach takes $N + 1$ forward runs. The overall agreement between the numerical and trajectory-based sensitivities demonstrates the general validity of the semi-analytic approach.

(a) Tracer sensitivity with respect to permeability

(b) Tracer sensitivity with respect to porosity

Figure 5.16 A comparison of sensitivities computed using a trajectory-based semi-analytic approach (left panels) and a numerical perturbation–based approach (Right panels) [from Vasco and Datta-Gupta (1999) and Datta-Gupta et al. (2002)]. Reprinted with permission. Copyright AGU.

5.5.3 Estimation of subsurface properties: tracer tomography

In this section we discuss the estimation of subsurface properties from experiments involving tracers. The properties of interest are three-dimensional distributions of permeability, porosity and, in case of a partitioning tracer, the saturation distribution of an in-situ fluid phase. In most cases, we are faced with an inverse problem in which remote observations are used to infer the internal properties of an object (Menke, 1989; Parker, 1994; Aster et al., 2013). As noted above, a trajectory-based approach is a particularly efficient means for formulating and solving many inverse problems. There are at least two attractive features to recommend it. First, as is evident from the section on the asymptotic approach, the forward problem of calculating the tracer concentration at an observation point partitions into two sub-problems: a travel time computation and the computation of the amplitude.

This partitioning provides additional flexibility in treating the inverse problem. Thus, one can first match tracer arrival or breakthrough times to produce initial estimates of flow properties. The advantage is the quasi-linearity of travel time tomography, leading to better convergence properties. Second, a trajectory-based approach provides semi-analytic expressions for sensitivities, simply requiring one set of forward runs for every model update.

Formulation: tracer travel times

As an illustration, consider an inverse problem based upon the use of tracer travel times, governed by Equation (5.67). Given a set of observed travel times T_{obs} and an initial estimate of the flow properties of the porous medium, we can use Equations (5.66) and (5.67) to calculate the expected travel times T_{cal}. One can relate the residual $\delta\tau_r = T_{obs} - T_{cal}$ to a model perturbation using Equations (5.71), (5.72), and (5.73)

$$\begin{aligned}\delta\tau_r &= -\int_{\Psi_o} \frac{\partial s(\mathbf{x})}{\partial k}\delta k(\mathbf{x})d\mathbf{x} - \int_{\Psi_o} \frac{\partial s(\mathbf{x})}{\partial \phi}\delta\phi(\mathbf{x})d\mathbf{x} \\ &= \int_{\Psi_o} s_o\frac{\delta k}{k}d\mathbf{x} - \int_{\Psi_o} s_o\frac{\delta\phi}{\phi}d\mathbf{x} \\ &= \int_{\Psi_o} s_o\left(\frac{\delta k}{k} + \frac{\delta\phi}{\phi}\right)d\mathbf{x}.\end{aligned} \tag{5.75}$$

where $s_o(\mathbf{x})$ is the slowness (5.66), calculated using the properties of the initial or background porous medium. From Equation (5.75) we conclude that one can only resolve an effective permeability, given by the linear combination

$$\frac{\delta k_e}{k_o} = \frac{\delta k}{k_o} - \frac{\delta\phi}{\phi} \tag{5.76}$$

from an inversion based solely upon tracer arrival times. There are a number of ways to deal with this limitation. One could augment the travel times with other observations to try and resolve the trade-off. One can reformulate the problem in terms of a better resolved variable such as the ratio k/ϕ. Or one can simply solve for effective permeability and interpret the results. Often, the relative variations in permeability are much larger than relative porosity variations. Also, the spatial variations in porosity can be better characterized using well and seismic data compared to permeability. Thus, the effective permeability might reflect the relative changes in permeability, particularly if we restrict the changes in porosity.

Models of porous media are usually defined over a network or a grid of cells. The grid may be structured, such as a finite difference grid, or an unstructured mesh, but it will be described by a finite set of discrete elements. Most commonly, each element has a constant set of properties such as porosity and permeability.

In that case, due to the linearity of integration, Equation (5.75) can be decomposed into a sum of integrations along segments of trajectories intersecting each grid element. Figure 1.7 in Chapter 1 gives an intuitive presentation of one simple discretization. If we denote the collection of elements or grid blocks intersected by the i-th trajectory by E_i, then the integral expression (5.75) may be rewritten as the sum (neglecting porosity variations)

$$\delta\tau_i = \sum_{j \in E_i} l_{ij}\bar{s}_j \frac{\delta k_j}{\bar{k}_j} \tag{5.77}$$

where l_{ij} is the length of the segment of the i-th trajectory intersecting the j-th element. The overbars in $\bar{s}_i$ and $\bar{k}_j$ indicate that the quantities are taken from the background model. With just a single breakthrough time, we would be rather limited in our ability to image flow properties. Fortunately there are instruments such as multi-level samplers, that provide much denser spatial sampling (Figure 1.10). In addition, there are multi-chemical tracer systems in which each well segment can be identified with a particular chemical signature. Coupled with horizontal wells, such a system can be used to identify flow patterns within a reservoir.

One can also couple tracer experiments with laboratory and geophysical imaging in order to monitor tracer migration in three dimensions. With automated systems, such as the multi-level continuous active source seismic monitoring (ML-CASSM) system it is possible to monitor the tracer migration with dense sampling in both time and space (Daley et al., 2011; Ajo-Franklin et al., 2012). We will examine one particular example, x-ray imaging of tracer migration through a core sample, in some detail (Figure 5.17). This experiment illustrates the advantages of the partitioning of the inverse problem into a phase or travel time matching component and an amplitude matching problem.

One of the issues associated with the use of time-lapse laboratory or geophysical imaging data is that it can be difficult to relate the magnitudes of changes in the observations to the state of the porous medium, for example to relate time-lapse seismic amplitude changes to changes in saturation and pressure. An alternative approach, that may be used when there are a sufficient number of time-lapse snapshots, avoids using the magnitudes of changes directly. Rather, the methodology makes use of the **onset time** of a change in an observation (Press et al., 1992; Vasco et al., 2014, 2015). The onset time is the time at which the measured quantity starts to deviate from its background or initial value. For example, in the volume elements (voxels) of the core cross-section plotted in Figure 5.17 the onset time is associated with time at which the attenuation changes from its initial value. There are a number of ways to define an onset time. For example, when the changes in attenuation exceed some threshold value, such as 5 percent of the peak value, or when the attenuation changes most rapidly. Because this time is sensitive to the arrival of the saline

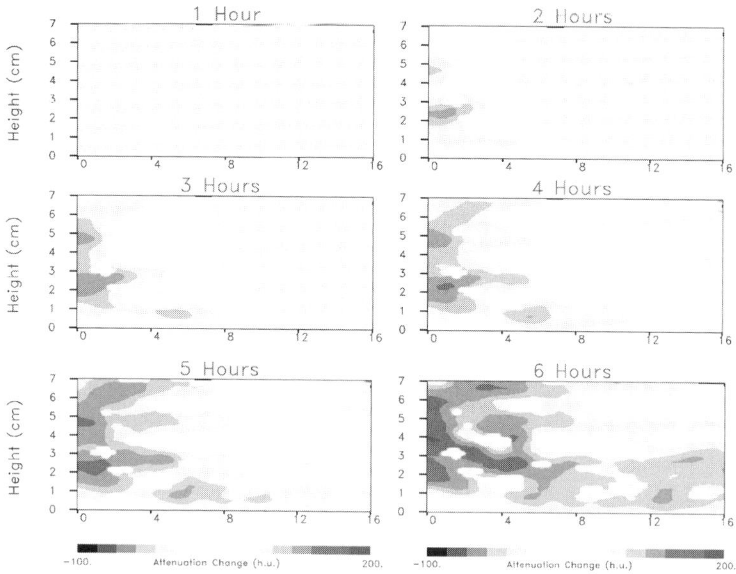

Figure 5.17 Tomographic x-ray attenuation data from the injection of a saline tracer into a sandstone core. For the colour version, please refer to the plate section.

tracer, and insensitive to coupling parameters that must be known when amplitudes are used, the onset time may be used as a measure of the tracer arrival time, T_{obs}, at each voxel. Onset time estimates are increasingly accurate with improved temporal sampling. However, as shown in Vasco et al. (2015), one can estimate large-scale variations in flow properties using surveys that are widely spaced in time. In Figure 5.17 the x-ray images are an hour apart but newer systems allow for much denser time sampling. In Figure 5.18 we plot the onset times extracted from the laboratory x-ray attenuation data.

Given a collection of x-ray onset or tracer arrival times, and the associated constraints of the form (5.77), we can define a system of equations constraining updates to a given permeability model

$$\delta\boldsymbol{\tau} = \mathbf{M}\delta\boldsymbol{\kappa}, \tag{5.78}$$

where $\delta\boldsymbol{\tau}$ is a vector of N travel time residuals and $\delta\boldsymbol{\kappa}$ is the vector of J unknown model parameter perturbations. The elements of the matrix are the combination of the parameters given in Equation (5.77), that is

$$M_{ij} = \frac{l_{ij}\bar{s}_j}{\bar{k}_j}.$$

The inverse problem is nonlinear because the trajectories depend upon the flow properties. Each time the model is updated the trajectories change and must be

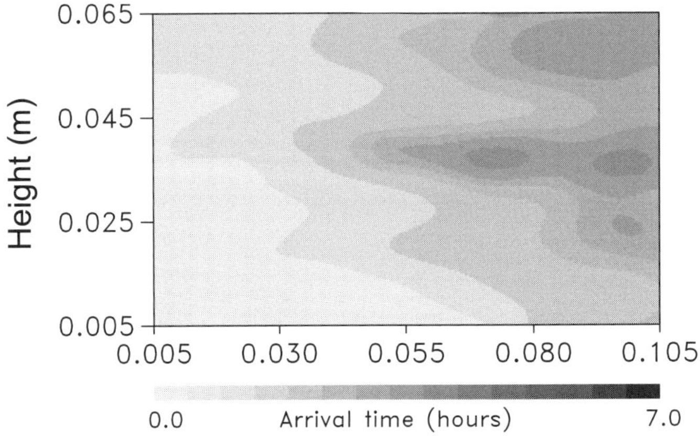

Figure 5.18 Onset time of the changes in x-ray attenuation from the background value.

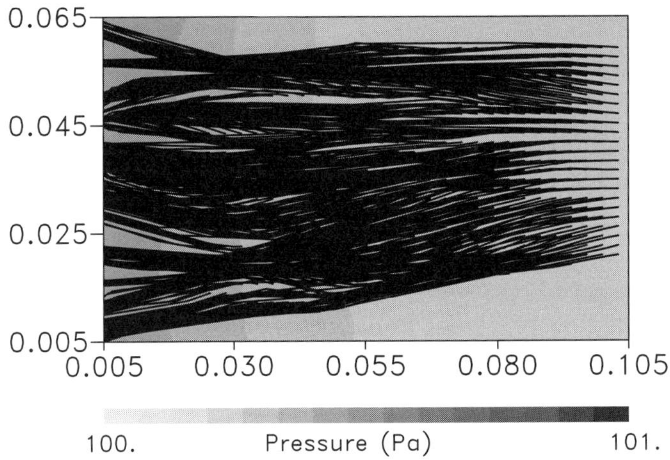

Figure 5.19 Trajectories associated with the final model update.

recomputed for the current porous medium. For example, in Figure 5.19 we plot the trajectories corresponding to the final update (50th iteration) of the effective permeability (Figure 5.20).

Because the model is an approximation of the actual porous medium and due to experimental limitations, we expect model and observational errors. Such errors preclude an exact solution to the system (5.78). In the face of such uncertainty we follow Gauss's lead and attempt to solve the system in a least-squares sense. That is, we seek to minimize the sum of the squares of the residuals

$$\mu_d^2 = (\delta\boldsymbol{\tau} - \mathbf{M}\delta\boldsymbol{\kappa})^t \cdot (\delta\boldsymbol{\tau} - \mathbf{M}\delta\boldsymbol{\kappa}) . \tag{5.79}$$

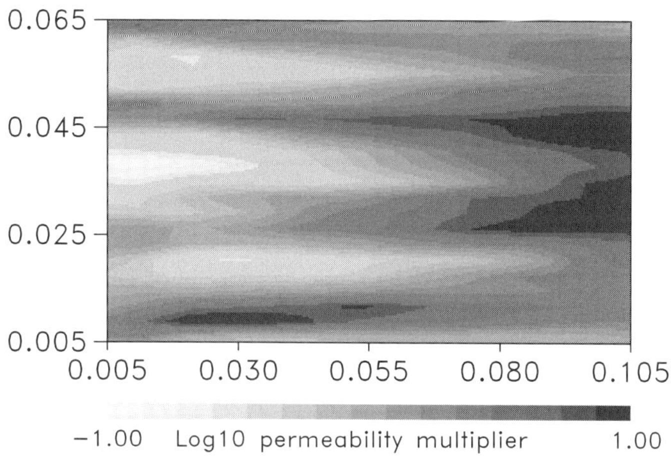

Figure 5.20 Final effective permeability model.

However, we are faced with an additional consideration because real-world applications very often involve a large number of unknown parameters and a limited number of measurements. Typically, there are many more unknowns than constraints, and the inverse problem tends to be ill-posed, leading to non-uniqueness and instability in the least-squares estimates (Parker, 1994). One remedy is to add penalty terms to the sum of the square of the residuals (5.79) and then minimize the total, a process known as **regularization** (Menke, 1989; Parker, 1994; Aster et al., 2013). Two common penalty functions are quadratic forms associated with the size and the roughness of the model updates. The norm penalty, given by

$$\mu_n^2 = \delta\boldsymbol{\kappa}^t \cdot \delta\boldsymbol{\kappa}, \tag{5.80}$$

ensures that the model perturbation is not large and the updated model does not stray too far from our current model. This makes intuitive sense, given that we are taking an iterative approach involving a local linearization. Too large a step might violate the conditions of the linearization. The roughness penalty

$$\mu_r^2 = (\mathbf{L}\delta\boldsymbol{\kappa})^t \cdot (\mathbf{L}\delta\boldsymbol{\kappa}), \tag{5.81}$$

accounts for the fact that each travel time datum is an integrated response over an entire trajectory and is incapable of resolving fine-scale features of the model. The multiplier $\mathbf{L}$ is a differencing matrix that provides a finite difference approximation to the spatial curvature of the model. For example, the j-th term of the vector $\mathbf{L}\delta\boldsymbol{\kappa}$ might be given by

$$\begin{aligned}(\mathbf{L}\delta\boldsymbol{\kappa})_j = 6\delta\kappa_j - \delta\kappa_{j+1} - \delta\kappa_{j-1} - \delta\kappa_{j+n} - \delta\kappa_{j-n} \\ -\delta\kappa_{j+m} - \delta\kappa_{j-m}\end{aligned} \tag{5.82}$$

where n and m are the offsets in the indices that are necessary to recover the elements adjacent to the j-th cell in the y and z directions, respectively. Rather than simply minimizing the data misfit μ_d^2, we now find the model update $\delta\boldsymbol{\kappa}$ that minimizes the augmented misfit function

$$\mu^2 = \mu_d^2 + w_n \mu_n^2 + w_r \mu_r^2, \tag{5.83}$$

where w_n and w_r are weighting coefficients that determine the importance of honoring the regularization relative to fitting the data. A necessary condition for a minimum of μ^2 is the vanishing of its gradient with respect to the model parameters:

$$\nabla_\kappa \mu^2 = -\mathbf{M}^t \delta\boldsymbol{\tau} + \left[\mathbf{M}^t\mathbf{M} + w_n\mathbf{I} + w_r\mathbf{L}^t\mathbf{L}\right]\delta\boldsymbol{\kappa} = 0 \tag{5.84}$$

or

$$\left[\mathbf{M}^t\mathbf{M} + w_n\mathbf{I} + w_r\mathbf{L}^t\mathbf{L}\right]\delta\boldsymbol{\kappa} = \mathbf{M}^t\delta\boldsymbol{\tau}. \tag{5.85}$$

These are the normal equations, first encountered in Chapter 4. The solution of the regularized inverse problem, our model update $\delta\boldsymbol{\kappa}$, is obtained by solving the system (5.85). At this point it is worth making a few observations regarding the nature of these equations. First, given the advancement of technology and the continuing integration of remote imaging techniques with fluid flow modeling and inversion, it is possible that the system (5.85) could be rather large. Second, each of the matrices in Equation (5.85), **M**, **I**, and **L**, is sparse. This is evident from Figure 5.19 because the non-zero elements of each row of the matrix are associated with the cells intersected by one particular trajectory. And a single trajectory only samples a small percentage of the total number of cells in the model. Thus, only a small percentage of the elements of each matrix are non-zero. Note that the same cannot be said about the matrix in the square brackets in Equation (5.85) because the matrix products, such as $\mathbf{M}^t\mathbf{M}$, are not necessarily sparse. These observations suggest taking advantage of sparse matrix methods in which we only store and manipulate the non-zero elements of the coefficient matrices (Saad, 2003). One particularly useful scheme for solving large, sparse, linear systems, the LSQR algorithm of Paige and Saunders (1982), was widely adopted for geophysical imaging. Approaches such as these have led to the efficiencies that are associated with tomographic imaging and which have proven to be so helpful in large-scale imaging problems (Nolet, 1987). Note that these sparse matrix techniques are also applicable to more sophisticated approaches for solving the nonlinear inverse problem, methods such as the conjugate gradient algorithm and limited-memory quasi-Newton methods. However, we limit ourselves to discussing the iterative updating scheme, we do not have space for even a cursory review of these other methods. Fortunately, there are several excellent texts that provide useful introductions to these valuable techniques

(Gill et al., 1982; Press et al., 1992; Fletcher, 2000; Nocedal and Wright, 2006). As a final note, we point out that we can use Equation (5.85) to derive a formal expression for our estimated model update

$$\begin{aligned}\delta\hat{\boldsymbol{\kappa}} &= \left[\mathbf{M}^t\mathbf{M} + w_n\mathbf{I} + w_r\mathbf{L}^t\mathbf{L}\right]^{-1}\mathbf{M}^t\delta\boldsymbol{\tau} \\ &= \mathbf{G}^{\dagger}\delta\boldsymbol{\tau},\end{aligned} \tag{5.86}$$

where

$$\mathbf{G}^{\dagger} = \left[\mathbf{M}^t\mathbf{M} + w_n\mathbf{I} + w_r\mathbf{L}^t\mathbf{L}\right]^{-1}\mathbf{M}^t \tag{5.87}$$

is known as the **generalized inverse**. This quantity is primarily useful for derivations and algorithm development. We are unlikely to ever try and form the generalized inverse explicitly, nor to even construct the matrix inside its square brackets, a potentially large, dense, and unwieldy array.

Assessing the estimates: model parameter covariance and resolution matrices

In solving Equation (5.85), perhaps using the LSQR algorithm (Paige and Saunders, 1982), we can obtain an estimate of our model update. But, we have already noted that the parameters are typically poorly constrained and that the data may not resolve anything but the large-scale spatial variations in properties. Even if we have dense spatial sampling, as provided by the array of multi-level samplers plotted in Figure 1.10 in Chapter 1, we may still have poor spatial resolution, due to the spatial averaging that is an inherent property of tracer data. For example, in a uniform flow field the trajectories can average over significant distances in particular directions (Figure 1.10). Also, our solution depends upon the regularization weighting parameters w_n and w_r. How does the solution change as these parameters are varied? What can we use to guide our choice of these weights? Because of these limitations, an important part of the solution of an inverse problem involves assessing the uncertainty and resolution associated with a particular estimate.

The uncertainty associated with the model parameter estimates is perhaps the easiest issue to deal with because it is almost always a part of any exercise in parameter estimation. Thus, we have a long history to draw from and help may be found in many quarters (Tarantola, 1987; Menke, 1989; Parker, 1994; Aster et al., 2013). The most important factors in calculating the model parameter uncertainties will be the nature of the errors contained in the data and the fact that, at least for a given model perturbation, the estimate is a linear function of the observations. The linear dependence of the estimate upon the data is clear from Equation (5.85). The nature of the errors contained within the data is less clear-cut. In order that this discussion not take us far-a-field we are going to assume that the errors follow a Gaussian distribution. This assumption is also compatible with our decision to use a least-squares approach, as it provides a maximum-likelihood estimate for the model parameters in the presence of Gaussian errors (Menke, 1989). Such

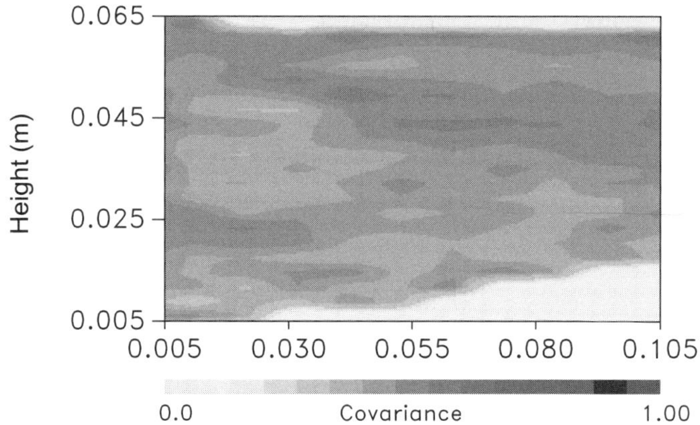

Figure 5.21 Model parameter standard errors obtained from the model parameter covariance matrix.

errors can be characterized by a data covariance matrix, denoted by $\mathbf{C}_{dd}$. If the data errors are Gaussian, characterized by a data covariance matrix, and the relationship between the model parameter estimates and the data is linear, then the posteriori model parameter **covariance matrix** associated with the model parameters, $\mathbf{C}_{mm}$, is given by

$$\mathbf{C}_{mm} = \mathbf{G}^{\dagger}\mathbf{C}_{dd}(\mathbf{G}^{\dagger})^{t} \tag{5.88}$$

(Tarantola, 1987). There may be some issues associated with computing $\mathbf{C}_{mm}$ as it requires computing the generalized inverse. It may be advantageous to compute the model parameter covariance matrix using a singular value decomposition (Menke, 1989). In Figure 5.21 we plot the model parameter standard errors, the square root of the model parameter variances (diagonal elements of the covariance matrix). Because the temporal sampling of the snapshots, plotted in Figure 5.17, is one hour, we assume that the uncertainty of the arrival times is 30 minutes. This results in a rather large model parameter uncertainty associated with the model parameter estimates plotted in Figure 5.20.

The question of model parameter resolution is not commonly dealt with in fields outside of geophysics. That is unfortunate because model parameter resolution allows one to judge the quality of the estimates in a manner that is not so dependent upon the assumed data errors. In order to derive the resolution matrix one assumes that there is an actual or true distribution of properties, the resolution matrix relates this true model to the estimated model parameters. If no errors are present, the relationship between the true model $\delta\boldsymbol{\kappa}$ and the error-free data is expressed by Equation (5.78). Substituting this expression for $\delta\boldsymbol{\tau}$ in Equation (5.86) results in

$$\widehat{\delta\boldsymbol{\kappa}} = \mathbf{G}^{\dagger}\mathbf{M}\delta\boldsymbol{\kappa}, \tag{5.89}$$

which relates the true model, if there were such a model, to our estimates of the model parameters. Therefore, the **resolution matrix** is given by

$$\mathbf{R} = \mathbf{G}^{\dagger}\mathbf{M} \tag{5.90}$$

and

$$\widehat{\delta\kappa} = \mathbf{R}\delta\kappa. \tag{5.91}$$

The resolution matrix $\mathbf{R}$ may be interpreted as a linear filter through which we view the actual spatial distribution of flow properties. The rows of the resolution matrix are averaging coefficients indicating the contribution of various other parameters to our estimate of a property in a given cell. In Figure 5.22 we plot elements from two rows of the resolution matrix. The fact that few other cells contribute to the

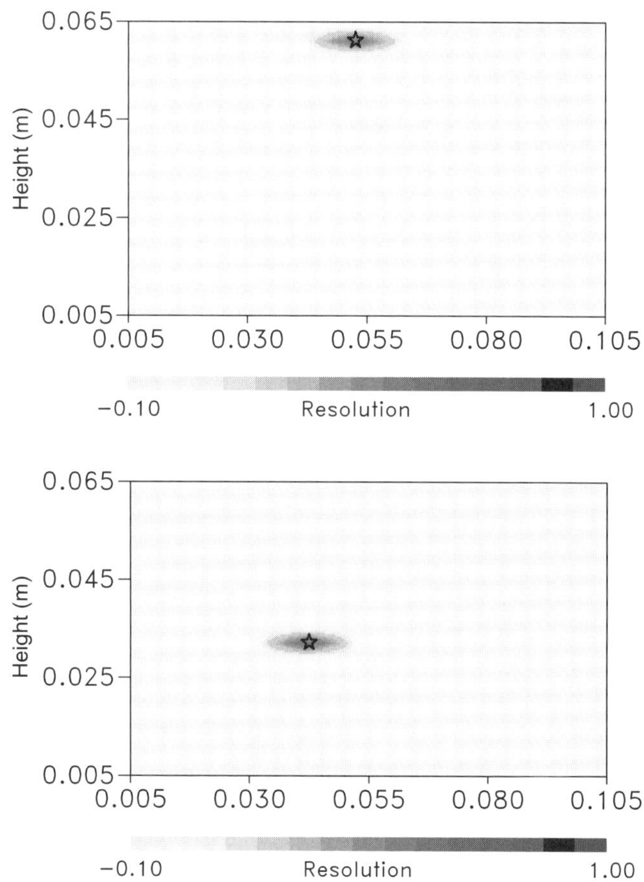

Figure 5.22 Two rows of the model parameter resolution matrix. The elements of each row are averaging kernels indicating the contribution of each parameter in the model to an estimate of the value in the cell marked by the unfilled star.

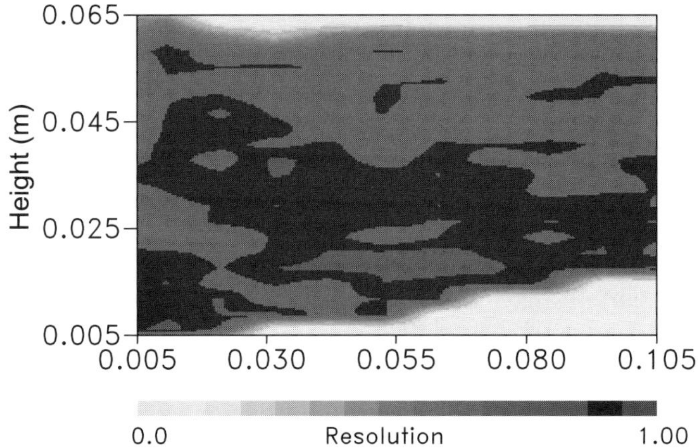

Figure 5.23 Estimated of model parameter resolution provided by the diagonal elements of the resolution matrix.

estimate indicates that it is fairly well resolved spatially. Thus, even though the trajectories are all sub-parallel to the length of the core, the temporal sampling provides resolution along each trajectory. In an ideal case the resolution matrix would be an identity matrix and we would resolve the properties of each cell perfectly, with no tradeoff between adjacent estimates nor with other classes of parameters. An estimate of the resolution of a model parameter is provided by the diagonal elements of the resolution matrix. If the diagonal value is close to 1 then the model parameter is well resolved but if it is near 0 then the parameter is poorly resolved. In Figure 5.23 we plot the diagonal elements of the resolution matrix associated with our x-ray imaging experiment. From this figure one can see that most of the model parameters are well resolved. The poorly resolved cells are located near the top and bottom of the core where there are few or no trajectories. One advantage of the resolution matrix is that it is independent of the data uncertainty and only depends on our sensitivities and the geometry of our experiment. The concept of model parameter resolution is particularly important for inverse modeling based upon a sparse distribution of data from widely spaced observation points.

There is a trade-off between model parameter resolution and model parameter covariance. That is, if one tries to improve the model parameter resolution, for example by reducing the model roughness w_r in the inversion, this will tend to increase the model parameter covariance. Thus, one can use this trade-off to determine those values of the weighting w_r that balance the desire to have reasonable error estimates against the goal of well or moderately resolved parameters. This balancing argument has been formalized into the idea of a trade-off curve, in which reasonable values of model parameter uncertainty and model parameter resolution are sought (Menke, 1989; Aster et al., 2013).

In closing, we must point out that there is an alternative probabilistic formulation of the inverse problem based on the well-known Bayes theorem which provides a convenient framework for combining the data and prior information (Tarantola, 1987). The Bayesian approach updates our prior knowledge of the parameter distribution by integrating data to derive a posterior distribution which represents solution to the inverse problem. Advanced sampling techniques such as Markov Chain Monte Carlo (MCMC) methods can be used to sample from the posterior distribution and to quantify the uncertainty in the parameter estimates. Some of the advantages of the trajectory-based methods in terms of fast forward modeling and sensitivity computations will also apply to Bayesian inversion and sampling (Ma et al., 2008). A huge body of literature exists on Bayesian inverse modeling and its link with the more deterministic framework discussed earlier. The reader is referred to the book by Tarantola (1987) for further information.

5.6 Summary and conclusions

An asymptotic formulation of a solution to the equation governing advection and transport leads to a trajectory-based approach. In the limit of no dispersion, the trajectory-based technique reduces to the streamline formulation for transport modeling. These methods provide insight and efficiency in both forward and inverse modeling. In particular, the techniques provide a foundation for tracer tomography and efficient imaging of flow properties.

6

Immiscible fluid flow

6.1 Introduction

In the Chapter 5 we considered the case in which transported material did not influence the fluid flow. Now we consider the situation in which at least two distinct fluids are present in the porous medium and the presence or absence of one of the fluids will have a significant impact on the flow of the other. In Chapter 2 we discussed the governing equations for the flow of such immiscible fluids, so we will limit our remarks to the key features of such flow. One restriction maintained in this chapter is that the mass transfer between fluid phases is negligible. Such a restriction is not necessary for the adoption of trajectory-based methods, but simply serves to limit the complexity of our treatment.

6.2 Governing equations for two-phase flow

A classic example of the type of immiscible fluid flow that we have in mind is the movement of oil and water within a porous material. In fact, this example illustrates some of the most important characteristics of multiphase fluid flow within a solid containing connected pore space. For example, it is often the case that the molecules of one fluid will have an affinity for the molecules of the solid matrix. This fluid tends to adhere to the solid surface of the pores, or to wet the solid. We will denote the saturation of this 'wetting' phase by S_w. The other fluid, simply referred to as the 'non-wetting' phase, has less attraction for the solid and tends to occupy the central regions of the pores. The saturation of the non-wetting phase is denoted by S_n. Since the pore volume of the medium is completely occupied by some combination of the two fluids, the saturations must sum to unity

$$S_w + S_n = 1. \tag{6.1}$$

The differential affinity for the solid has another consequence, leading to surface tension and curvature in the boundary separating the two fluids. The result is greater pressure in the non-wetting fluid relative to the pressure in the wetting fluid. This pressure difference is termed the capillary pressure, p_c, and is most frequently assumed to be a unique function of the fluid saturation:

$$p_n - p_w = p_c(S_w). \tag{6.2}$$

The capillary pressure curve $p_c(S_w)$ is an important aspect of multiphase fluid flow and a key constitutive equation. However, the difference in capillary pressure is often much smaller than the average fluid pressure. As pointed out by Peaceman (1977), and shown below, this fact can be used to derive a governing equation for the average fluid pressure.

As noted in Chapter 2, the governing equations for multiphase flow are rooted in the conservation of mass for the l-th fluid phase, as expressed by the equation,

$$\frac{\partial\left(\varphi\rho_l S_l\right)}{\partial t} + \nabla\cdot\left(\rho_l \mathbf{q}_l\right) = Q_l, \tag{6.3}$$

where l denotes either w or n, φ is the porosity, ρ_l is the density of the l-th fluid. The source term on the right-hand-side, $Q_l(\mathbf{x}, t)$, represents either the injection or withdrawal of the l-th fluid at one or more locations. The vector $\mathbf{q}_l$ is the superficial flow velocity, given by the multiphase extension of Darcy's law, Equation (2.128) in Chapter 2, attributing the fluid flow velocity to two forces: the fluid pressure gradient and the gravitational attraction g acting in the direction of the vector $\mathbf{z}$. For the wetting phase Darcy's law takes the particular form

$$\mathbf{q}_w = -\lambda_w\left(\nabla p_w - \rho_w g\mathbf{z}\right) \tag{6.4}$$

where $\mathbf{q}_w$ is the velocity of the wetting phase, and we have defined the **phase mobility** of the wetting fluid

$$\lambda_w = \frac{k_{rl}(S_w)k}{\mu_w} \tag{6.5}$$

k being the absolute permeability, $k_{rw}(S_w)$ is the relative permeability function, an additional constitutive equation. One can think of the relative permeability function as a reduction factor in the permeability due to the presence and interference of the other fluid. That is, if the medium was fully saturated by the wetting phase, $k_{rw}(S_w)$ equals 1 and the product $k_{rw}(S_w)k$ equals the absolute permeability k. For S_w less than 1 the function $k_{rw}(S_w)$ becomes successively smaller, approaching zero as S_w approaches small values. The final constitutive equation relates the density of the fluid phases to the fluid pressure, $\rho_l(p_l)$ and is related to the compressibilities of the fluids. In fact, we can formally define the fluid compressibilities as

$$c_l(p_l) = \frac{1}{\rho_l}\frac{d\rho_l}{dp_l}. \tag{6.6}$$

The expression (6.3) represents a system of two equations in the two unknown fluid saturations S_w and S_n. The equations are linked by the fact that the saturations sum to unity, as expressed by (6.1). While we can combine the set of equations with appropriate constitutive equations to construct a solvable system, there is some advantage and additional physical insight if we reformulate them. In particular, we can derive a pressure equation and a saturation equation. This is advantageous because of the very different nature of the pressure variation in comparison to the saturation variation. For example, the saturation changes propagate with the same velocity, coupled by the constraint (6.1), while the pressure changes can propagate at a substantially different velocity (Vasco, 2011). This fact has implications for the modeling of multiphase flow, allowing for the time separation that is used to ones advantage in streamline modeling, as described later in this chapter. Thus, the saturation equation can be solved over long time intervals, during which the average fluid pressure is largely constant

6.2.1 A pressure equation

The pressure differential equation is obtained from the Equations (6.3) by eliminating the time derivatives of the saturation. If we expand the time derivative in each of these equations we have

$$\rho_l S_l \frac{\partial \varphi}{\partial t} + \varphi S_l \frac{d\rho_l}{dp_l}\frac{\partial p_l}{\partial t} + \varphi \rho_l \frac{\partial S_l}{\partial t} + \nabla \cdot (\rho_l \mathbf{q}_l) = Q_l. \tag{6.7}$$

Dividing each equation by the appropriate density ρ_l, adding them together, and noting that $S_w + S_n = 1$, a constant, we obtain a single equation

$$\frac{\partial \varphi}{\partial t} + \varphi S_w c_w \frac{\partial p_w}{\partial t} + \varphi S_n c_n \frac{\partial p_n}{\partial t} + \frac{1}{\rho_w}\nabla \cdot (\rho_w \mathbf{q}_w) + \frac{1}{\rho_n}\nabla \cdot (\rho_n \mathbf{q}_n) = \hat{Q}_t, \tag{6.8}$$

where we have used the definition (6.6) and defined the weighted source term representing the total flow rate

$$\hat{Q}_t = \frac{Q_w}{\rho_w} + \frac{Q_n}{\rho_n}, \tag{6.9}$$

in terms of injected volume rather than mass. If the rock and fluids are incompressible then the compressibilities are zero and any spatial or temporal derivatives of the densities vanish as well. Therefore, Equation (6.8) takes a particularly simple form:

$$\nabla \cdot \mathbf{q}_t = \hat{Q}_t \tag{6.10}$$

where

$$\mathbf{q}_t = \mathbf{q}_w + \mathbf{q}_n \tag{6.11}$$

is the total velocity

Incompressible fluids are something of an idealization, and the effect of compressibility typically must be taken into account in any realistic model of fluid flow. However, it is usually the case that the capillary pressure is very small in comparison to the average fluid pressure given by

$$\bar{p} = \frac{p_w + p_n}{2}. \tag{6.12}$$

In the box below we rewrite the pressure Equation (6.8) in terms of the average and capillary pressures. Note that, because the capillary pressure is considered to be a function of the saturation [see Equation (6.2)], the final equation in Box 6.1 is really an equation in terms of the average fluid pressure and the saturation. Often the capillary pressure terms are negligible in comparison to the average pressure terms (Peaceman, 1977, p. 18) and the final pressure equation in Box 6.1 reduces to an equation for $\bar{p}$,

$$\varphi\bar{c}\frac{\partial\bar{p}}{\partial t} - \frac{1}{\rho_w}\nabla\cdot[\rho_w\lambda_w\nabla\bar{p}] - \frac{1}{\rho_n}\nabla\cdot[\rho_n\lambda_n\nabla\bar{p}] = \hat{\hat{Q}}. \tag{6.13}$$

where the coefficient of the time derivative is

$$\bar{c} = \frac{1}{\varphi}\frac{d\varphi}{d\bar{p}} + S_w c_w + S_n c_n, \tag{6.14}$$

and the modified source term

$$\hat{\hat{Q}} = \hat{Q}_t - \frac{1}{\rho_w}\nabla\cdot\left(\rho_w^2\lambda_w g\mathbf{z}\right) - \frac{1}{\rho_n}\nabla\cdot\left(\rho_n^2\lambda_n g\mathbf{z}\right) \tag{6.15}$$

contains the effects of gravitational forces as well as those due to injection and withdrawal.

Box 6.1 **The pressure equation in terms of the average and capillary pressures**

In this box we rewrite the pressure Equation (6.8) in terms of the average $\bar{p}$ and capillary p_c pressures. First, the wetting and non-wetting fluid phase pressures are written as

$$p_w = \bar{p} - \frac{1}{2}p_c$$

and

$$p_n = \bar{p} + \frac{1}{2} p_c.$$

These expansions are then substituted into Equation (6.8), taking into account the generalized Darcy's law, given by the expression (6.4) for the wetting phase. The resulting equation is

$$\frac{\partial \varphi}{\partial t} + \varphi S_w c_w \frac{\partial \bar{p}}{\partial t} - \frac{1}{2} \varphi S_w c_w \frac{\partial p_c}{\partial t} + \varphi S_n c_n \frac{\partial \bar{p}}{\partial t} + \frac{1}{2} \varphi S_n c_n \frac{\partial p_c}{\partial t}$$

$$- \frac{1}{\rho_w} \nabla \cdot \left[\rho_w \lambda_w \left(\nabla \bar{p} - \frac{1}{2} \nabla p_c - \rho_w g \mathbf{z} \right) \right]$$

$$- \frac{1}{\rho_n} \nabla \cdot \left[\rho_n \lambda_n \left(\nabla \bar{p} + \frac{1}{2} \nabla p_c - \rho_n g \mathbf{z} \right) \right] = \hat{Q}_t.$$

We can recast this as an equation in $\bar{p}$ and p_c, with coefficients that depend upon the saturations and other parameters if we rewrite the time derivative of porosity, φ, in terms of the time derivative of the average fluid pressure. That is, we assume that the porosity is a function of the average fluid pressure, $\varphi(\bar{p})$, so that the first time derivative in the equation above may be written as

$$\frac{\partial \varphi}{\partial t} = \frac{d\varphi}{d\bar{p}} \frac{\partial \bar{p}}{\partial t}.$$

Now we can regroup terms, collecting the coefficients of each time derivative. Furthermore we can move all the gravitational terms to the right-hand-side, and define a modified source term that includes these gravitational effects. Thus, the source term will now act over all regions of the model due to the pervasive nature of the gravitational force. The resulting pressure equation is

$$\varphi \bar{c} \frac{\partial \bar{p}}{\partial t} + \varphi c_c \frac{\partial p_c}{\partial t} - \frac{1}{\rho_w} \nabla \cdot \left[\rho_w \lambda_w \left(\nabla \bar{p} - \frac{1}{2} \nabla p_c \right) \right] - \frac{1}{\rho_n} \nabla \cdot \left[\rho_n \lambda_n \left(\nabla \bar{p} + \frac{1}{2} \nabla p_c \right) \right] = \hat{\hat{Q}},$$

where we have defined the coefficients

$$\bar{c} = \frac{1}{\varphi} \frac{d\varphi}{d\bar{p}} + S_w c_w + S_n c_n,$$

$$c_c = \frac{S_n c_n - S_w c_w}{2},$$

and the modified source term

$$\hat{\hat{Q}} = \hat{Q}_t - \frac{1}{\rho_w} \nabla \cdot \left(\rho_w^2 \lambda_w g \mathbf{z} \right) - \frac{1}{\rho_n} \nabla \cdot \left(\rho_n^2 \lambda_n g \mathbf{z} \right).$$

The coupling to saturation changes occur in several locations in Equation (6.13) and the equation is non-linear because the densities are functions of pressure. For incompressible fluids the coefficients c_w and c_n vanish, and the densities on the left-hand-side drop out. In that case, the coupling to saturation is primarily through the saturation-dependence of the **total phase mobility**, $\lambda_t = \lambda_w + \lambda_n$, and the source term $\bar{Q}$. Furthermore, the equation is nearly linear in the average pressure.

6.2.2 A saturation equation

We still need to model the evolution of the fluid saturation. In many ways, this is our ultimate goal and the fluid pressure is just an intermediate step in that direction. Furthermore, even if capillary pressure effects are negligible and we can use Equation (6.13) to calculate the average pressure, the coefficients still depend upon the saturation, the coupling is still present. Therefore, we shall need one additional equation in order to specify the physical system completely, and to determine the pressure and saturation. Starting from one of the conservation Equations (6.3), we can derive the necessary equation for the saturation of one of the fluid phases. The other saturation may be determined from the constraint (6.1). For the sake of illustration, we will consider the wetting phase saturation S_w and begin with the equation

$$\frac{\partial\left(\varphi\rho_w S_w\right)}{\partial t} + \nabla\cdot\left(\rho_w \mathbf{q}_w\right) = Q_w. \tag{6.16}$$

Let us see what we can do to rewrite Equation (6.16) in terms of the total fluid velocity, $\mathbf{q}_t$, and S_w. So we will need to express $\mathbf{q}_w$ in terms of these quantities. Taking the gradient of Equation (6.2), defining capillary pressure, gives

$$\nabla p_c = \nabla p_n - \nabla p_w. \tag{6.17}$$

Using Equation (6.4) and its non-wetting counterpart, we can write ∇p_c in terms of the fluid velocities:

$$\nabla p_c = \frac{\mathbf{q}_w}{\lambda_w} - \frac{\mathbf{q}_n}{\lambda_n} + \left(\rho_n - \rho_w\right) g\mathbf{z}. \tag{6.18}$$

Using Equation (6.11) to write $\mathbf{q}_n$ in terms of the total velocity and the velocity of the wetting phase, and the fact that $p_c(S_w)$ is actually a function of the saturation of the wetting phase, we can rewrite and rearrange Equation (6.18) to provide an expression for $\mathbf{q}_w$, we have

$$\mathbf{q}_w = f_w\mathbf{q}_t + \lambda_n f_w\left(\rho_n - \rho_w\right) g\mathbf{z} - h_w\nabla S_w, \tag{6.19}$$

after multiplying by $\lambda_w\lambda_n/(\lambda_w + \lambda_n)$. In Equation (6.19) $f_w(S_w)$ is the **fractional flow curve**, or the flux function,

$$f_w(S_w) = \frac{\lambda_w}{\lambda_w + \lambda_n} \tag{6.20}$$

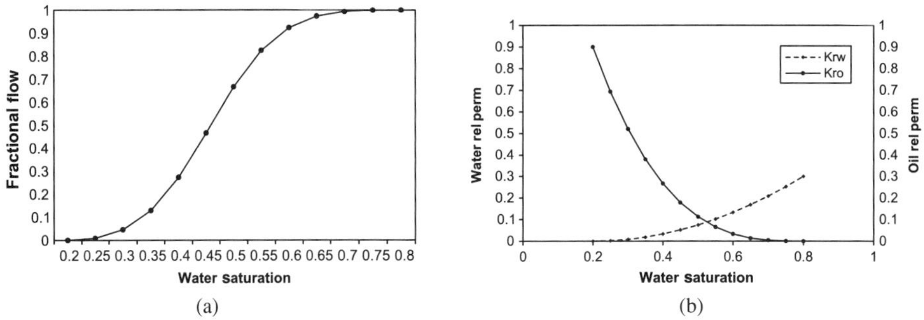

Figure 6.1 Example of relative permeability (a) for two-phase oil-water displacement (b) and water fractional flow.

that depends upon the relative permeabilities and the viscosities of the fluids. Interestingly, the fractional flow f_w does not depend upon the absolute permeability because k divides out when the ratio is formed. The fractional flow curve is an extremely important characteristic of a porous medium and the two fluids that it contains.A typical example of relative permeability and fractional flow curves for a two-phase oil-water system is shown in Figure 6.1. The function $h_w(S_w)$ accounts for the dependence of the capillary pressure upon the fluid saturation

$$h_w(S_w) = -\frac{\lambda_w \lambda_n}{\lambda_w + \lambda_n} \frac{dp_c}{dS_w} \tag{6.21}$$

and vanishes when the capillary pressure effect may be neglected or is constant. Substituting expression (6.19) into the governing Equation (6.16) gives a partial differential equation for the saturation of the wetting phase

$$\frac{\partial (\varphi \rho_w S_w)}{\partial t} + \nabla \cdot \rho_w f_w [\mathbf{q}_t + \lambda_n (\rho_w - \rho_n) g\mathbf{z}] - \nabla \cdot \rho_w h_w \nabla S_w = Q_w. \tag{6.22}$$

Equation (6.22) is of mixed character, depending upon the relative magnitude of the coefficients. Generally, if capillary effects are important, the equation is parabolic (diffusive) in nature. If capillary effects can be neglected, or when the convective term containing the total velocity vector $\mathbf{q}_t$ dominates, the equation is hyperbolic (wave-like), and reduces to

$$\frac{\partial (\varphi \rho_w S_w)}{\partial t} + \nabla \cdot \rho_w f_w [\mathbf{q}_t + \lambda_n (\rho_w - \rho_n) g\mathbf{z}] = Q_w. \tag{6.23}$$

6.3 An asymptotic approach

Trajectory-based methods for modeling the flow of immiscible fluids have a long history. The earliest work is rooted in the method of characteristics and will be discussed later. This includes the classic work of Buckley and Leverett (1942) on one-dimensional two-phase fluid displacement as well as multi-dimensional generalizations. Later efforts branched into streamtube and streamline-based modeling of two-phase flow and transport (Datta-Gupta and King, 2007). Asymptotic techniques for modeling immiscible fluid flow in a heterogeneous medium are a more recent development, although trajectory-based methods have been use to model non-linear fluid flow for some time (Courant and Friedrichs, 1948; Keller, 1954).

Because fluid velocities typically increase as a function of saturation, multiphase fluid flow often leads to shelf-sharpening fronts, particularly when capillary effects are not significant. Such sharp fronts are quite compatible with the central tenet of the asymptotic approach: that the length scale of the propagating front, denoted by l, is much smaller than the length scale of the heterogeneity (L) within a given porous layer. We shall discuss this in more detail shortly. But first, we have to settle on the governing equation that will underlie our analysis.

6.3.1 The governing equation

It is certainly possible to apply the asymptotic approach to the more general governing Equation (6.22) or indeed, even to the original system (6.3) (Taniuti and Nishihara, 1983; Vasco, 2004a; Debnath, 2005; Vasco, 2011, p. 95) but the result is too complicated to serve as an introduction to the technique. In order to keep the complexity to a minimum and maintain continuity with the streamline developments to follow, we will make some simplifying assumptions. First, there is the issue of the total fluid velocity, $\mathbf{q}_t$, or fluid pressure to deal with. That is, Equation (6.22) contains the vector $\mathbf{q}_t$, that follows from the solution of the pressure Equation (6.13). Because the fluid pressure variations are transient and decrease rapidly with distance from the well, one can often use the steady-state pressure associated with a particular set of boundary conditions. The upshot is, one can often solve the pressure equation in a piecewise fashion in time. When the boundary conditions change, for example due to the placement of a new injection or production well, the pressure is recalculated to capture the response due to the new configuration. The resulting quasi-static pressure field is used in Equation (6.22). Similarly, the pressure is recomputed when the flow rate is adjusted, or a well is taken off-line. Second, we will adopt the convection-dominated Equation (6.23), neglecting the effects of capillary pressure. Third, we will assume that the compressibility of the wetting phase can be neglected so that we shall not have to deal with the complications of a varying aqueous fluid density, ρ_w. In many applications the wetting phase is water, a fairly incompressible liquid. The density may often be treated as constant for this fluid phase and we may factor it out of Equation (6.23). Note that we will allow for a compressible non-wetting phase, and hence ρ_n may vary spatially and temporally. Finally, since we will only be dealing with the wetting phase, and there is no danger of confusion, we shall designate the saturation by S in order to save on subscripts. Based upon these considerations, we adopt Equation (6.23) and factor out the density ρ_w, giving

$$\varphi \frac{\partial S}{\partial t} + \nabla \cdot [f_w \mathbf{q}_t + g_w \mathbf{z}] = \hat{Q}_w, \tag{6.24}$$

where

$$\hat{Q}_w = \frac{Q_w}{\rho_w}. \tag{6.25}$$

is the volume flux at the source, and

$$g_w = \lambda_n (\rho_w - \rho_n) g f_w \tag{6.26}$$

is a coefficient related to the gravitational forces. We are also assuming that the porosity does not change in response to the pressure and saturation change and so we can pull it out from under the derivative.

6.3.2 An asymptotic equation for the phase

Recall from the previous chapters that the asymptotic solution is tailored to a porous medium containing smoothly-varying properties between known interfaces. Because the characteristic length-scale of the heterogeneity, L is assumed to be much larger than the characteristic length-scale of the fluid front, l, we can define the ratio

$$\varepsilon = \frac{l}{L}. \tag{6.27}$$

where $\varepsilon \ll 1$. Because the heterogeneity varies slowly in comparison to the saturation change across the front, many of the coefficients vary in a similar fashion. Therefore, it makes sense to write Equation (6.24) in terms of slowly varying temporal and spatial coordinates, denoted by T and $\mathbf{X}$, respectively. The slow coordinates may be defined in terms of ε:

$$T = \varepsilon t \tag{6.28}$$

and

$$\mathbf{X} = \varepsilon \mathbf{x}. \tag{6.29}$$

Our representation of $S(\mathbf{X}, T)$ will take advantage of the fact that ε is small, representing the solution as the power series

$$S(\mathbf{X}, T) = S_0(\mathbf{X}, T) + \sum_{i=1}^{\infty} \varepsilon^i S_i(\mathbf{X}, T, \theta) \tag{6.30}$$

where $\theta(\mathbf{x}, t)$ represents the phase associated with the propagating two-phase front. The presence of the phase indicates that we are treating the saturation change as a propagating front and θ is related to the propagation time of the front to locations within the reservoir. The function $\theta(\mathbf{x}, t)$ controls the kinetics of the front and is a key variable. Note that we have pulled out the $S_0(\mathbf{X}, T)$ term from the sum because it represents the slowly-varying background saturation variation and does not depend upon the phase. For $\varepsilon \ll 1$ the series (6.30) is dominated by it's first few terms. In many cases only the first one or two terms are of interest.

If we are going to rewrite Equation (6.24) in terms of the slow coordinates then we must rewrite the temporal and spatial derivatives in terms of T and $\mathbf{X}$. In addition, we must account for the presence of the phase function $\theta(\mathbf{x}, t)$, and it's implicit dependence upon $\mathbf{x}$ and t. The resulting expression for the time derivative is

$$\begin{aligned}\frac{\partial S}{\partial t} &= \frac{\partial T}{\partial t}\frac{\partial S}{\partial T} + \frac{\partial \theta}{\partial t}\frac{\partial S}{\partial \theta} \\ &= \varepsilon\frac{\partial S}{\partial T} + \frac{\partial \theta}{\partial t}\frac{\partial S}{\partial \theta}\end{aligned} \tag{6.31}$$

where we have made use of the relation (6.28) between t and T. A similar expression,

$$\nabla S = \varepsilon\overline{\nabla}S + \nabla\theta\frac{\partial S}{\partial \theta} \tag{6.32}$$

holds for the spatial derivatives, where $\overline{\nabla}$ denotes that the gradient involves spatial derivatives with respect to the slow coordinates. If we substitute these expressions for the derivatives in Equation (6.24) we end up with the expression

$$\varphi\left(\varepsilon\frac{\partial S}{\partial T} + \frac{\partial \theta}{\partial T}\frac{\partial S}{\partial \theta}\right) + \left(\varepsilon\overline{\nabla} + \nabla\theta\frac{\partial}{\partial \theta}\right)\cdot[f_w\mathbf{q}_t + g_w\mathbf{k}\cdot\mathbf{z}] = \varepsilon\frac{\partial m_w}{\partial T} \tag{6.33}$$

where we have used the definition (6.25) of $\hat{Q}_w$ and the fact that the flow rate is the mass flux, the rate of change of the fluid mass, $m_w(t)$, at the borehole.

Many of the coefficients in Equation (6.33), such as f_w, $\mathbf{q}_t$, and g_w depend upon the saturation S. To account for this dependence, we expand these functions as power series in S and substitute the representation (6.30) for S, resulting in

$$\begin{aligned}f_w(S) &= f_w(S_o) + \varepsilon\frac{\partial f_w}{\partial S}S_1 + O(\varepsilon^2), \\ \mathbf{q}_t(S) &= \mathbf{q}_t(S_o) + \varepsilon\frac{\partial \mathbf{q}_t}{\partial S}S_1 + O(\varepsilon^2), \\ g_w(S) &= g_w(S_o) + \varepsilon\frac{\partial g_w}{\partial S}S_1 + O(\varepsilon^2),\end{aligned} \tag{6.34}$$

where the derivatives are evaluated at the saturation S_0. We substitute these expansions and the asymptotic power series (6.30) for S into Equation (6.33). The resulting equation contains an infinite series of terms, each one containing ε to some power. Because ε is small, we shall be concerned with the terms of the lowest order in ε. When all is said and done, and we have made all the substitutions and power series expansions, retaining only terms of lowest order (ε), we are left with the equation

$$\varphi\frac{\partial S_o}{\partial T}+\overline{\nabla}\cdot[f_w(S_o)\mathbf{q}_t(S_o)+g_w(S_o)\mathbf{k}\cdot\mathbf{z}]+\left[\varphi\frac{\partial\theta}{\partial t}+\mathbf{v}\cdot\nabla\theta\right]\frac{\partial S_1}{\partial\theta}=\frac{\partial m_w}{\partial T}, \tag{6.35}$$

where we have defined the vector

$$\mathbf{v}=\frac{\partial f_w}{\partial S}\mathbf{q}_t+f_w\frac{\partial\mathbf{q}_t}{\partial S}+\frac{\partial g_w}{\partial S}\mathbf{k}\cdot\mathbf{z}. \tag{6.36}$$

In many cases the total flow velocity $\mathbf{q}_t$ is not a strong function of saturation and we may neglect its derivative with respect to S, resulting in the reduced expression for $\mathbf{v}$

$$\mathbf{v}=\frac{\partial f_w}{\partial S}\mathbf{q}_t+\frac{\partial g_w}{\partial S}\mathbf{k}\cdot\mathbf{z}. \tag{6.37}$$

Because the background saturation satisfies the governing Equation (6.24) the first two sets of terms and the right-hand-side of equation (6.35) vanish. We assume that S_1 does in fact depend upon the phase. As a consequence, its partial derivative with respect to θ is never zero. Therefore, the quantity in the second set of square brackets in Equation (6.35) must vanish. This constraint produces a linear, first-order partial differential equation for $\theta(\mathbf{x},t)$

$$\varphi\frac{\partial\theta}{\partial t}+\mathbf{v}\cdot\nabla\theta=0. \tag{6.38}$$

Defining the slowness vector

$$\mathbf{p}=\nabla\theta \tag{6.39}$$

one can write the equation for θ as the Hamilton–Jacobi equation

$$\varphi\frac{\partial\theta}{\partial t}+\mathbf{v}\cdot\mathbf{p}=0, \tag{6.40}$$

one of the most important classes of equations in mathematical physics, among other fields (Courant and Hilbert, 1962). Such equations have been studied extensively in mathematics and geometrically inspired solutions, based upon the method of characteristics, are available [see Chapter 3]. This approach leads to a trajectory-based solution, as we shall now see. One can also solve Equation (6.38) or Equation (6.40) using purely numerical techniques such as level set methods and fast marching methods (Sethian, 1999; Osher and Fedkiw, 2003). The methods are based on the viscosity solution of a Hamilton–Jacobi equation and have become popular for calculating wave propagation times.

6.3.3 The method of characteristics, trajectories, and travel times

In Section 3.5 of Chapter 3 we discussed a method for solving first-order equations such as (6.40), the method of characteristics. The approach is fairly general and is applicable to linear, quasi-linear, and fully non-linear first-order partial differential equations. Thus, we are on fairly firm ground in applying the technique to Equation (6.40). In fact, because the governing Equation (6.24) itself is a quasi-linear, first-order equation, we could have applied the technique directly, without the intervention of an asymptotic representation. Such an approach has a long history, from the work of Buckley and Leverett (1942) to studies of compressible flow related to gas dynamics (Courant and Friedrichs, 1948). The real advantage of asymptotic and related WKB techniques, methods that partition the solution into amplitude and phase functions, is that they can convert second-order partial differential equations to first-order, non-linear partial differential equations. Such equations are ripe for the application of the method of characteristics. We will not need to take advantage of that aspect here, since our equation is already first-order, rather we will be focused on the semi-analytic, trajectory-based nature of the solution. In particular, the partitioning of the equation into an equation for the phase, or travel time, and an equation for the amplitude, offers some insight into the nature of the propagation of the two-phase front. We remind the reader that the full governing Equation (6.22), containing capillary effects, is actually a second-order partial differential equation. An asymptotic approach is applicable to this more general equation (Vasco, 2011).

Let us return to the method of characteristics and its application to Equation (6.40). As in Chapter 3 we will consider the partial differential equation to be a function in the two sets of variables $\mathbf{x}$ and $\mathbf{p}$. Note the presence of the time derivative in Equation (6.40). This is not a serious issue because, as shown in Chapter 3, we can just add time to the set of independent variables $\mathbf{x}$ and add its partial derivative, denoted by q is Chapter 3, as a component of the vector of derivatives, $\mathbf{p}$. However, we adopt a simpler approach if we take advantage of the mathematical structure of the equation. First, note that the equation is linear in the derivatives. Second, if we divide through by φ then Equation (6.40) is separable and the spatial derivatives and spatially dependent coefficients may be moved to the right-hand-side, isolating the time derivative. Thus, we expect that the solution is also separable, motivating us to attempt a solution of the form

$$\theta(\mathbf{x}, t) = \vartheta(\mathbf{x}) - t. \tag{6.41}$$

We will use this form of the solution in the derivation that follows.

Equation (6.40) may be interpreted as the vanishing of a function in **x** and **p**, given by

$$F(\mathbf{x}, \mathbf{p}) = \mathbf{v} \cdot \mathbf{p} - \varphi = 0, \tag{6.42}$$

a linear partial differential equation for $\vartheta(\mathbf{x})$, the spatial component of the phase. As explained in Chapter 3, the partial differential Equation (6.42) is equivalent to the system of ordinary differential equations [see Box 3.1 for the full set of ray equations]

$$\begin{aligned} \frac{d\mathbf{x}}{ds} &= \nabla_p F = \mathbf{v} \\ \frac{d\vartheta}{ds} &= \mathbf{p} \cdot \nabla_p F = \mathbf{p} \cdot \mathbf{v} = \varphi, \end{aligned} \tag{6.43}$$

where s is a variable signifying position along the trajectory, and we have used Equation (6.42) to replace $\mathbf{p} \cdot \mathbf{v}$ by φ. Note that, because the governing equation is of first-order, the ray equations uncouple, and the equation for **p** does not really play a role in our trajectory calculations.

We can combine the first ray Equation (6.43) with the definition of **v**, Equation (6.37) to get a defining expression for the trajectory

$$\frac{d\mathbf{x}}{ds} = \frac{df_w}{dS}\mathbf{q}_t + \frac{\partial g_w}{\partial S}\mathbf{z} \tag{6.44}$$

that includes both the influence of the flow field as well as the effects of gravity. However, if the non-wetting phase is very compressible and of a sufficiently different density then one will have to update the pressure field and the saturation changes periodically. However, for incompressible constituents with moderate density contrasts, the flow field may change slowly once the transients due to changes in the well conditions have dissipated. Thus, we would only need to update the pressure field infrequently and can solve for saturation changes along the trajectories $\mathbf{x}(s)$. Therefore, Equation (6.44) serves as our entry point into the world of streamline simulation and trajectory-based modeling in the next section.

We can combine the two ray Equations (6.43) to derive an explicit expression for $\vartheta(\mathbf{x})$, a measure of the traveltime of the two-phase front, and for the phase $\theta(\mathbf{x}, t)$, a reduced time. As a first step, note that

$$\frac{d\mathbf{x}}{ds} \cdot \frac{d\mathbf{x}}{ds} = \left(\frac{dx}{ds}\right)^2 = \mathbf{v} \cdot \mathbf{v} = \mathrm{v}^2$$

where v is the magnitude of the vector **v**. Integrating Equation (6.43) for the spatial component of the phase produces the expression

$$\vartheta(\mathbf{x}) = \int_{\mathbf{x}} \varphi ds = \int_{\mathbf{x}} \frac{\varphi}{\mathrm{v}} dx.$$

The velocity $\mathbf{v}$ is given in Equation (6.37):

$$\mathbf{v} = \frac{df_w}{dS}\mathbf{q}_t + \frac{\partial g_w}{\partial S}\mathbf{z}.$$

In situations in which gravity is not important the travel time of the saturation front is given by

$$\vartheta(\mathbf{x}) = \int_{\mathbf{x}} \frac{\varphi}{\mathrm{v}} dx = \int_{\mathbf{x}} \left(\frac{df_w}{dS}\right)^{-1} \frac{\varphi}{|\mathbf{q}_t|} dx, \tag{6.45}$$

an expression that we shall meet again later in this chapter, and the phase is

$$\theta(\mathbf{x}, t) = \int_{\mathbf{x}} \left(\frac{df_w}{dS}\right)^{-1} \frac{\varphi}{|\mathbf{q}_t|} dx - t.$$

6.4 Streamline modeling of immiscible fluid flow

The asymptotic methodology produced an equation for the trajectory $\mathbf{x}(s)$ and for the phase $\vartheta(\mathbf{x})$ without too much effort. And, after examining the preceding chapters, the reader can probably fill in the steps needed to derive an equation for the zeroth-order amplitude S_0. However, at this juncture we will go in a different direction and consider a streamline approach for computing the evolution of the saturation. In streamline simulation the solution of the saturation Equation (6.22), or one of it's simplifications, is partitioned into a family of one-dimensional calculations along pre-determined trajectories. In many situations the streamlines coincide with the trajectories of the preceding asymptotic analysis. In streamline simulation one calculates the evolution of the saturation amplitude along the streamline using analytical or numerical technique such as finite-differences. Because the calculation is dominantly one-dimensional, the technique scales well with problem size and efficiencies ensue for large three-dimensional problems.

Our starting point is the governing Equation (6.24), neglecting gravitational forces. Moving the velocity divergence term to the right-hand-side where it can be treated as a source term, one has

$$\varphi\frac{\partial S}{\partial t} + \frac{df_w}{dS}\nabla S \cdot \mathbf{q}_t = \hat{Q}_w - f_w \overline{\nabla} \cdot \mathbf{q}_t. \tag{6.46}$$

It is not essential to neglect gravity. In fact, Equation (6.44) provides a trajectory valid in the presence of gravitational forces. Streamline approaches typically account for gravity via a numerical technique known as operator splitting, as noted below. The right-hand-side of Equation (6.46) may be thought of as an effective source term, accounting for the effects of compressibility and also flow into and out of wells.

6.4.1 Transforming the saturation equation into characteristic or streamline coordinates

In streamline simulation, and in most other numerical techniques, there is an underlying finite-difference or finite-element grid for the pressure calculations where Equation (6.13) is solved. However, the saturation calculations are, in a sense, free of such a grid. That is, these calculations take full advantage of the characteristic coordinates and characteristic variables defined below. In particular, the streamlines are intimately related to $\mathbf{q}_t$, the total flow velocity. In streamline simulation one transforms the saturation, defined in the three-dimensional space of the finite-difference grid into characteristic coordinates. In the transformation of Equation (6.46) into characteristic coordinates we shall assume that we are not at a well nor in a region where compressibility effects are important. Thus, the source terms on the right side of Equation (6.46) are zero. Generalization to compressible flow will be discussed later. Also, things are a bit more compact if we consider the partial differential equation in terms of the derivative of the fractional flow function

$$\varphi \frac{\partial S}{\partial t} + \mathbf{q}_t \cdot \nabla f_w = 0. \tag{6.47}$$

We may rewrite this equation in terms of the vector

$$\boldsymbol{v} = \frac{1}{\varphi} \mathbf{q}_t \tag{6.48}$$

to arrive at an equation governing the evolution of the saturation in space and time:

$$\frac{\partial S}{\partial t} + \boldsymbol{v} \cdot \nabla f_w = 0. \tag{6.49}$$

It is evident from Equation (6.48) that the transformation and the streamline coordinates are computed in terms of the total phase velocity $\mathbf{q}_t$ of the fluid mixture. This is important because, away from sources and sinks, the total velocity field is continuous. This may not be the case for the velocity field of individual fluids, as a particular phase can appear or disappear, depending upon the conditions within the porous medium. The development of streamline coordinates using total phase velocity follows the same procedure described in Chapter 5 and is valid for both compressible and incompressible flow (Datta-Gupta and King, 2007).

Our goal is to express Equation (6.49) in terms of a set of trajectory-based characteristic variables (King and Datta-Gupta, 1998). To this end, we define a local coordinate system oriented with respect to the vector field $\boldsymbol{v}$. We specify the coordinate system by the functions (s, ψ, ϕ) where s is oriented in the direction of $\boldsymbol{v}$, while ψ and ϕ are associated with coordinate axes orthogonal to $\boldsymbol{v}$. The gradient operator in physical space is decomposed into components in the new coordinate directions

$$\nabla = \nabla s \frac{\partial}{\partial s} + \nabla \psi \frac{\partial}{\partial \psi} + \nabla \phi \frac{\partial}{\partial \phi}. \tag{6.50}$$

The new spatial variables s, ψ, and ϕ provide a physically based coordinate system in which to model the evolving saturation front. Because of the orthogonality of the ψ and ϕ axes to $\boldsymbol{\nu}$ we have

$$\boldsymbol{\nu} \cdot \nabla = \boldsymbol{\nu} \cdot \nabla s \frac{\partial}{\partial s} = \nu_s \frac{\partial}{\partial s}, \tag{6.51}$$

where ν_s denotes the component of $\boldsymbol{\nu}$ along the trajectory. Thus, we can write Equation (6.49) solely in terms of the time, t, and the position along the trajectory, s,

$$\frac{\partial S}{\partial t} + \nu_s \frac{\partial f_w}{\partial s} = 0. \tag{6.52}$$

Equation (6.52) takes a simpler form if we rewrite it in terms of a new variable

$$\tau(s) = \int_{\mathbf{x}(0)}^{\mathbf{x}(s)} \frac{1}{\nu_s} d\mathbf{x} = \int_{\mathbf{x}(0)}^{\mathbf{x}(s)} \frac{\varphi}{|\mathbf{q}_t|} d\mathbf{x} \tag{6.53}$$

where the integral is taken along the trajectory $\mathbf{x}(s)$ from the source to the observation point at a distance s along the trajectory. In adopting τ as a spatial variable we are accounting for the variations in the velocity field which in turn encompasses geologic heterogeneity. Implicit in this procedure is the assumption that the total mobility $\lambda_t = \lambda_w + \lambda_n$ does not vary strongly as a function of saturation. If λ_t is significantly influenced by the saturation then one must recalculate $\mathbf{q}_t$ at frequent intervals and regenerate the streamlines. That is, the time interval of interest must be sub-divided into enough increments over which the pressure field is stable. The variable τ is the multiphase equivalent of the **time of flight** introduced in Chapter 5. In terms of the new independent variable, (6.52) simplifies to

$$\frac{\partial S}{\partial t} + \frac{\partial f_w}{\partial \tau} = 0, \tag{6.54}$$

a quasi-linear hyperbolic equation in a single spatial variable, τ, and time. This is the **Buckley–Leverett equation** expressed in streamline coordinates (Buckley and Leverett, 1942). Because it is quasi-linear, we can solve Equation (6.54) using the method of characteristics as discussed next. Alternatively, due to the reduction from three spatial variables to a single variable τ, Equation (6.54) can be solved efficiently using a numerical technique such as finite-differences.

6.4.2 Solving for the saturation along the streamlines

A semi-analytic expression for the saturation

We reiterate that, as a quasi-linear equation, we can apply the method of characteristics directly to the saturation equation. However, rather then apply it to the original Equation (6.49) in three spatial dimensions, we will treat Equation (6.54), written in characteristic or streamline coordinates,

$$\frac{\partial S}{\partial t} + c(S)\frac{\partial S}{\partial \tau} = 0, \tag{6.55}$$

where c, sometimes referred to as the velocity function, is the derivative of the fractional flow curve, a function of $S(\tau, t)$,

$$c(S) = \frac{df_w}{dS}, \tag{6.56}$$

making Equation (6.55) decidedly non-linear.

In order to complete the formulation of the problem we need to specify initial and boundary conditions. For the **Riemann problem** we envision flow from left to right along a horizontal τ (spatial) axis, and consider an initial saturation distribution in the form of a step:

$$S(\tau, 0) = \begin{cases} S_l, & \tau \leq 0 \\ S_r, & \tau > 0, \end{cases}, \tag{6.57}$$

where $S_r < S_l$. Here S_r denotes the initial saturation to the right, or 'downstream', of the front while S_l is the saturation behind, or 'upstream', of the front. For the boundary conditions we will assume that the far-field values of saturation are $S(-\infty, t) = S_l$ and $S(\infty, t) = S_r$. While the specification (6.57) is a simple step, it covers some important situations, such as the injection of a fluid into a porous medium containing a different fluid. It can also represent the advection of an existing fluid interface, such as a gas-water boundary. Finally, the solution to the Riemann problem provides a basis for the Lagrangian front-tracking algorithm outlined below.

In method of characteristics, rather than allowing the variables τ and t to be completely independent, we parameterize the problem by the time t. That is, we consider the variation in $S(\tau, t)$ along a curve $\tau(t)$. Some insight is provided by considering the total derivative of $S(\tau, t)$, bearing in mind that by assumption, τ depends upon t and the function $\tau(t)$ is differentiable,

$$\frac{dS}{dt} = \frac{\partial S}{\partial t} + \frac{d\tau}{dt}\frac{\partial S}{\partial \tau}. \tag{6.58}$$

Comparing this expression for the total derivative with the governing partial differential Equation (6.55), we observe that if we stipulate that the path $\tau(t)$, the characteristic curve, satisfies

$$\frac{d\tau}{dt} = c(S), \tag{6.59}$$

then Equation (6.55) states that the total derivative of S along $\tau(t)$ vanishes

$$\frac{dS(\tau(t),t)}{dt} = \frac{\partial S}{\partial t} + c(S)\frac{\partial S}{\partial \tau} = 0. \tag{6.60}$$

Let us denote the value of τ at the initial time $t = 0$ by τ_0. Equation (6.60) implies that S is constant along $\tau(t)$ [we denote this constant value by S_0], it follows that $c(S)$ is also constant along the path. Equation (6.59) implies that the derivative of $\tau(t)$ with respect to t is also constant along the path. Integrating Equation (6.59) with respect to t we find that the curve $\tau(t)$ is a linear function of t, a straight line in (τ, t)-space,

$$\tau(t) = \tau_0 + c(S_0)t. \tag{6.61}$$

We can solve Equation (6.61) for $\tau_0 = \tau - c(S_0)t$, allowing us to construct the solution at any given time and location by starting with the initial conditions, $(\tau_0, 0, S_0)$, and traveling along the characteristics with speed $c(S_0)$. Because the saturation is constant along the characteristic curve, we have the representation of $S(\tau, t)$

$$S(\tau(t), t) = S(\tau_0, 0) = S\left[\tau - c(S_0)t, 0\right]. \tag{6.62}$$

The feasibility of this approach depends upon the nature of the flux function. An example of a well-behaved flux function $F = f_w(S)$ is shown in Figure 6.2, along with velocity function $c(S)$, the saturation characteristics, and the corresponding saturation solution $S(\tau(t), t)$. Because the initial saturations are constant on each side of the origin, the characteristic lines within each of the two sets are parallel. With the initial saturation distribution given by the step (6.57) at $\tau = 0$, the characteristic lines intersecting the t-axis to the left of the origin, corresponding to S_l, are steeper than those intersecting the t-axis to the right of the origin [see panel (c) in Figure 6.2]. Thus, the characteristic lines for the two sets diverge and do not intersect. There is a 'fan' of trajectories lying between the two families of characteristics. This expansive fan defines a third family of characteristics, representing a **rarefaction** wave. Each characteristic line within the fan is associated with a value of saturation contained in the initial jump from S_r to S_l found at $\tau = 0$. Thus, the angular spread of the fan is bounded by slopes associated with the limits of the step. Physically the rarefaction wave represents the spreading out of the step over time, with the toe of the saturation front, corresponding to the lowest saturations, propagating the fastest [see panel (b) in Figure 6.2], and moving away from regions of higher saturation. It is evident from Figure 6.2 that every point in the $\tau - t$ plane, other then the origin, lies on a unique characteristic line. The origin is a singular point associated with the saturation jump at $t = 0$.

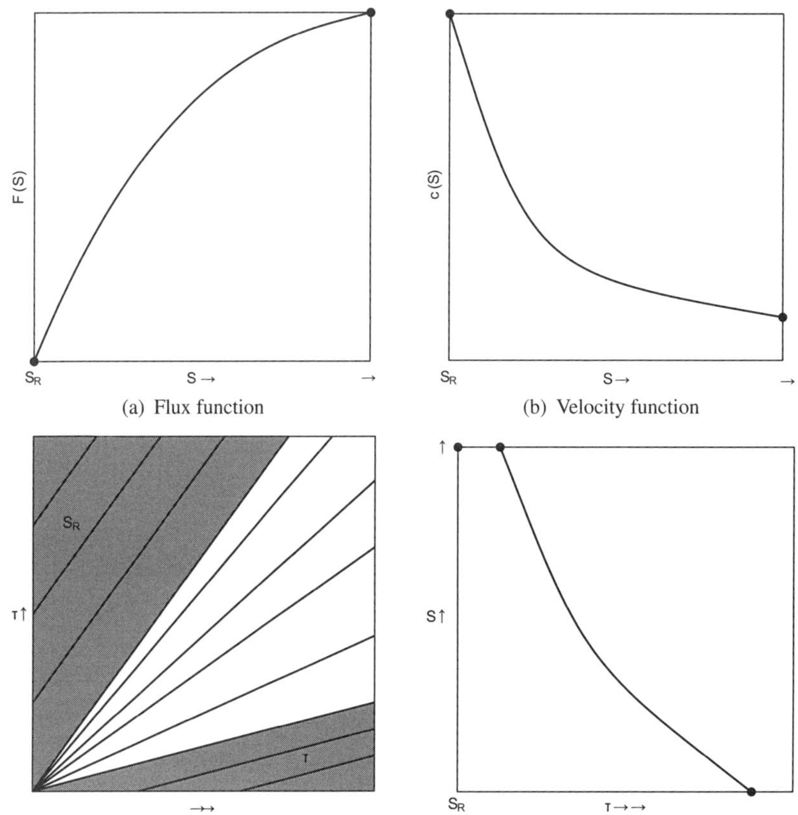

(a) Flux function (b) Velocity function (c) Saturation characteristics (d) Saturation solution

Figure 6.2 An illustration of the solution of Equation (6.55) for a convex flux function.

The geometrical construction given above is not possible when the characteristics intersect in the (τ, t) plane, as in Figure 6.3, resulting in two values of saturation at a particular location. This is because the intermediate saturations moving at a higher speed overtake the lower saturations ahead. Preventing this behavior requires the flux function to be convex, resulting in the saturation velocity increasing from left (upstream) to right. In multiphase flow, the above condition is often violated because the flux function, $f_w(S)$ may be non-convex resulting in a non-monotonic velocity profile. Such a situation is illustrated in Figure 6.3 for a typical quadratic flux function commonly encountered in water-oil displacement. The velocity function in Figure 6.3 does not satisfy the condition given above, and the intermediate saturations move at a faster speed. This leads to the intersection of characteristics and physically unrealistic multivalued solutions as shown in Figure 6.3d.

In reality, the difficulties noted above simply highlight the limitations of our modeling. To begin with, one could call into question the very use of a partial

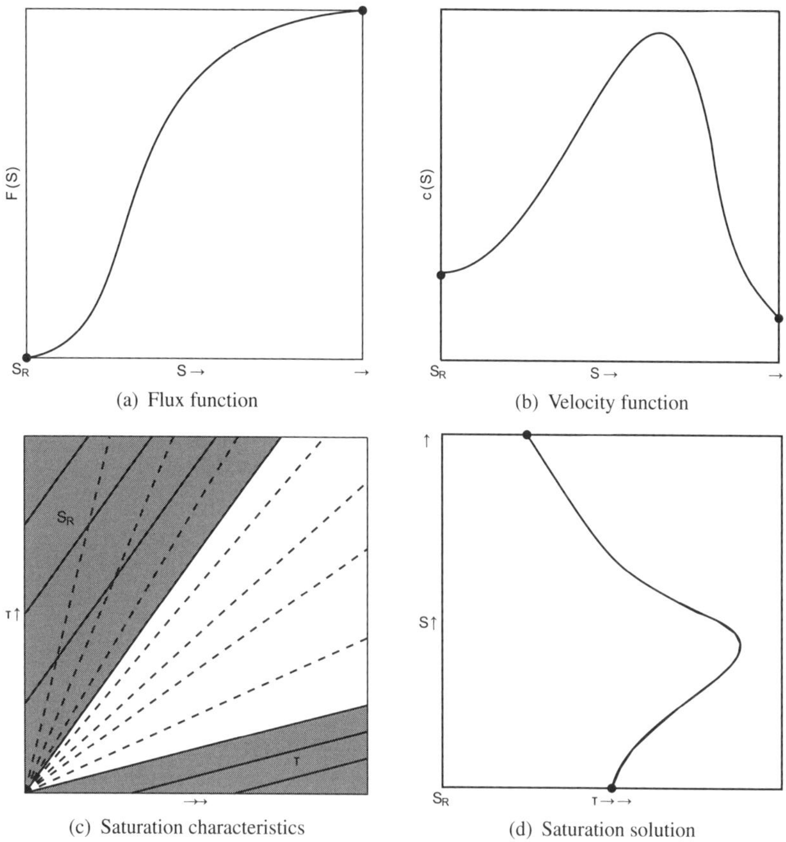

(a) Flux function (b) Velocity function

(c) Saturation characteristics (d) Saturation solution

Figure 6.3 An illustration of the solution of Equation (6.55) with a non-convex flux function, leading to multivalued solution.

differential equation to describe a process that often results in solutions containing discontinuities. Even if the initial solution contained the appropriate continuity and differentiability, a discontinuity may appear due to non-linear steepening. This has prompted the development of approaches that allow for functions lacking derivatives at some points. The simplest approach involves reformulating the original differential Equation (6.54) in an integral form, either from first principles or by simply integrating with respect to τ between two fixed points a and b

$$\int_a^b \frac{\partial S}{\partial t} d\tau + \int_a^b \frac{\partial f_w}{\partial \tau} d\tau = 0 \tag{6.63}$$

or, if S is sufficiently continuous,

$$\frac{d}{dt} \int_a^b S d\tau + f_w(S(b,t)) - f_w(S(a,t)) = 0, \tag{6.64}$$

an expression that does not involve derivatives of S. Techniques such as integral finite differences, that discretize a volume into elemental subdomains and evaluate the mass balance for each element (MacNeal, 1953; Todd, 1959; Narasimhan and Witherspoon, 1976; Pruess and Narasimhan, 1982) take advantage of such formulations and avoid local derivatives and their associated difficulties.

An alternative approach is motivated by rewriting Equation (6.55) in terms of $f_w(S)$, multiplying the resulting equation by a smooth function $\Upsilon(\tau, t)$ and integrating over a region R in the $\tau - t$ plane:

$$\int\int_R \left[\frac{\partial S}{\partial t} + \frac{\partial f_w}{\partial \tau}\right] \Upsilon d\tau dt = 0.$$

Integration by parts, and stipulating that the function $\Upsilon(\tau, t)$ vanishes on the boundary of the region of interest R, produces an expression

$$\int\int_R \left[S\frac{\partial \Upsilon}{\partial t} + f_w(S)\frac{\partial \Upsilon}{\partial \tau}\right] d\tau dt = 0, \tag{6.65}$$

that is equivalent to (6.55) but does not involve any derivatives of the saturation. Solutions $S(\tau, t)$ that satisfy Equation (6.65) for all arbitrary continuous functions $\Upsilon(\tau, t)$ are said to be **weak solutions** of the differential Equation (6.54) or (6.55). Again, the essential idea is to go back to an integral formulation of the governing equations, as this form may retain its validity even when there is not enough differentiability to justify a differential formulation. One can show that the weak solution actually solves the integral formulation of the governing equations (Chorin and Marsden, 1993). Mathematical formulations in terms of weak solutions provide a basis for some important numerical methods, such as finite element analysis (Belytschko et al., 2014).

Often the breakdown of the solution is due to approximations in the modeling itself. For example, one might recall that the original saturation Equation (6.22), written here in an abbreviated form, contained second-order spatial derivatives and a coefficient $h_w(S)$ representing capillary forces,

$$\varphi\frac{\partial S}{\partial t} + \nabla \cdot \mathbf{F}_w(S) - \nabla \cdot h_w(S)\nabla S = 0, \tag{6.66}$$

where we have assumed an incompressible wetting phase, defined the fractional flow vector

$$\mathbf{F}_w = \rho_w f_w \left[\mathbf{q}_t + \lambda_n (\rho_w - \rho_n) g\mathbf{z}\right],$$

and are considering a source-free region. The second derivative term represents dissipation and/or dispersion effects that become important when the saturation starts to varying rapidly in space. Thus, it will have the greatest influence in the regions where the curvature of the saturation distribution is the largest, such as the regions where the saturation variation is steepening due to non-linearity. In fact, once the transient effects have died away, a stable solution emerges, representing a balance between the non-linear steepening and the dissipation due to capillary forces (Whitham, 1974, p. 33). Second derivative, or 'viscosity', terms have been used to great effect in stabilizing numerical methods for the solution of non-linear conservation and Hamilton–Jacobi equations (Sethian, 1999; Holden and Risebro, 2002; Osher and Fedkiw, 2003). In this procedure, known as **viscous regularization**, one perturbs the original Equation (6.55) by adding a diffusive term:

$$\frac{\partial S}{\partial t} + c(S)\frac{\partial S}{\partial \tau} + \epsilon \frac{\partial^2 S}{\partial \tau^2} = 0, \tag{6.67}$$

and considers the limit as $\epsilon \to 0$. One can show that the solution of this equation in this limit is a weak solution of the original governing Equation (6.55) (Chorin and Marsden, 1993). As an aside, we note that when $c(S) = S$ in Equation (6.67), it can be transformed into the linear heat equation and an analytic solution follows (Hopf, 1950; Cole, 1951). This particular form is known as Burgers equation (Whitham, 1974; Karpman, 1975; Sachdev, 2000) and, as one of the best known examples of a solvable non-linear equation, has been extensively studied.

In spite of the limitations of our modeling, we can actually construct a relatively useful and instructive solution to the Riemann problem, based upon a few simple assumptions. First, returning to Equation (6.54), we note that the expression is self-similar and admits a symmetry group. Because the flux function $f_w(S)$ is given by (6.20), and the phase mobilities are given by definitions such as (6.5), we have

$$f_w(S) = \frac{\lambda_w}{\lambda_w + \lambda_n} = \frac{k_{rw}}{k_{rw} + \mu_w k_{rn}/\mu_n}. \tag{6.68}$$

The important point is that the absolute permeability $k(x)$, a quantity that appears in the phase mobilities λ_w and λ_n and typically varies strongly with x, divides out when the ratio of phase mobilities is formed. Note that the spatially-varying properties, such as φ and $\mathbf{q}_t$, are now contained within the new spatial variable τ. Thus, for a given porous material, and in a pressure range limited by our chosen time interval, we may assume that the fractional flow function f_w only depends upon the saturation S. For a fractional flow function of that nature, both the governing Equation (6.54) and the initial conditions (6.57) are invariant with respect to the transformations from the scaling group

$$t' = \alpha t \tag{6.69}$$

$$\tau' = \alpha \tau \tag{6.70}$$

where α is a constant. That is, if the independent variables are transformed according to (6.69) and (6.70), both the governing equation and the initial conditions retain their form. Furthermore, the ratio

$$z = \frac{\tau}{t} = \frac{\tau'}{t'} \tag{6.71}$$

is invariant with respect to the scaling group as well. We expect the solution to display the same symmetry as it's defining equation and boundary conditions. One way to ensure this symmetry is to write the solution as a function of the invariant or self-similar variable z. Thus, we assume that the solution has the form

$$S(\tau, t) = S\left(\frac{\tau}{t}\right) = S(z) \tag{6.72}$$

and substitute this form into the governing Equation (6.54) or, equivalently, Equation (6.55). Substituting $S(z)$ into Equation (6.55) produces the ordinary differential equation

$$\frac{1}{t}\frac{df_w}{dS}\frac{dS}{dz} - \frac{\tau}{t^2}\frac{dS}{dz} = 0. \tag{6.73}$$

Under the assumption that $0 < t < \infty$ we may multiply through by t, and write equation entirely in terms of z:

$$\left[\frac{df_w}{dS} - z\right]\frac{dS}{dz} = 0. \tag{6.74}$$

Equation (6.74) implies that either $S(z)$ is constant over some interval of z, or that

$$z = \frac{df_w}{dS}(S). \tag{6.75}$$

If the derivative of $f_w(S)$ is a monotonic function of s then we can invert it for S as a function of z. That is, we can solve Equation (6.75) for $S(z)$,

$$S(z) = \frac{df_w}{dS}^{-1}(z) = c^{-1}(z). \tag{6.76}$$

Unfortunately, while $f_w(S)$ itself is typically monotonic, its derivative with respect to S, represented by the function $c(S)$, is usually non-monotonic. Figure 6.2 shows an example of a monotonic function $c(S)$, while Figure 6.3 shows a non-monotonic example. The non-linear Equation (6.74), in conjunction with a non-monotonic flux function typically has multiple solutions. The nature of the solutions to this equation will depend on the flux function as well as on the initial or boundary conditions at hand.

Given a multitude of possible solutions we need an additional condition to extract the one that makes physical sense. This is given by the so called **entropy condition** (Oleninik, 1957), which simply states that of all possible solutions, the correct one has the fastest speed. The name originated from applications in gas dynamics, an area that generated much of the early work in non-linear fluid flow and shock formation [see Courant and Friedrichs (1948)]. We can motivate this criterion through a more in-depth analysis of the Riemann problem with the troublesome non-convex flux function, plotted in Figure 6.3. It should be noted that the entropy condition can take various forms, and readers may encounter different definitions in other texts. As noted in Logan (2008, p. 145), for our particular governing Equation (6.54), it can be shown that an alternative form of the entropy condition is the inequality

$$\frac{df_w}{dS}(S_r) > c(S) > \frac{df_w}{dS}(S_l),$$

providing bounds on the speed of a propagating disturbance.

Before we dive into our analysis, it would benefit us to try and imagine how the initial step in saturation,

$$S(\tau, 0) = \begin{cases} S_l, & \tau \leq 0 \\ S_r, & \tau > 0, \end{cases},$$

subject to the flux function shown in Figure 6.3, would evolve both in time and space. Even if we cannot calculate it's shape accurately without the mathematics, we should be able to discern the main features that a stable solution might possess. First, the saturation jump propagates with a finite velocity, so that ahead of the front the saturation will be at the initial value S_r. Second, the initial discontinuity is likely to persist for some length of time so we shall have to deal with functions that are not continuous. Behind the front the saturation distribution will evolve, perhaps leading to smoothly-varying behavior. At a sufficient distance behind the front the solution should approach the limiting value S_l. From the nature of the flux function, and the associated velocity function $c(S)$, plotted in Figure 6.3, it is clear that some intermediate saturation will propagate the fastest, eventually leading to a jump to that value as the front propagates. The higher values of saturation, with lower velocities, will lag behind, while the lower and slower values of saturation will be overrun by the fluid that propagates the fastest. Therefore, the entropy condition stipulating that the correct solution has the fastest speed, is in accordance with our intuition.

To further our understanding we require a little mathematics. In particular, to construct a solution we shall need a few specific quantities, such as the speed of the front and the jump in saturation across it. Because our focus is on the saturation front itself, we cannot simply gloss over the loss of continuity and the breakdown of the governing differential Equation (6.55). At the very least we must entertain the possibility of discontinuous solutions, allowing for weak solutions of the governing equations. One could approach the problem formally, starting with the weak formulation (6.65), dividing the integral into two regions where the solution is continuous and a boundary dividing the regions. Integration by parts then gives the governing equation in each region, that vanishes, and leads to a condition on the jumps in S and $f_w(S)$ across the boundary. Our derivation will be a bit more informal, based upon the idea of viscous regularization, introduced a few pages back. To this end, we perturb the governing equation by adding a diffusive term

$$\frac{\partial S}{\partial t} + c(S)\frac{\partial S}{\partial \tau} + \epsilon \frac{\partial^2 S}{\partial \tau^2} = 0, \tag{6.77}$$

and consider what happens in the limit as $\epsilon \rightarrow 0$. The approach is a variation of an argument presented in Whitham (1974, p. 33). While the non-linear term in (6.77) promotes steepening, the second derivative term leads to diffusive smoothing. Because the saturation downstream, to the right of the saturation front, is constant, and we are interested in a stable front in which the transient effects have died away, we shall assume that the front is propagating with a constant velocity V. Therefore, consider a solution in the form of a stable profile, denoted by the function Π, moving with velocity V

$$S = \Pi(\xi) \tag{6.78}$$

where

$$\xi = \tau - Vt. \tag{6.79}$$

The reader may wonder why that this solution is not in the self-similar form (6.72). The second derivative term in Equation (6.77) changes the equation and its corresponding symmetries. Also, one can relate the translational symmetry of (6.78) to the self-similarity of Equation (6.72) by a logarithmic change in independent and dependent variables (Barenblatt, 1979; Sachdev, 2000). Substituting (6.78) into Equation (6.77) gives the differential equation

$$[c(S) - V]\frac{d\Pi}{d\xi} = \epsilon \frac{d^2\Pi}{d\xi^2}. \tag{6.80}$$

Integrating this equation with respect to ξ gives

$$f_w(S) - VS = \epsilon \frac{d\Pi}{d\xi}(S) + A \tag{6.81}$$

where A is a constant of integration. Consider this expression for points to the right of the front, where $S = S_r$

$$f_w(S_r) - VS_r = \epsilon \frac{d\Pi}{d\xi}(S_r) + A \tag{6.82}$$

and for a point on the front where the saturation is S^*:

$$f_w(S^*) - VS^* = \epsilon \frac{d\Pi}{d\xi}(S^*) + A. \tag{6.83}$$

Note that, in taking the derivative in Equation (6.83) we may have to take the limit from the left side of the front. Subtracting Equation (6.82) from Equation (6.83), taking the limit as $\epsilon \to 0$, and solving for V,

$$V = \frac{f_w(S^*) - f_w(S_r)}{S^* - S_r}, \tag{6.84}$$

an expression for the velocity of the front in terms of the changes in S and f_w across the front.

Only one task remains, to find the jump in saturation across the front, that is to determine S^*. Here we used the fact that, for a stable front, the velocity of the front due to the change in saturation from S_r to S^*, given by Equation (6.84), should equal the velocity of the peak saturation obtained within the front, which is S^*. If this were not the case we would observe transient effects as the saturation distribution rearranged itself. Now the velocity associated with the saturation is $c(S^*)$. Thus, we have the equality

$$c(S^*) = \frac{df_w}{dS}(S^*) = \frac{f_w(S^*) - f_w(S_r)}{S^* - S_r}, \tag{6.85}$$

a non-linear equation that can be solved for S^* using an iterative technique such as Newton's method. In Figure 6.4a we plot S^*, the point at which the cord from S_r to S^*, the right-hand-side of Equation (6.85), equals that tangent df_w/dS on the left-hand-side of the equation. This is consistent with the condition that of all possible saturation jumps, the physically correct one moves with the fastest speed. Thus, the chord from S_r to S^* must lie above all other chords drawn from S_r to any point on the flux function as shown in Figure 6.3(a). That is the geometric content of Equation (6.85).

Putting it altogether, we can now present a complete solution for the saturation as a function of t and τ. The central idea is to replace df_w/dS by a monotonic function. The particular monotonic function incorporates all of the ideas given above. In particular, the saturation ahead of the front is S_r. The front consists of a jump from S_r to S^*, traveling at velocity $c(S^*)$, given by Equation (6.85). The monotonic function is constructed using the 'concave envelope' of the flux function rather than the original function, as illustrated in Figure 6.4c. The segment of the fractional flow curve between S_r and S^* is replaced by the cord, the straight line segment from S_r to S^*. As indicated in Equation (6.85) this cord is tangent to the fractional flow curve at S^*. Because the segment of the fractional flow curve from S^* to S_l is already concave, we retain that portion of the curve without any changes. The modified, concave fractional flow curve is denoted by a tilde, $\tilde{f}_w$. We can now write

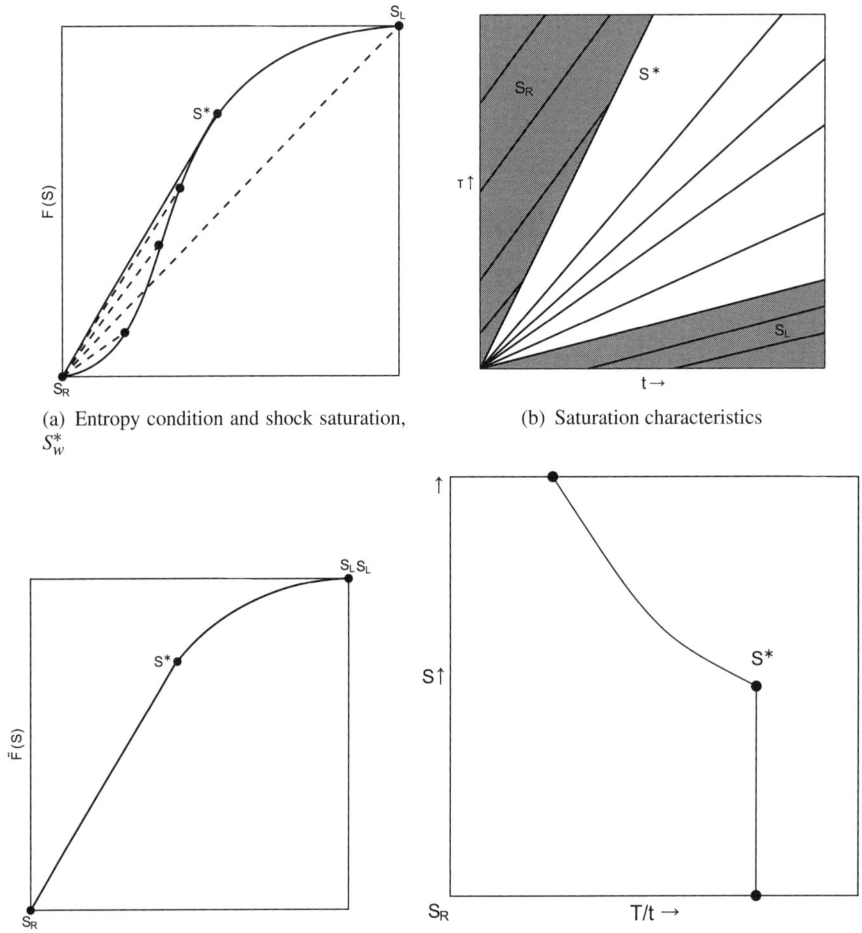

(a) Entropy condition and shock saturation, S_w^*

(b) Saturation characteristics

(c) Concave envelope

(d) Saturation solution

Figure 6.4 An illustration of the entropy condition and identification of the correct solution.

down the general solution of the Buckley–Leverett equation with a flux function that changes from convex to concave

$$S(\tau,t) = \begin{cases} S_l, & \dfrac{\tau}{t} < \dfrac{d\tilde{f}_w}{dS}(S_l) \\ \left[\dfrac{d\tilde{f}_w}{dS}\right]^{-1}\left(\dfrac{\tau}{t}\right), & \dfrac{d\tilde{f}_w}{dS}(S_l) \le \dfrac{\tau}{t} \le \dfrac{d\tilde{f}_w}{dS}(S_r) \\ S_r, & \dfrac{\tau}{t} > \dfrac{d\tilde{f}_w}{dS}(S_r) \end{cases} \tag{6.86}$$

with a particular realization given in Figure 6.4d.

The solution (6.86) tells us that after start of the injection of the aqueous phase only the in situ fluid is produced until the front arrives. The saturation of the aqueous phase in the produced fluid rises sharply at the beginning and continues to rise at a slower rate later. The arrival time of the front at location x is given by

$$T = \left[\frac{df_w}{dS}(S^*) \right]^{-1} \tau(x), \tag{6.87}$$

where $\tau(x)$ is given by Equation (6.53), so that

$$T = \left[\frac{df_w}{dS}(S^*) \right]^{-1} \int_{\mathbf{x}(0)}^{\mathbf{x}(s)} \frac{\varphi}{|\mathbf{q}_t|} d\mathbf{x}. \tag{6.88}$$

In the solution (6.86) the jump at the leading edge of the front occurs when $\tau/T = df_w/dS(S_r)$.

Box 6.2 **The Riemann problem for three fluids**

In many situations three fluids, such as water, oil, and gas, are present in a porous medium and are free to flow. It is important to understand the nature of such three phase flow and in this box we outline how the techniques that we have just described extend to this case. The topic has been studied by Pope (1980), Helfferich (1981), Shearer and Trangenstein (1989), Guzman and Fayers (1997), Juanes and Patzek (2004), and Lie and Juanes (2005). In addition to the saturations of the wetting and non-wetting phases, denoted by S_w and S_n, respectively, we have a phase with saturation S_g, perhaps representing the presence of gas. The saturations of the three phases must sum to unity

$$S_w + S_n + S_g = 1.$$

Using this relationship, we can reduce the equations for the conservation of mass, the equivalent of Equation (6.55), to

$$\frac{\partial S_w}{\partial t} + \frac{\partial f_w}{\partial S_w}\frac{\partial S_w}{\partial \tau} + \frac{\partial f_w}{\partial S_g}\frac{\partial S_g}{\partial \tau} = 0$$

$$\frac{\partial S_g}{\partial t} + \frac{\partial f_g}{\partial S_w}\frac{\partial S_w}{\partial \tau} + \frac{\partial f_g}{\partial S_g}\frac{\partial S_g}{\partial \tau} = 0,$$

where the fractional flow for each phase, $f_w(S_w, S_g)$ and $f_g(S_w, S_g)$ depends upon the two independent saturations. We may write the system of equations in vector-matrix form

$$\frac{\partial \mathbf{S}}{\partial t} + \mathbf{A}(\mathbf{S})\frac{\partial \mathbf{S}}{\partial \tau} = 0$$

where $\mathbf{S}$ is the vector

$$\mathbf{S} = \begin{bmatrix} S_w \\ S_g \end{bmatrix}$$

and $\mathbf{A}(\mathbf{S})$ is the Jacobian matrix of partial derivatives

$$\mathbf{A}(\mathbf{S}) = \begin{bmatrix} f_{ww} & f_{wg} \\ f_{gw} & f_{gg} \end{bmatrix}.$$

For brevity, we are using a subscript notation for the partial derivatives of the fractional flow functions

$$f_{ij} = \frac{\partial f_i}{\partial S_j}.$$

The properties of the system of equations, and the nature of its solutions, will depend upon the eigenvalues and eigenvectors of the Jacobian matrix $\mathbf{A}(\mathbf{S})$. For example, if the eigenvalues are real and distinct then the system is termed hyperbolic and the solution has the characteristics of a propagating wave. If the eigenvalues are real but equal, the system is parabolic and the solution is diffusive in nature. The two eigenvalues of the matrix $\mathbf{A}(\mathbf{S})$ are given by

$$e_{\pm} = \frac{1}{2}\left[f_{gg} + f_{ww} \pm \sqrt{(f_{ww} - f_{gg})^2 + 4f_{gw}f_{wg}}\right].$$

The eigenvalues of the matrix $\mathbf{A}(\mathbf{S})$ depend directly on the three phase relative permeability functions. Some commonly used relative permeability functions contain regions in saturation space in which the equations are not hyperbolic (Bell et al., 1986), perhaps an indication that they are incomplete.

For the three-phase Riemann problem we seek a solution of the governing system of partial differential equations, subject to the initial condition

$$\mathbf{S}(\tau, 0) = \begin{cases} \mathbf{S}_l & \tau < 0 \\ \mathbf{S}_r & \tau > 0 \end{cases},$$

where we refer to $\mathbf{S}_l$ and $\mathbf{S}_r$ as the left and right saturation states, respectively. Note that the system of equations and the accompanying initial conditions admit the same scaling symmetry group

$$t' = \alpha t$$
$$\tau' = \alpha \tau$$

encountered in two-phase flow. In order to ensure that the solution maintains this symmetry, we write it in terms of the invariant variable

$$z = \frac{\tau}{t}.$$

Substituting a solution of the form $\mathbf{S}(z)$ results in the system of ordinary differential equations

$$[\mathbf{A}(\mathbf{S}) - z\mathbf{I}]\frac{d\mathbf{S}}{dz} = 0,$$

where $\mathbf{I}$ is the identity matrix. If this system of equations is to have a non-trival solution $\mathbf{S}$, the coefficient matrix must be singular and so the determinant of the quantity within the square brackets must vanish. In fact, this equation is an eigenvalue problem, where z is an eigenvalue of $\mathbf{A}(\mathbf{S})$, given above as $e_{\pm}$, and $d\mathbf{S}/dz$ is a right eigenvector. The eigenvectors indicate directions of admissible changes in saturation, given by

$$\frac{dS_g}{dz} = \frac{1}{2}\left[f_{gg} - f_{ww} \pm \sqrt{\left(f_{ww} - f_{gg}\right)^2 + 4f_{gw}f_{wg}}\right]\frac{dS_w}{dz}$$

(Helfferich, 1981). There are two families of saturation paths, the fast and slow paths, corresponding to the larger and smaller eigenvalues, respectively. Along these paths the saturation travels with a characteristic velocity given by the eigenvalues. For the three-phase Riemann problem we construct a saturation path that connects the left ($\mathbf{S}_l$) and right ($\mathbf{S}_r$) states. As shown in (Juanes and Patzek, 2004), the path can be constructed by combining specific fast and slow integral curves of the eigenvector equation emerging from the left and right states and intersecting in an intermediate state $\mathbf{S}_m$. The determination of the intermediate state differentiates the three-phase problem from the two-phase problem. Falls and Schulte (1992) lay out the principles providing a straight-forward procedure for determining the saturation routes.

While we have focused on two fluids, the solution for a step in saturation, the Riemann problem, can be extended to three fluids. Box 6.2 outlines how this extension may be carried out. When it is applicable, an analytic solution of the saturation equation leads to a substantial reduction in computation time and can provide valuable insight. For example, we have learned that the onset of the saturation change is determined by τ, defined the integral of $\varphi/|\mathbf{q}_t|$ over the streamline $\mathbf{x}(s)$ [see Equation (6.53)]. Furthermore, from the form of the solution for saturation, Equation (6.86), we observed that the magnitude of the saturation change depends upon the fractional flow curve. As we have seen, the fractional flow curve does not depend directly on the permeability $k(\mathbf{x})$. Thus, while the arrival time of the saturation depends upon the permeability, through $\mathbf{q}_t$, the size of the saturation change along a trajectory only depends upon the relative permeabilities and the fluid viscosities, as evident in Equation (6.68).

In deriving the solution of the Riemann problem, we have assumed that the saturation distribution ahead of the front is uniform. In general, such an assumption is not valid when the streamlines need to be updated and the saturations need to be remapped onto a new set of trajectories. The formulation developed for the Riemann problem can be extended to a heterogeneous saturation distribution using a piecewise constant approach originally developed by Dafermos (1972, 2000) and extended by Holden et al. (1988) and Bratvedt et al. (1996). This hybrid scheme provides a numerical method that is described in the next section, along with more conventional finite-difference algorithms.

Exercise 6.1. Use the software 'TRACE3D' to simulate water injection into an oil reservoir with uniform permeability and a five-spot well configuration (a central injector and four surrounding producers). Use the data files in the folder 'Examples/Simulation/5-Spot-Hom' by creating a project in the same folder and running the software. Use analytic solution option (already selected) for two-phase water-oil displacement calculations along streamlines. Visualize the pressure distribution, streamline trajectories, time of flight from the injector/producer, partitioning by well regions, well allocation, water saturation distribution, oil-water production and recovery history. Examine the sensitivity of the waterflood performance to the fractional flow curve by changing the water/oil viscosities and the relative permeability curves.

Exercise 6.2. Repeat the Exercise 6.1 for a heterogeneous permeability field (in 'Examples/Simulation/5-Spot-Het'). View the heterogeneous permeability field and the streamlines side by side to examine how the trajectories are impacted by the heterogeneity. Notice how the streamlines tend to cluster along the high permeability regions and the time of flight reflects the preferential fluid movement. Also, notice how the oil recovery is impacted by heterogeneity.

Numerical techniques for calculating saturation

The most general approach for calculating the evolving saturation distribution is one that is based upon numerical methods such as finite differences. The chief drawback of numerical techniques is the computational burden that they impose. Fortunately, in the trajectory-based approach, we have reduced the computational undertaking considerably, from a partial differential equation in four independent variables, Equation (6.47), to an equation in τ and t, Equation (6.54). We will have to solve a sequence of such problems, one for each trajectory or streamline. But the decomposition provides added flexibility, enabling us to pick the particular set of trajectories that are the most important. There are several texts on numerical methods for solving the equations governing multiphase flow and front propagation (Holden and Risebro, 2002; Osher and Fedkiw, 2003; Datta-Gupta and King, 2007), and we defer to them for the details of such computations. However, let us take a moment to outline what these calculations involve.

In general, the numerical techniques can be divided into Lagrangian, or front tracking methods, and Eulerian, or grid-based methods. The nomenclature is related to the two possible descriptions of material deformation or motion, given in Section 2.3 of Chapter 2. In an Eulerian approach, $\mathbf{x}$ and t are independent variables and the saturation is a dependent quantity, the numerical algorithm solves Equation (6.54):

$$\frac{\partial S}{\partial t} + \frac{\partial f_w}{\partial \tau} = 0, \tag{6.89}$$

directly on a fixed grid. A method such as finite differences is a typical example of an Eulerian algorithm and these methods have a long history and are widely used in modeling multiphase fluid flow (Peaceman, 1977; Aziz and Settari, 1979). Eulerian approaches work well for dispersive transport but have difficulties with advection-dominated flow and sharp fronts. In the Lagrangian front tracking approach, one follows the propagation of quantities of interest, such as a jump in saturation at a front. As in the Lagrangian description of motion, the spatial coordinate (in this case τ) is the dependent variable, governed by

$$\frac{\partial \tau}{\partial t} = \frac{df_w}{dS}. \tag{6.90}$$

The equation is a generalization of Equation (6.59), accounting for the dependence of τ on S. The Lagrangian approach works best when modeling the propagation of sharp fronts and its performance can degrade in areas of smooth saturation variation (Datta-Gupta and King, 2007). We should note that the distinction between Eulerian and Lagrangian techniques hold for both the global problem of solving the original governing equations, such as Equation (6.3), and also for solving the saturation equation in $\tau - t$ coordinates along the streamline, Equation (6.54). In the latter case, we are adopting a trajectory-based approach, following the fluids and not operating on a grid, but we solve the reduced saturation Equation (6.89) on the trajectory, using either a Lagrangian or an Eulerian algorithm. It is this $\tau - t$ case that we shall describe in this section.

Because the Lagrangian front-tracking approach relies upon the solution of the Riemann problem, which was just discussed, we consider it first. In the Lagrangian-based algorithm each 'particle' that we are tracking is a pre-defined jump in saturation, let us say with a typical incremental resolution of δS. Intuitively, we approximate a general saturation variation by a sequence of small steps, and represent the fractional flow curve by a piecewise-linear construction (Holden and Risebro, 2002; Datta-Gupta and King, 2007). A saturation profile then appears as an arrangement of regions of constant saturation. The front tracking method is

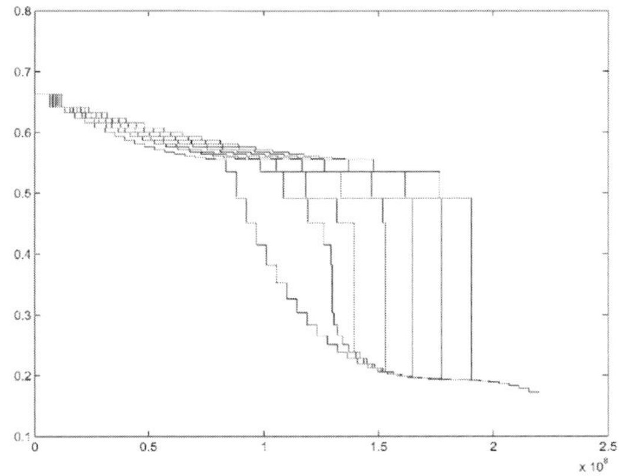

(a)

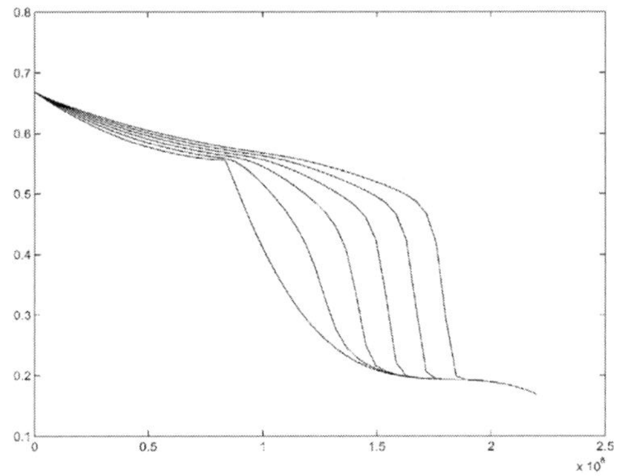

(b)

Figure 6.5 Streamline transport calculations (a) Lagrangian front tracking method, (b) Eulerian method with single point upwinding (from SPE 79693). Reprinted with Permission. Copyright SPE.

based on the observation that if $f_w(S)$ is approximated as a piecewise linear function and the initial condition is approximated as piecewise constant, then the solution is also piecewise constant (Dafermos, 1972). The speed of each increment can be obtained from the jump condition, Equation (6.85), where S_r changes for each jump considered. Combining these operations produces a semi-analytic technique for determining the saturation evolution along a streamline. The results of this approach are illustrated in Figure 6.5a. More increments, each with a smaller saturation jump, will improve the spatial resolution of the saturation profile. As the fronts propagate they may interact and collide, leading to shock formation. If the incremental fronts spread further apart, a rarefaction wave results. More details on front tracking in general, along with specific algorithms, can be found in the book by Holden and Risebro [2002].

The second class of numerical approximations, Eulerian approaches, represent $S(\tau, t)$ at discrete points on a numerical grid. For simplicity we shall assume that the grid spacing is uniform in each direction, perhaps after reinterpolation from an irregular grid, with grid spacings of $\Delta\tau$ and Δt. We shall denote the saturation at grid point (i, n), that is $S(i\Delta\tau, n\Delta t)$, by the shorthand S_i^n. One can use standard finite-difference techniques to discretize Equation (6.89)

$$\frac{S_i^{n+1} - S_i^n}{\Delta t} + \frac{f_{i+1/2} - f_{i-1/2}}{\Delta\tau} = 0, \tag{6.91}$$

where $f_{i+1/2}$ is the flux at the boundary between nodes i and $i + 1$, and n is the timestep counter. In order to minimize the number of subscripts we have changed the name of the flux function from f_w to simply f. As might be imagined, one must use care when applying finite difference algorithms to non-linear equations. Difficulties, including instabilities and convergence to the wrong solution, are possible. However, one is at least guaranteed immunity from such incorrect convergence if the finite difference method used to solve Equation (6.89) can be written in **conservation form** (Le Veque 1990, p. 124; Sethian 1999, p. 55), a numerical approximation of a conservation equation,

$$\frac{S_i^{n+1} - S_i^n}{\Delta t} + \frac{F(S_{i+1}^n, S_i^n) - F(S_i^n, S_{i-1}^n)}{\Delta\tau} = 0, \tag{6.92}$$

or, solving for S_i^{n+1},

$$S_i^{n+1} = S_i^n - \frac{\Delta t}{\Delta\tau}\left[F(S_{i+1}^n, S_i^n) - F(S_i^n, S_{i-1}^n)\right], \tag{6.93}$$

where $F(S_{i+1}^n, S_i^n)$ and $F(S_i^n, S_{i-1}^n)$ are **numerical flux functions** (Osher and Fedkiw, 2003, p. 157). The flux functions are approximations for $f_{i+1/2}$ and $f_{i-1/2}$ and the various possible choices for these functions lead to the wide variety of finite difference algorithms that are now available. For simplicity, we have assumed that the flux functions only depend upon pairs of nodal saturations. More generally, the numerical flux functions can depend upon an extended set of nodal saturations [see Le Veque (1990, p. 124)]. If we abstract things a bit, and consider the right-hand-side of Equation (6.93) as a general function

$$S_i^{n+1} = M(S_{i-1}^n, S_i^n, S_{i+1}^n) \tag{6.94}$$

of the three nodal saturations, then the function $M(S_{i-1}^n, S_i^n, S_{i+1}^n)$ is said to be **monotone** if it is non-decreasing function of all of its arguments (Crandall and Majda, 1980). A monotone scheme in conservative form produces a solution that satisfies the entropy condition (Le Veque, 1990).

There are two guiding principles that help in designing finite difference algorithms for modeling immiscible fluid flow. One is adherence to the physics of the process. For example, the flow of material should be from the 'upwind' direction, from a higher potential to a lower potential. Another guiding principle is perhaps better described as a cautionary note. The algorithms should avoid methods that are unstable in the face of steep gradients, as found in a propagating saturation front. Any higher-order polynomial approximation methods must be used with great care. Thus, in the essentially nonoscillatory (ENO) method one uses adaptive polynomial interpolation designed to avoid steep gradients (Harten and Osher, 1987; Osher and Fedkiw, 2003, p. 155). Higher order methods, based upon additional terms in a Taylor series expansion of $S(\tau, t + \Delta t)$, are more accurate (as expected), but the price we pay is oscillatory behavior near discontinuities. Thanks to non-linearity, such oscillations may grow over time.

One approach to maintaining stability is simply to use a low-order approximation for the numerical flux function, such as single-point upstream weighting, where

$$F(S_{i-1}^n, S_i^n) = f(S_{i-1}^n). \tag{6.95}$$

This is the **upwind scheme**, first-order accurate but highly diffusive, rapidly smoothing any discontinuities that might be present (Press et al., 1992, p. 832). In Figure 6.5b we plot an application of this approach to the Riemann problem of oil-water displacement. Godunov (1959) came up with a rather clever first-order approach, based upon a solution of the Riemann problem for each finite difference interval. That is, for a given interval between $i - 1$ and i, let $\tilde{S}_i$ be the solution of the Riemann problem with the initial data

$$\tilde{S}_i(x, 0) = \begin{cases} S_{i-1}^n, & x \leq 0 \\ S_i^n, & x > 0 \end{cases}$$

where x is a local coordinate system with the origin at the left end of the interval and S_i^n and S_{i+1}^n are the nodal saturations at the n-th time step. The numerical flux function is defined in terms of the solution of the Riemann problem

$$F(S_{i-1}^n, S_i^n) = f[\tilde{S}_{i-1}(0, \Delta t)]. \tag{6.96}$$

The solution at the $n + 1$ time step and the location $i - 1/2$ is either a rarefaction or a shock, depending on the relative size of S_{i-1}^n and S_i^n. The global solution is a piecing together of all the intervals. In order to avoid interference from neighboring intervals, the time step is limited in size by

$$\frac{df}{dS}\Delta t \leq \Delta \tau, \tag{6.97}$$

the **Courant-Fredrichs-Levy** (CFL) condition (Holden and Risebro, 2002, p. 66). If the characteristic speeds are all non-negative then Godunov's method is equivalent to single-point upstream weighting.

Even though steep gradients are present at saturation fronts, much of the remainder of a solution is smooth. Thus, there is a strong temptation to use higher-order schemes, with their more rapid convergence to the smoothly-varying component of the solution. This has led to the use of hybrid methods, typically consisting of the linear combination of a lower-order and a higher-order method (Holden and Risebro, 2002, p. 66). The numerical flux function is the linear combination

$$F(S_{i-1}^n, S_i^n) = \alpha F_l(S_{i-1}^n, S_i^n) + (1-\alpha)\, F_h(S_{i-1}^n, S_i^n) \tag{6.98}$$

where F_l is a low-order flux function, such as the single-point upstream flux function (6.95) or the Godunov flux function (6.96). Examples of higher-order flux functions F_h can be found in Holden and Risebro (2002) and Datta-Gupta and King (2007). The weighting function $\alpha(S_{i-1}^n S_i^n)$ is close to zero when the saturation distribution is smooth and approaches 1 near discontinuities. The general form (6.98) has led to a great many finite difference algorithms. Perhaps the most successful class of algorithms are the total variation diminishing (TVD) schemes (Harten, 1983; Le Veque, 1990; Datta-Gupta and King, 2007), used in conjunction with flux limiters. The total variation at time n is defined as the sum over all grid blocks:

$$TV(S^n) = \sum |S_{i+1}^n - S_i^n|, \tag{6.99}$$

a measure of the roughness of the saturation distribution along the streamline. A TVD scheme has the property that the roughness never increases over time, $TV(S^{n+1}) \le TV(S^n)$. A consequence of this, first observed by Harten (1983), is that the algorithm preserves monotonicity. The basic procedure for constructing a TVD scheme is to combine lower-order and higher-order fluxes, according to Equation (6.98), and then to impose limiter functions on the higher-order fluxes. Sweby (1984) derived properties on the limiter functions that guarantee monotonicity and prevent extraneous oscillations in the solution (Datta-Gupta and King, 2007).

It is worth pointing out the relative advantages of the Lagrangian and the Eulerian approaches. The biggest advantage of the Lagrangian front tracking method is that it is unconditionally stable and thus, has no intrinsic time step limitations. However, the applicability of the method is limited by the class of problems with available Riemann solutions and complexities such as multicomponent flow can present difficulties. The advantage of the Eulerian method is its generality and ability to model complex physical processes. However, the time step size in Equation (6.92) will be limited by the Courant-Fredrichs-Levy condition, Equation (6.97), where df/dS is the maximum speed over the saturation range for an adjacent pair of cells. Clearly small $\Delta\tau$ will limit the allowable time step size. However, an important advantage of the trajectory-based calculations is that the one-dimensional solution can be optimized independently along each streamline. Unlike conventional finite-differences, the method is not limited by any global Courant-Fredrichs-Levy stability criterion, that is, one imposed over the entire three-dimensional simulation grid.

Exercise 6.3. This is a continuation of Exercise 6.1. Use the software 'TRACE3D' to simulate water injection into an oil reservoir with uniform permeability and a five-spot well configuration (a central injector and four surrounding producers). Use numerical solution option (already selected) for two-phase water-oil displacement calculations along streamlines and compare the results with the analytic solution in Exercise 6.1. Use the data files in the folder 'Examples/Simulation/5-Spot-Hom-Num' by creating a project in the same folder and running the software. Notice that the numerical solution, while more general, requires specification of the time step via CFL number. Visualize the pressure distribution, streamline trajectories, time of flight from the injector/producer, partitioning by well regions, well allocation, water saturation distribution, oil-water production and recovery history. Notice the effects of numerical dispersion on the water saturation distribution as compared to the analytic solution.

Exercise 6.4. Repeat the Exercise 6.3 for a heterogeneous permeability field (in 'Examples/Simulation/5-Spot-Het-Num'). Compare the results with the analytic solution in Exercise 6.2.

Exercise 6.5. One of the ways to counter the adverse impact of heterogeneity on oil recovery is through optimal well placement in relation to the underlying heterogeneity. Repeat the Exercise 6.4 for a different well configuration (in 'Examples/Simulation/5-Spot-Het-Num-W'). Visualize the pressure distribution, streamline trajectories, time of flight from the injector/producer, water saturation distribution to see how optimal well placement counteracts heterogeneity effects, resulting in improved recovery. Notice in the input file 'WELL.DATA' how well locations have been changed.

6.4.3 Modifications for gravitational, capillary, and compressional effects

In Chapter 5 we discussed trajectory-based modeling of single phase transport under steady-state conditions. For multiphase flow, the velocity field may change over time due to a variety of factors. As we have seen, the saturation and pressure equations are truly coupled, so temporal variations in saturation can lead to variations in the flow field. In addition, due to time-dependent flow rates at existing wells and/or the introduction or removal of a well, the boundary conditions can change. Under such changes the trajectories will move, generating three-dimensional fluxes orthogonal to the original flow directions. Time-varying velocity fields can often be approximated as a sequence of steady velocity fields. Thus, for unsteady state an additional complexity is introduced in terms of updating the pressure fields and the streamlines at selected time intervals. The choice of the time interval should be such that the variations in velocity, measured in a quantitative sense, are kept small.

Time-varying velocity fields are but one means of introducing fluxes across streamlines, gravitational and capillary forces are another. Such fluxes can be handled via a numerical technique known as **operator splitting**. For example, consider the case in which gravity is important but capillary effects may be neglected. The time evolution of saturation in Equation (6.23) can be split into two components, convective and gravitational. Thus, one solves for the convective part of the flow for a brief time increment

$$\frac{\partial(\varphi S)}{\partial t} + \nabla \cdot f_w \mathbf{q}_t = 0 \tag{6.100}$$

and then updates the saturation for the gravitational component of the flow

$$\frac{\partial(\varphi S)}{\partial t} + \nabla \cdot f_w \lambda_n (\rho_w - \rho_n) g\mathbf{z} = 0. \tag{6.101}$$

In these equations we are assuming that the aqueous phase is incompressible. When gravity is accounted for, the individual phase velocities will no longer be aligned with the total velocity. We can still use the streamline time of flight based on the total velocity as a spatial coordinate. Thus, the convective component can be solved along the streamline coordinate using the transformations given above to derive the reduced Buckley–Leverett Equation (6.54)

$$\frac{\partial S}{\partial t} + \frac{\partial f_w}{\partial \tau} = 0. \tag{6.102}$$

In a similar fashion, Equation (6.101), representing cross streamline mechanisms such as gravity can be reduced to an equation in time t and a coordinate ζ in the direction, $\mathbf{z}$, of the gravity field

$$\frac{\partial S}{\partial t} + \frac{1}{\phi}\frac{\partial}{\partial \zeta}\left[f_w \lambda_n (\rho_w - \rho_n) g\right] = 0. \tag{6.103}$$

When it is necessary to update the streamlines to account for changing velocity field, an important consideration is the remapping and resampling of saturation. The remapping of the saturations is also crucial for a correct gravity update.

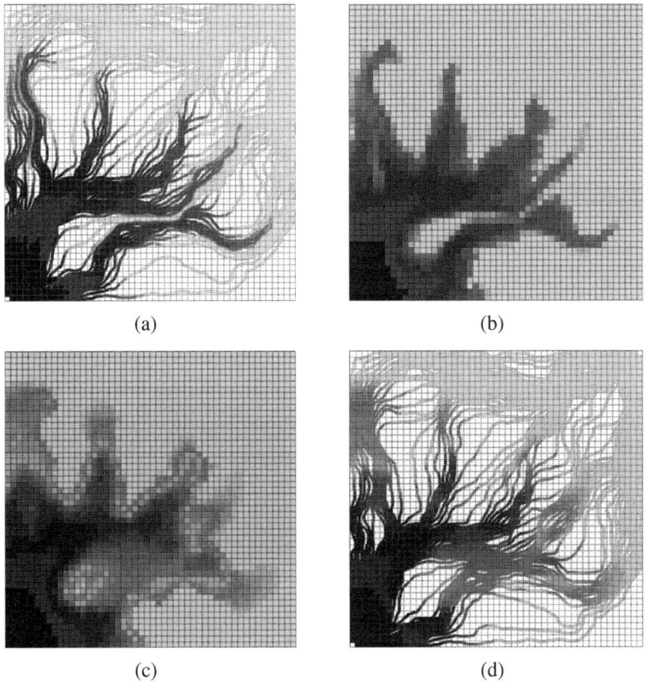

Figure 6.6 Stepwise illustration of the multiphase streamline approach in a quarter five spot with a single injector and a producer. (a) saturation transport along streamlines, (b) saturation mapping on to the grid (note that more streamlines were used for the mapping than are displayed here), (c) accounting for cross-streamline mechanisms (capillarity) on the grid, and (d) resampling of saturation on to the streamline trajectories for the next time interval calculations. For the colour version, please refer to the plate section.

Two distinct approaches have been used for mapping of saturations (Bratvedt et al., 1996; Datta-Gupta and King, 2007). In the first approach, a direct streamline-to-streamline resampling is performed in three dimensions. With a sufficient density of streamlines, the line to line approach can preserve the saturation fronts. A simpler alternative is to map the changes onto the underlying finite-difference grid used for the pressure calculations. In this second approach, the saturations are averaged within the grid blocks intersected by streamlines. One disadvantage of this technique is a loss of resolution, resulting in a smearing of the saturation front. Figure 6.6 shows the steps for multiphase transport calculations. Those include solution of the 1-D saturation transport equations and accounting for cross-streamline mechanisms on the grid via saturation mapping and resampling.

Although the operator splitting method has been successfully applied to account for fluxes transverse to streamlines, it can require frequent pressure updates to achieve convergence, particularly when the gravitational flux is significant. This can

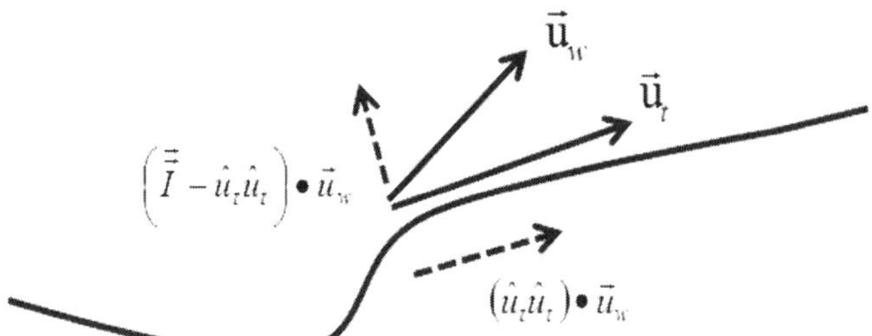

Figure 6.7 Under the influence of gravity, the total velocity and phase velocities may not be aligned. The components of the water phase flux parallel and orthogonal to the total velocity are shown above. The parallel component can be solved along streamline trajectories which are based on total velocity for multiphase flow. The orthogonal component is accounted for using operator splitting. Reprinted with permission. Copyright SPE. [From Tanaka et al. (2013)]

reduce some of the computational advantages of the streamline approach compared to more traditional finite difference method. To circumvent this, variations of the operator splitting approach have been proposed such as orthogonal projection of the water flux into components parallel and perpendicular to the streamlines (Tanaka et al., 2013), in order to minimize grid updates, as shown in Figure 6.7.

While capillary pressure effects may be incorporated via operator splitting, there are also semi-analytic solutions to draw upon. To date, most semi-analytical studies have assumed that the fluids are incompressible. At the very least, we will take the aqueous phase to be incompressible, so that the governing Equation (6.22) reduces to

$$\frac{(\varphi \partial S)}{\partial t} + \nabla \cdot [f_w \mathbf{q}_t + g_w \mathbf{z} - h_w \nabla S] = 0, \tag{6.104}$$

where, we are denoting the saturation of the wetting phase by S, g_w contains terms related to gravitational forces

$$g_w = f_w \lambda_n (\rho_w - \rho_n) g, \tag{6.105}$$

and we are considering a region away from any sources or sinks. In addition it is typically assumed that the flow is one dimensional, either radial or longitudinal (in the direction of $\mathbf{q}_t$). For example, consider a coordinate system in which the coordinate x is along the flow direction and the saturation variation is dominantly in that direction. Rewriting Equation (6.104) solely in terms of t and x,

$$\frac{\partial (\varphi S)}{\partial t} + \frac{\partial}{\partial x} \left[f_w q_t + g_w \sin\alpha - h_w \frac{\partial S}{\partial x} \right] = 0, \tag{6.106}$$

where q_t is the magnitude of the flow vector, and α is the angle between the flow direction and the vertical direction. Early work (Fokas and Yortsos, 1982; Yortsos and Fokas, 1983) produced an exact analytic solution in the case of a horizontal reservoir and a constant flow rate with specific capillary pressure and fractional flow functions. The specific functional forms are not general enough to model realistic fluid flow but the solution can still be useful for testing numerical algorithms. Chen (1988) reviewed a series of published solutions, including a self-similar solution for the case of axisymmetric flow and linear, one-dimensional flow. One of the self-similar solutions, one-dimensional flow with the specific boundary condition $Q_w(0,t) = Q_0 t^{-1/2}$, was presented in McWhorter (1971) and elaborated upon by McWhorter and Sunada (1990). This solution was recently used to derive a semi-analytic solution for immiscible two-phase flow with a dispersing component within one of the phases (Schmid et al., 2011). However, the specific form of the boundary condition is not commonly used in actual applications, limiting the usefulness of the analytic solution.

Chen (1988) also discussed a solution corresponding to the propagation of a stable front that appears when the flow rate is constant. We can derive the governing equations for the front velocity and structure, if we adopt a solution of the form $S = S(\xi)$ where

$$\xi = x - Vt, \tag{6.107}$$

reducing Equation (6.106) to an ordinary differential equation,

$$\frac{d}{d\xi}\left[f_w q_t + g_w \sin\alpha - V\varphi S - h_w \frac{dS}{d\xi}\right] = 0. \tag{6.108}$$

Note that we used this technique in our study of the Riemann problem [see Equations (6.78) and (6.79)]. Integrating this expression once with respect to ξ results in a first-order differential equation for S

$$h_w \frac{dS}{d\xi} + V\varphi S - q_t \Psi(S) = C, \tag{6.109}$$

where C is a constant of integration. In Equation (6.109) we have followed Chen [1988] and, for the sake of brevity, defined

$$\Psi(S) = f_w - \frac{1}{q_t} g_w \sin\alpha, \tag{6.110}$$

assuming that the flow velocity does not vanish. Boundary conditions can be used to specify C and to determine the velocity of the propagating front. As in the Riemann problem, we assume that the saturation well ahead of the front is S_r. The saturation some distance behind the front is taken to be S_l. Therefore $S \to S_l$, a constant value, as $\xi \to -\infty$, and hence

$$\frac{dS}{d\xi} \to 0$$

in this limit. Thus, in the limit of large negative values of ξ, Equation (6.109) allows us to determine that $C = V\varphi S_l - q_t\Psi(S_l)$ and hence the first-order equation may be written

$$h_w \frac{dS}{d\xi} + V\varphi\,(S_l - S) - q_t\,[\Psi(S_l) - \Psi(S)] = 0. \tag{6.111}$$

Equation (6.111) may be solved using numerical methods. We may use the second boundary condition, specifying that the saturation approaches a constant value S_r, and that $dS/d\xi \to 0$ as $\xi \to \infty$, to produce an expression for the front velocity V

$$V = \frac{q_t}{\varphi}\frac{\Psi(S_l) - \Psi(S_r)}{S_l - S_r}. \tag{6.112}$$

From this expression we deduce that the velocity of the front is not influenced by the capillary pressure parameters contained in $h_w(S)$. It appears that the effect of capillary pressure is to alter the shape of the two-phase front but not its speed. This independence of the front speed from the diffusion coefficient is also observed in the well-known non-linear diffusion equation known as Burgers equation, where an exact solution is available (Whitham, 1974, p. 101; Sachdev, 2000, p. 131). Note that when gravitational forces are negligible, this expression reduces to the velocity of the front for the Buckley–Leverett equation, if we account for the definition (6.53) of τ and the fact that we have formulated the governing equation in terms of the spatial variable x. As indicated in Box 6.3, the effect of fluid compressibility is included within the numerical calculation of pressures and, with a few modifications, can be included in the corresponding streamline calculations.

Box 6.3 **Including compressibility**

Because the average fluid pressure $\bar{p}$ is calculated numerically, by solving Equation (6.13), the effects of compressibility are already accounted for. For example, the fluid compressibilities c_l, defined by Equation (6.6), are contained in the equation for average pressure. A key element in the modeling of compressible flow is the inclusion of appropriate equations-of-state relating densities and average fluid pressure. The pressure-dependence of the density does add to the non-linearity of the differential equation but is not a barrier to its solution by numerical methods. One can modify the streamline formulation, given primarily in Section 5.4 of Chapter 5, to incorporate effective density changes. In this box we touch upon the modifications necessary to accurately trace the trajectories. Our starting point is Equation (6.46), rearranged to isolate the source term on the right-hand-side

$$\varphi\frac{\partial S}{\partial t} + \nabla f_w \cdot \mathbf{q}_t + f_w \nabla \cdot \mathbf{q}_t = \hat{Q}_w.$$

Compressibility effects can be accounted for by noting that the tangents to the trajectories are still parallel to the velocity field generated by the solution of the pressure Equation (6.13). However, for the flow of multiple compressible fluids it is the total mass flux that is conserved. Thus, our streamline coordinate system is still defined with one axis along the total velocity vector $\mathbf{q}_t$ and the other two axes, with coordinate functions ψ and χ, in the plane perpendicular to $\mathbf{q}_t$. But, rather than Equation (5.46) relating $\mathbf{q}_t$ to the gradients of ψ and χ, we have (Bear, 1972)

$$\rho\mathbf{q}_t = \nabla\psi \times \nabla\chi,$$

where ρ denotes the effective density. Taking the divergence of both sides of this equation gives

$$\nabla \cdot \rho\mathbf{q}_t = \mathbf{q}_t \cdot \nabla\rho + \rho\nabla \cdot \mathbf{q}_t = \nabla \cdot (\nabla\psi \times \nabla\chi) = 0,$$

the last equality is a known identity from vector analysis. For a given flow field $\mathbf{q}_t$ this expression provides a differential equation for ρ. Combining Equations (6.51), (6.53), and (6.48), we may write the above equation as

$$\varphi\frac{\partial\rho}{\partial\tau} + \rho\nabla \cdot \mathbf{q}_t = 0,$$

or, upon rearrangement,

$$\frac{\partial \ln \rho}{\partial\tau} = -\frac{1}{\varphi}\nabla \cdot \mathbf{q}_t.$$

If one considers a small enough time increment so that the flow field does not change significantly, say the time increment Δt required for the fluid to traverse a grid block of a reservoir model, then the right-hand-side of the equation does not depend upon time and we have

$$\rho = \rho_o e^{-\frac{1}{\varphi}\nabla\cdot\mathbf{q}_t\tau}.$$

One can combine these expressions with the streamline tracing algorithms, such as that of Pollock [1988], where the cell velocities within each grid block of a finite-difference model are given by locally linear expressions (Equations (5.53))

$$(q_t)_i = (q_1)_i + a_i\left[x_i - (x_1)_i\right],$$

where a_i are local coefficients and $\mathbf{x}_1$ is the entry point of the trajectory into the grid block and $\mathbf{q}_1$ is the velocity at that entry point. Using this expression we find that

$$\nabla \cdot \mathbf{q}_t = \sum_{i=1}^{3} a_i$$

and

$$\rho = \rho_o e^{-\frac{1}{\varphi}\sum_i a_i \tau}.$$

Note that for incompressible flow, the velocity field is divergence-free and $\sum_{i=0}^{3} a_i = 0$.

Returning to our initial governing equation, we can transform it into streamline coordinates

$$\frac{\partial S}{\partial t} + \frac{\partial f_w}{\partial \tau} + \frac{f_w}{\varphi}\nabla \cdot \mathbf{q}_t = \frac{\hat{Q}_w}{\varphi},$$

and use the expression above to write it as

$$\frac{\partial S}{\partial t} + \frac{\partial f_w}{\partial \tau} = \frac{\hat{Q}_w}{\varphi} + f_w \frac{\partial \ln \rho}{\partial \tau}.$$

The change in effective density acts as a source term. Thus, streamline techniques are applicable in the presence of compressible flow with a few modifications, such as the computation of effective density and the inclusion of an appropriate source term.

Exercise 6.6. Polymer flooding is an improved oil recovery scheme designed to counter the adverse impact of heterogeneity on oil recovery. It involves increasing the viscosity of the injected water by dissolving polymer in it. The increased water viscosity steepens the fractional flow curve, resulting in a more favorable (piston-like) displacement. Repeat the Exercise 6.4 for a polymer flood modeled by increasing water viscosity (in 'Examples/Simulation/5-Spot-Het-Num-P'). Visualize the pressure distribution, streamline trajectories, time of flight from the injector/producer, water saturation distribution to see how polymer flood counteracts heterogeneity effects, resulting in improved recovery. Notice in the input file 'SIMULATION.DATA' that the pressure field needs to be updated and streamlines regenerated in this example to account for the effects of viscosity contrast on the pressure distribution and the resulting changes in velocity field.

Exercise 6.7. This is a continuation of exercise 6.4. Here four additional producing wells are introduced after some time to recover the unswept oil in the reservoir (in 'Examples/Simulation/5-Spot-Het-Num-Infill'). This kind of development strategy is commonly referred to as **infill drilling**. Visualize the pressure distribution, streamline trajectories, time of flight from the injector/producer, water saturation distribution to see how infill drilling counteracts heterogeneity effects, resulting in improved recovery. The changing well configuration is given in the input file 'WELL.DATA'. Whenever the well configuration is changed and/or the streamlines are updated, the saturation distribution must be mapped from the old set of streamlines to the new set of streamlines. This mapping can be a potential source of error in streamline-based flow simulation.

Exercise 6.8. Use the change in variables

$$\xi = x - Vt,$$

to reduce the partial differential equation

$$\frac{\partial\left(\varphi S_w\right)}{\partial t} + \frac{\partial}{\partial x}\left[f_w q_t + g_w \sin\alpha - h_w \frac{\partial S_w}{\partial x}\right] = 0,$$

to the ordinary differential equation for front propagation with capillary forces,

$$\frac{d}{d\xi}\left[f_w q_t + g_w \sin\alpha - V\varphi S_w - h_w \frac{dS_w}{d\xi}\right] = 0.$$

6.5 Applications

6.5.1 Flow visualization

In streamline simulation the trajectories are computed using total fluid flux, that is by summing the fluxes of all the phases. Streamlines based on total flux are continuous and allow the use of time of flight as a spatial coordinate. We have seen in Chapter 5 that these streamlines are convenient for visualizing the impact of heterogeneity on the flow field. For example, streamlines tend to cluster in regions of high permeability and spread in regions of low permeability. The visualization of total flux streamlines also indicates the movement of tracer and water flood fronts, injector-producer relationship, swept volumes for injectors and drainage volumes for producers. However, streamlines based upon the total flux can mask important aspects of reservoir flow embedded in the individual phase fluxes. These include reservoir dynamics, such as phase distribution, appearance and disappearance of phases and reservoir drive mechanisms. The phase streamlines can be computed in the same manner as the total fluid streamlines, but using the individual phase flux instead of total flux. These trajectories, though discontinuous, provide insight none-the-less.

Figure 6.8 displays a snapshot of the phase streamlines for waterflooding in a heterogeneous reservoir. The permeability distribution is the same as shown in Figure 5.11; however, in this case three phases are present: water, oil, and gas.

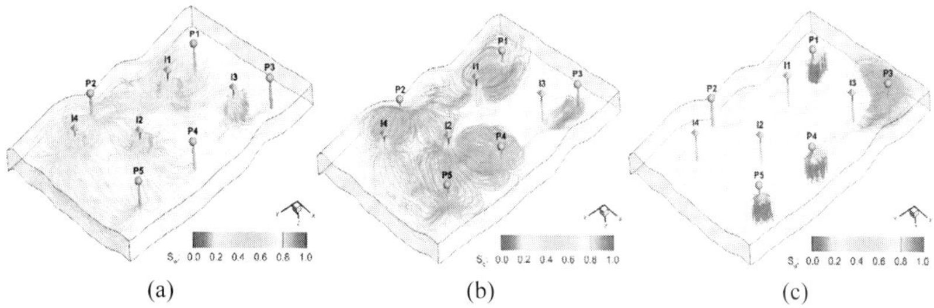

Figure 6.8 Visualization of phase streamlines. (a) Water phase streamlines showing movement of the injected water, (b) oil phase streamlines showing mobile oil being drained by producers, and (c) gas phase streamlines showing regions where depletion drive is active. For the colour version, please refer to the plate section.

The water phase streamlines have been traced, starting from the injectors, indicating the movement of the injected water and the region in which the waterflood is active. The oil and gas phase streamlines have been traced starting from the producers. The oil phase streamlines display the mobile oil regions that is being accessed by individual producers and provide guidelines for placement of new producing wells. The gas streamlines show the regions where gas is evolving because of pressure depletion, promoting primary recovery and potential loss of reservoir energy. Thus, for multiphase flow the visualization of the phase streamlines can provide some unique insights in addition to the visualization of the total streamlines discussed in the Chapter 5.

6.5.2 Sensitivity computations

In this section we shall derive the sensitivity coefficients associated with model parameter perturbations. That is, we shall determine how small variations in reservoir flow properties introduce small changes in observable quantities, such as the multi-phase front arrival times. We shall make use of our time of flight estimates, and the expression (6.88) for the phase travel time. In addition to the arrival time, we shall consider saturation histories, or the magnitude of the saturation changes, governed by Equation (6.86). For clarity, we shall neglect both capillary and gravitational effects. Towards the end of the section we discuss how to include the effects of gravity and temporal variations in the trajectories.

Arrival time sensitivities

Arrival time sensitivities follow from either of the two semi-analytical expressions, (6.45) or (6.88), in conjunction with an equation for the total velocity $\mathbf{q}_t$. A little work is required in order to pull out the important parameters contained within the flow vector $\mathbf{q}_t$. Specifically, we have that

$$\begin{aligned}\mathbf{q}_t &= \mathbf{q}_w + \mathbf{q}_n \\ &= -\lambda_w \nabla p_w - \lambda_n \nabla p_n \\ &= -\left(\lambda_w + \lambda_n\right) \nabla p \end{aligned} \tag{6.113}$$

where the last equality follows from the assumption that we can neglect capillary pressure, so that p_w equals p_n. Using the definition (6.5) of the phase mobilities λ_w and λ_n

$$\begin{aligned}\mathbf{q}_t &= -\left(\frac{k_{rw}}{\mu_w} + \frac{k_{rn}}{\mu_n}\right) k \nabla p \\ &= -\kappa k \nabla p \end{aligned} \tag{6.114}$$

where we have defined

$$\kappa = \frac{k_{rw}}{\mu_w} + \frac{k_{rn}}{\mu_n}, \tag{6.115}$$

the total fluid mobility. We shall assume that the total fluid mobility is a weak function of saturation.

We now consider how a perturbation in the properties of a porous medium, such as porosity and permeability, relates to a perturbation in the arrival time of the saturation front at an observation point. Because the goal is to derive sensitivities in order to update our model of a porous medium, we shall consider perturbations in the properties of individual grid blocks of the model. Thus, our perturbations will be localized and of limited extent. Let us denote the reservoir properties associated with the background model by a subscript o. Substituting the expression (6.114) for $\mathbf{q}_t$ into Equation (6.88) we find that the travel time in the background model is given by

$$T_o = \int_{\mathbf{x}_o(s)} \frac{1}{\kappa_o} \left(\frac{df_w}{dS}\right)^{-1} \frac{\varphi_o}{k_o |\nabla p_o|} ds \tag{6.116}$$

where $\mathbf{x}_o(s)$ denotes the trajectory in the background medium and we are allowing for a heterogeneous background model. We have moved the function $(df_w/dS)^{-1}$ inside the integral to account for variations in the background saturation along the trajectory. Consider a perturbation in the background porosity, $\varphi(\mathbf{x}) = \varphi_0(\mathbf{x}) + \delta\varphi(\mathbf{x})$, and a corresponding change in the travel time to an observation point:

$$T = \int_{\mathbf{x}_o(s)} \frac{1}{\kappa_o} \left(\frac{df_w}{dS}\right)^{-1} \frac{\varphi_o + \delta\varphi}{k_o |\nabla p_o|} ds \tag{6.117}$$

for a flow field that is assumed to be constant. In Equation (6.117) we are integrating along the trajectory $\mathbf{x}_o(s)$ calculated using this background model rather than the trajectory in the perturbed model. It has been shown (Nolet, 1987) that the difference in trajectories leads to contributions to the travel time that are of second order in $\delta\varphi$ and can be neglected in our first-order sensitivity estimates. Note that if the flow field is time-varying, we may have to sub-divide the total time into increments over which the pressure field is nearly constant. The change in travel time is the difference $\delta T = T - T_0$ given by subtracting the integral expression (6.116) from (6.117)

$$\begin{aligned}\delta T &= \int_{\mathbf{x}_o(s)} \frac{1}{\kappa_o}\left(\frac{df_w}{dS}\right)^{-1} \frac{1}{k_o|\nabla p_o|}\delta\varphi ds \\ &= \int_{\mathbf{x}_o(s)} \frac{1}{|\mathbf{v}_o|}\frac{\delta\varphi}{\varphi_o} ds \end{aligned} \tag{6.118}$$

where $\mathbf{v}_o$ is the flow velocity of the water, calculated using the background model of the porous medium

$$\mathbf{v}_o = \kappa_o \frac{df_w}{dS}\frac{k_o}{\varphi_o}\nabla p_o. \tag{6.119}$$

One can calculate $\mathbf{v}_o$ either by combining the individual contributions, according to Equation (6.119), or by direct numerical computation, extracting the phase velocity from the output of a numerical simulator. Sensitivities are functions relating perturbations in model parameters, such as $\delta\varphi(\mathbf{x})$, to changes in observable quantities, such as the front arrival time T. Formally, the sensitivity is the quantity that multiplies $\delta\varphi$ in the integrand of the expression (6.118) for δT. From Equation (6.118) we see that sensitivity associated with perturbations in the porosity, φ, is given by

$$\frac{\partial T}{\partial \varphi} = \frac{1}{\varphi_o|\mathbf{v}_o|}. \tag{6.120}$$

We may follow a similar procedure with respect to the permeability $k(\mathbf{x})$ and the pressure gradient magnitude $|\nabla p|$ to determine the sensitivities

$$\frac{\partial T}{\partial k} = -\frac{1}{k_o|\mathbf{v}_o|} \tag{6.121}$$

and

$$\frac{\partial T}{|\nabla p|} = -\frac{1}{|\nabla p||\mathbf{v}_o|} \tag{6.122}$$

respectively. Equations (6.120), (6.121), and (6.122) account for the explicit dependence of T on φ, k, and $|\nabla p|$. There is also an implicit dependence of $|\nabla p|$ field on both φ and k, through the pressure equation. We could account for this dependence explicitly by calculating the partial of the pressure gradient magnitude with respect to φ and k and substituting into (6.121) to obtain additional terms for the sensitivities. However, this approach leads to complicated expressions. Our preferred approach is to take advantage of the fact that we recompute the pressure field after each update of the model of the porous medium. Thus, after each change in the flow properties we conduct a numerical simulation and obtain a new estimate of the pressure field, p, and its gradient, accounting for the implicit dependence. We can use the travel time sensitivities, as well as the trajectories, for the efficient imaging of flow properties, following a procedure akin to geophysical tomography.

Exercise 6.9. Following the procedure used to determine the porosity sensitivity (6.120), derive the sensitivities for the permeability $k(\mathbf{x})$ and the pressure gradient magnitude $|\nabla p|$, given by

$$\frac{\partial T}{\partial k} = -\frac{1}{k_o|\mathbf{v}_o|}$$

and

$$\frac{\partial T}{|\nabla p|} = -\frac{1}{|\nabla p||\mathbf{v}_o|}$$

respectively.

Exercise 6.10. Use the software 'TRACE3D' to update an initial permeability distribution (PERMX-INIT.DATA) by matching (via inversion) the water-cut response at the producing wells (WELL-OBSERVED.DATA) in a 9-spot water flood (a central injector with eight surrounding producers). Use the data files in the folder 'Examples/Inversion/9-Spot-Het' by creating a project in the same folder and running the software. Choose the travel time inversion option (already selected). View the results to examine the improvements in the travel time and amplitude match. Compare the initial and the updated permeability fields and also view the changes made to initial permeability distribution during inversion. Also compare the updated permeability field with the 'reference' permeability field that was used to generate the data ('reference' permeability field available in the folder 'Examples/Inversion/9-Spot-Het').

Exercise 6.11. This is a continuation of exercise 6.10. Update the initial permeability field via a joint inversion of water-cut response at the producing wells and bottom hole pressure (BHP) data at the injection well. Use the data files in the folder 'Examples/Inversion/9-Spot-Het-BHP' by creating a project in the same folder and running the software.

Saturation amplitude sensitivities

Travel times provide useful information related to the hydraulic conductivity within a porous medium. However, the arrival time provides just a single datum for a given source and observation point. Furthermore, the first arriving saturation at a particular location most likely took the fastest path and does not provide information on the flow properties away from that path. Finally, the initial portion of the breakthrough curve might be poorly sampled, or not sampled at all, and we may miss the initial arrival at a well altogether. For these reasons, it is desirable to use as much of the breakthrough curve as possible in order to infer flow properties between a set of wells. It this section we describe a simple method to accomplish this (Vasco et al., 1999; Vasco and Datta-Gupta, 2001a), extending the results of the previous section. Often it will be possible to conduct a travel time inversion first, adjusting an initial model to fit a set of observed arrival times. We can then use the amplitude data to further refine the model. This approach is often taken in geophysical waveform inversion.

The sensitivities will be based upon the self-similar form (6.72) for the saturation amplitude. The Buckley–Leverett solution (6.86) provided one example of such a solution. One may recall that adoption of this form allows for spatially varying $\mathbf{q}_t$ and flow properties, but does require the assumption the saturation ahead of the front is uniform. We will adopt this solution in a local piecewise sense, assuming that the saturation within grid blocks of the reservoir model have a uniform average saturation, though the saturation may vary slowly along a path traversing many blocks. Thus, we adopt the self-similar form of the solution when considering local propagation across a single grid block, and perturbations to that grid block alone.

We will begin by building upon the results of the previous section and consider how a perturbation in the time of flight introduces a small deviation in the saturation amplitude. Once we establish this relationship, we can follow the approach taken in the last section to relate the perturbation in the time of flight to the properties of the porous medium. Consider a perturbation in the properties along a trajectory, or flow path, $\mathbf{x}_o(s)$ within the porous medium, and the resulting perturbation in the time of flight, given by

$$\tau = \tau_o + \delta\tau \tag{6.123}$$

where τ_o is the value computed using the unperturbed porous structure. There will be a corresponding perturbation in the saturation at the observation point

$$S\left(\frac{\tau}{t}\right) = S_o\left(\frac{\tau}{t}\right) + \delta S\left(\frac{\tau}{t}\right) \tag{6.124}$$

where $S_o(\tau/t)$ is the saturation history in the unperturbed medium. As noted above, we are assuming that the saturation within the grid block, and ahead of the propagating saturation front, is uniform, at least locally, so that the self-similar form may be used. Expanding $S(\tau/t)$ in a Taylor series about the point τ_o/t we find that

$$S\left(\frac{\tau}{t}\right) = S\left(\frac{\tau_o+\delta\tau}{t}\right)$$
$$= S_o\left(\frac{\tau^0}{t}\right) + \frac{1}{t}S_o{}'\left(\frac{\tau_o}{t}\right)\delta\tau \qquad (6.125)$$

to first order in the time of flight perturbation, $\delta\tau$, where the prime denotes the derivative with respect to τ. From Equation (6.125) we find that the perturbation in saturation is given by

$$\delta S\left(\frac{\tau}{t}\right) = \frac{1}{t}S_o{}'\left(\frac{\tau}{t}\right)\delta\tau. \qquad (6.126)$$

We are almost in a position to relate variations in the saturation amplitude to perturbations in the properties φ, k, and $|\nabla p|$. We will need the relationship between the time of flight τ and flow properties, given by Equation (6.53) and repeated here for convenience,

$$\tau(s) = \int_{\mathbf{x}(0)}^{\mathbf{x}(s)} \frac{\varphi}{|\mathbf{q}_t|} d\mathbf{x}$$

where $\mathbf{q}_t$ is related by κ, k, and ∇p, according to Equation (6.114). Between this sub-section and the previous one, everything is in place to relate a perturbation in the amplitude of a saturation change to perturbations in φ, k , and $|\nabla p|$. We leave it as an exercise for the reader, to show that the sensitivities are given by

$$\frac{\partial S}{\partial \varphi} = \frac{1}{t}S_o{}'\frac{1}{\varphi_o|\mathbf{q}_o|} \qquad (6.127)$$
$$\frac{\partial S}{\partial k} = -\frac{1}{t}S_o{}'\frac{1}{k_o|\mathbf{q}_o|} \qquad (6.128)$$
$$\frac{\partial S}{\partial |\nabla p|} = -\frac{1}{t}S_o{}'\frac{1}{|\nabla p||\mathbf{q}_o|} \qquad (6.129)$$

We can provide a physical interpretation of these expressions. As an example, consider the porosity sensitivity, given by Equation (6.127). The fractional component $1/\varphi_o|\mathbf{q}_o|$ is the time of flight sensitivity associated with a perturbation in φ. The local slope of the saturation curve, S_o', converts the time-shift sensitivity to an amplitude-shift sensitivity. For example, steeper slopes, or larger values of S_o', mean that a given time-shift leads to larger amplitude shifts. Note that the full sensitivity is obtained by summing over all paths that contribute to the saturation history at a given observation point. A multitude of paths can provide sensitivity to an extended region between a set of source and observation wells.

Although we have derived the sensitivities in terms of saturation, we can convert them to fractional flow sensitivities by multiplying each sensitivity given above by df_w/dS. The derivative of the fractional flow function is computed numerically, based upon the relative permeability curves. We must emphasize that, even though $\delta|\nabla p|$ is treated as an unknown parameter in our derivation it functions more as a correction term. That is, given estimates of flow properties, boundary conditions, and flow rates we can compute the pressure field and the pressure gradient. And we do calculate the pressure field numerically for each iteration of our inversion. However, in solving for reservoir properties we do not know the pressure field exactly. The pressure gradient perturbation $\delta|\nabla p|$ is a correction term that should minimize the mapping of deviations in the pressure gradient into reservoir properties. At each iteration of the inversion we recalculate the pressure field given our current estimates of reservoir structure and overwrite $\delta|\nabla p|$ with the calculated value. If the trajectories are not significantly perturbed by the passage of the saturation front then the perturbed trajectory $\mathbf{x}(s)$ may be replaced by the unperturbed trajectory $\mathbf{x}_o(s)$.

Exercise 6.12. Starting with expression (6.125), show that the saturation amplitude sensitivities are given by

$$\frac{\partial S}{\partial \varphi} = \frac{1}{t} S_o{}' \frac{1}{\varphi_o |\mathbf{q}_o|}$$

$$\frac{\partial S}{\partial k} = -\frac{1}{t} S_o{}' \frac{1}{k_o |\mathbf{q}_o|}$$

$$\frac{\partial S}{\partial |\nabla p|} = -\frac{1}{t} S_o{}' \frac{1}{|\nabla p||\mathbf{q}_o|}$$

Sensitivity estimates and their accuracy

In this section, we explore the accuracy of the trajectory-based approach. Specifically, we compare the semi-analytic approach with numerical sensitivity estimates. In a purely numerical approach, conducted using a reservoir simulator, a given cell permeability is perturbed by a small amount and an entire flow simulation is run. Changes in the predicted observable quantities, such as the aqueous fluid saturation at each well, are then calculated. The sensitivity associated with a particular observation, due to changes in a particular cell, is then the ratio of the saturation change to the permeability change. While this approach is flexible and accurate it generally requires extensive computation, a complete simulation for each parameter, limiting the number of parameters that can be estimated. Contrast this with the trajectory-based sensitivities, expressed in Equations (6.127), (6.128), and (6.129), based upon a single reservoir simulation.

As an illustration, consider the sensitivities associated with the j-th water-cut estimate, the fraction of water extracted from a given well. First, a forward simulation is conducted using an initial model of the porous medium. This initial model would typically be constructed using geographically dispersed data, such as well-logs or cores. The change in fractional flow at an observation point due to perturbations in flow properties, are based upon the sensitivities (6.127), (6.128), and (6.129). If we simply consider perturbations in permeability, then the change in the j-th value of water-cut is

$$\delta f_j = -\sum_{n=1}^{N_j} \frac{1}{t} \frac{df_w}{dS} S_o{}' \left(\frac{\tau}{t}\right) \int_{\mathbf{x}} \frac{1}{k_o |\mathbf{q}_o|} \delta k dr, \tag{6.130}$$

where the sum is over all N_j trajectories contributing to this particular observation, and the integral is along each trajectory within the sum. The derivative df_w/dS appears because we are considering observations of fractional flow rather than saturation. Equation (6.130) is an expression for a change in fractional flow due to all permeability variations influencing a given observation. In order to isolate the change due to a variation in a particular cell of the model, say grid-block i, we simply consider the increments of the trajectory which pass through that block. Denoting the segment of the path within grid block i by ν_i, we isolate that part of Equation (6.130) that is sensitive to changes within the cell,

$$\frac{\delta f_j}{\delta k_i} = -\sum_{n=1}^{N_j} \frac{1}{t} \frac{df_w}{dS} S_o{}' \left(\frac{\tau}{t}\right) \int_{\nu_i} \frac{1}{k_o |\mathbf{q}_o|} dr \tag{6.131}$$

using the fact that the perturbation δk_i is constant within the grid block. As stated above, all the quantities appearing in Equation (6.131) are contained in the existing model or produced by a single numerical simulation.

To verify the accuracy of the asymptotic approach we compare it to a purely numerical sensitivity calculation. In this example we consider water injection in a quarter five-spot well configuration with a single injector-producer pair in opposite corners of the reservoir grid. Trajectories, representing flow between the two wells, are shown in Figure 6.9, along with the fraction of water appearing at the producing well as a function of time. In this case the numerical and asymptotic sensitivities associated with perturbations in porosity, shown in Figure 6.10, are essentially identical. Because the steady state velocity field is insensitive to porosity variations, the trajectories do not shift as a result of perturbations in porosity, and the asymptotic sensitivities are exact.

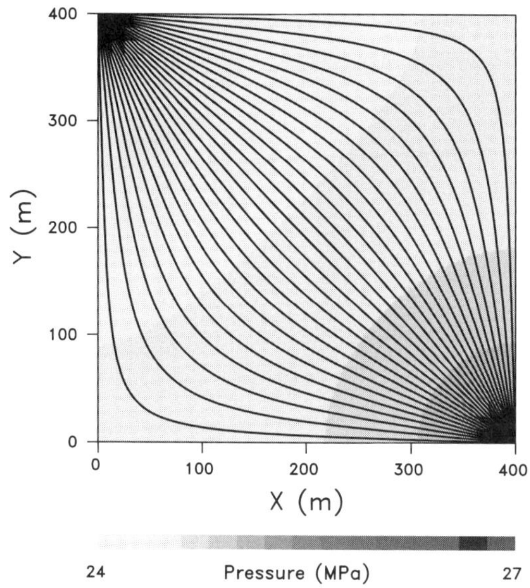

(a)

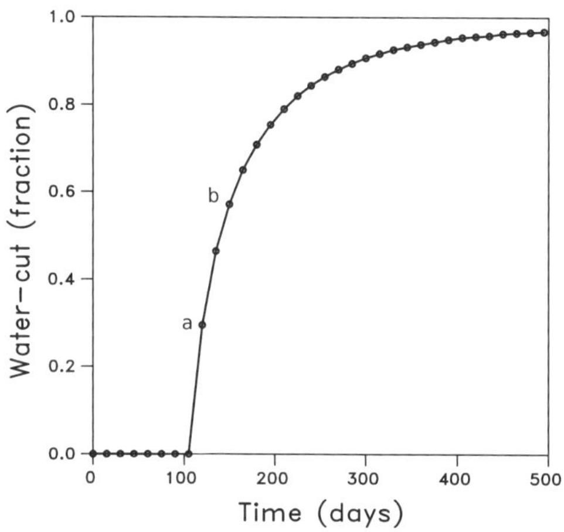

(b)

Figure 6.9 Water injection in a quarter five spot (a) streamline trajectories and (b) Fractional water production.

Sensitivities associated with changes in reservoir permeability, computed using Equation (6.131) are plotted in Figure 6.11. The numerical and asymptotic sensitivity coefficients are again very similar. However, there are small but detectable differences between the sensitivities. These differences are mainly due to the fact that, unlike porosity, the velocity field is influenced by the permeability variations. Thus, the assumption of static trajectories is only an approximation. The patterns are essentially the same, indicating that deviations in fractional flow data

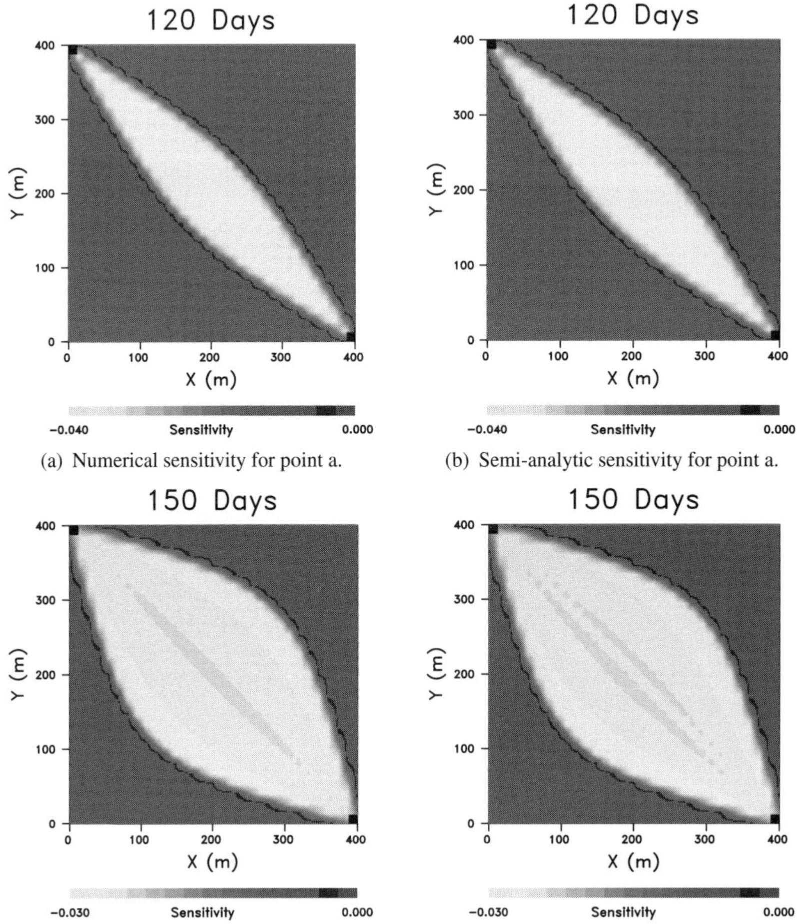

(a) Numerical sensitivity for point a.

(b) Semi-analytic sensitivity for point a.

(c) Numerical sensitivity for point b.

(d) Semi-analytic sensitivity for point b.

Figure 6.10 Comparison of trajectory-based semi-analytic sensitivities for porosity with numerical perturbation at two different times.

will be mapped into similar permeability changes by the numerical and asymptotic approaches.

Accounting for gravity and changing trajectories

So far, the trajectory-based analytic sensitivity computations assumed static trajectories. Many situations can lead to changes in the flow field and the corresponding streamlines. These include the appearance of new wells, the removal of existing wells, or isolating various reservoir zones to enhance production. Gravitational effects, and other cross streamline mechanisms, can also lead to changing streamlines. Trajectory-based sensitivity computations have been generalized to

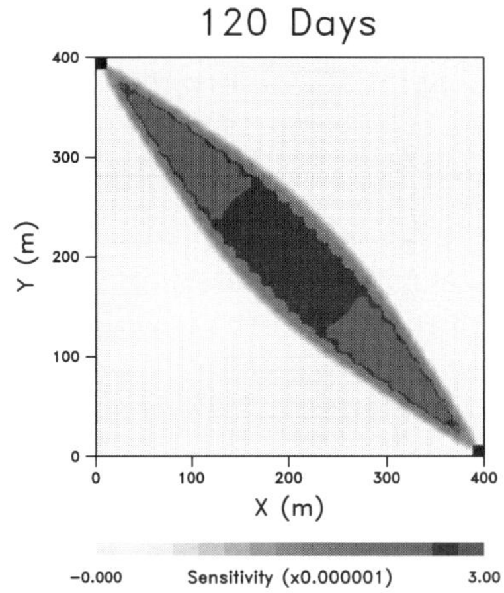

(a) Numerical sensitivity for point a.

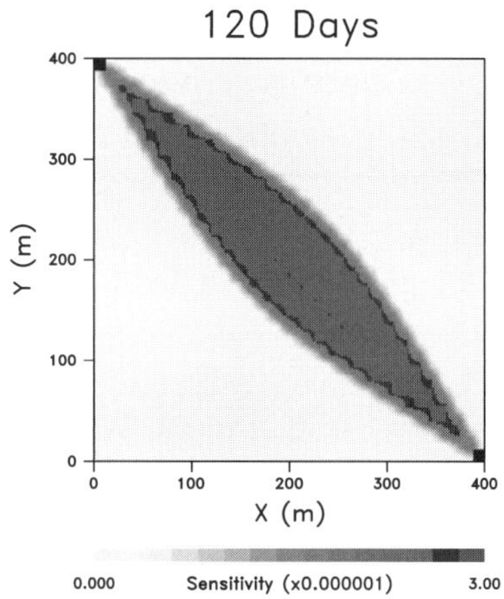

(b) Semi-analytic sensitivity for point a.

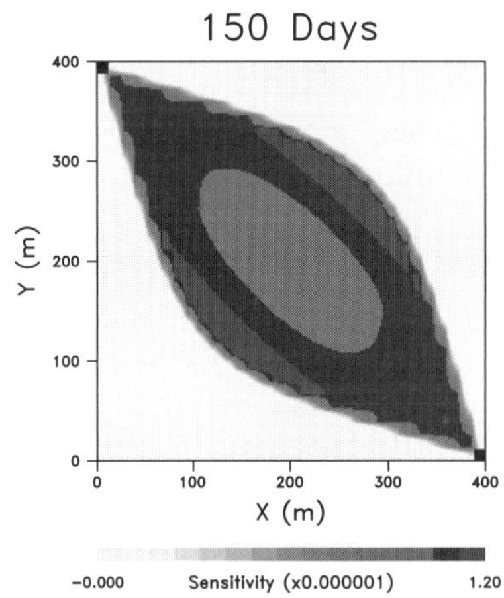

(c) Numerical sensitivity for point b.

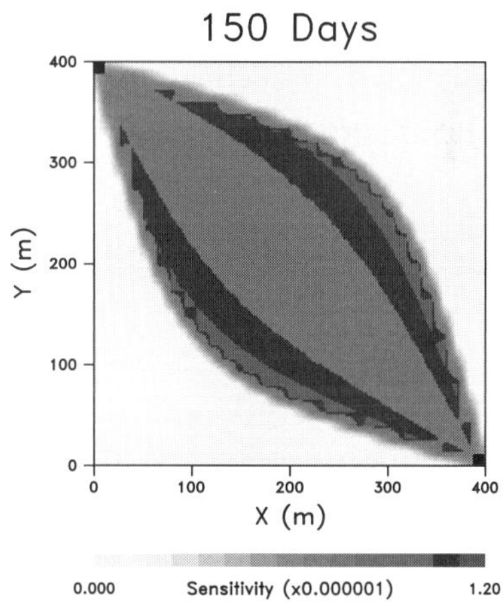

(d) Semi-analytic sensitivity for point b.

Figure 6.11 Comparison of trajectory-based analytic sensitivities for permeability with numerical perturbation at two different times specified by *a* and *b* in Figure 6.9.

handle such situations, albeit in an approximate manner. Two approaches have been proposed. In the first approach, He et al. [2002] propose a 'generalized travel time' inversion. It involves minimizing a time shift that is obtained by maximizing the cross-correlation between observed and computed production response. The sensitivities of the time shift can be obtained analytically, utilizing the arrival time sensitivities discussed before. The approach combines travel time and amplitude sensitivities and is analogous to seismic waveform inversion (Zhou et al., 1995). The generalized travel time inversion has been applied successfully to field applications (Kim et al., 2010). The other approach involves numerical solution of the sensitivity equation along one-dimensional trajectories [Gautier et al., 2001]. The sensitivity equation is derived by differentiating the finite-difference form of the transport equation along streamline trajectories.

Exercise 6.13. Use the software 'TRACE3D' to update an initial permeability distribution (PERMX-INIT.DATA) by matching (via inversion) the water-cut response at the producing wells (WELL-OBSERVED.DATA) in water flooding under changing well conditions. Use the data files in the folder 'Examples/Inversion/9-Spot-Het-Infill' by creating a project in the same folder and running the software. The water flooding starts with 4 producing wells and 4 additional producing wells are introduced after 1000 days (see Well.DAT). Choose the generalized travel time inversion option (already selected). View the results to examine the improvements in the travel time and amplitude match. Compare the initial and the updated permeability fields and also view the changes made to initial permeability distribution during inversion. Also compare the updated permeability field with the 'reference' permeability field that was used to generate the data ('reference' permeability field available in the folder 'Examples/Inversion/9-Spot-Het-Infill').

Exercise 6.14. This is a continuation of exercise 6.13. Update the initial permeability field via a joint inversion of water-cut response at the producing wells and bottom hole pressure (BHP) data at the injection well. Turn on the BHP matching option in the data file 'SIMULATION.DATA'.

6.5.3 Estimation of subsurface properties: multiphase flow and transport tomography

The advantages of the semi-analytic sensitivities provided by the trajectory-based approach might best be illustrated in applications involving geophysical observations. The wealth of data supplied by geophysical time-lapse monitoring can be a boon to reservoir characterization and the imaging of flow-related properties such as permeability. As noted by Vasco et al. (2004, 2008b) the computational efficiency associated with the semi-analytic sensitivities facilitates reservoir characterization based upon geophysical time-lapse data. Follow-up work has supported this conclusion (Rey et al., 2012; Watanabe et al., 2014), and new data sets with improved temporal resolution can potentially provide robust reservoir characterization based upon the onset of changes in geophysical parameters (Vasco et al., 2014, 2015). Here we illustrate the application of the trajectory-based sensitivity computation for reservoir model updating by integrating 4D seismic

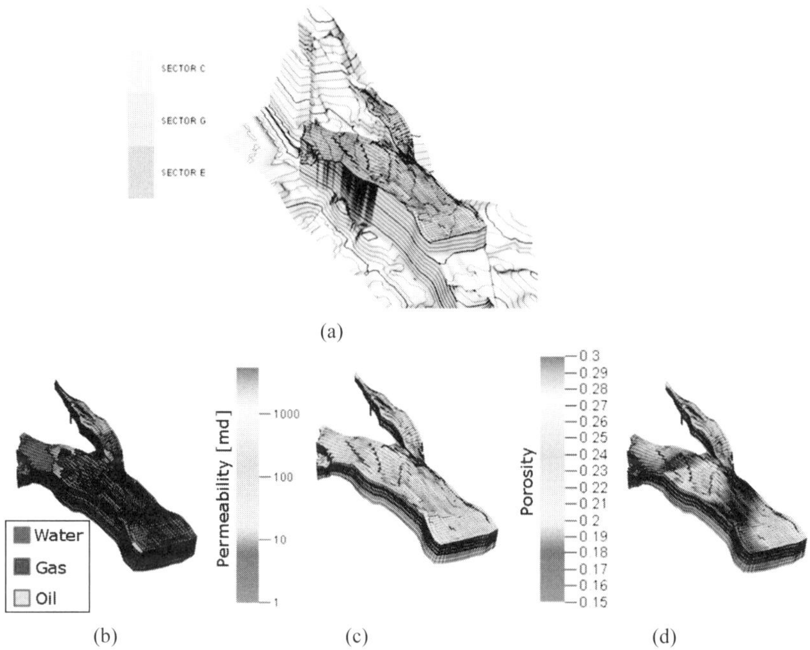

Figure 6.12 (a) Top surface of the reservoir indicating fault blocks, (b) initial fluid phase distribution, (c) initial horizontal permeability distribution, and (d) initial porosity distribution [from Rey et al. (2012)]. Reprinted with Permission. Copyright SEG. For the colour version, please refer to the plate section.

surveys with well production data. The approach is particularly well-suited for the calibration of high-resolution reservoir properties such as permeability, because the field-scale seismic data are areally dense while the production data effectively average properties between the wells.

The example discussed here involves waterflooding of a North Sea reservoir with multiple seismic surveys (Rey et al., 2012; Watanabe et al., 2014). The reservoir rocks are lower to middle Jurassic sandstones of a high quality, with an average porosity of 25 percent and permeability in the range of 200–2,000 mD. Figure 6.12(a) shows the top horizon of the reservoir, the respective segments (C, D, E, G) and the wells. Figures 6.12 (b)–(d) show the initial phase distribution accompanied by the porosity and permeability cell properties in the initial model. The production data include water, oil and gas rates in addition to bottom-hole pressure. These data are provided exclusively for the wells located in the E-segment; therefore, we will concentrate the calibration of the reservoir model over this segment. Figure 6.13 shows the location of the three producers (E-3H, E-2H, E-3AH) and two injectors (F-1H, F-3H), together with their hydraulic connectivity depicted by streamline trajectories that are colored per injector-producer

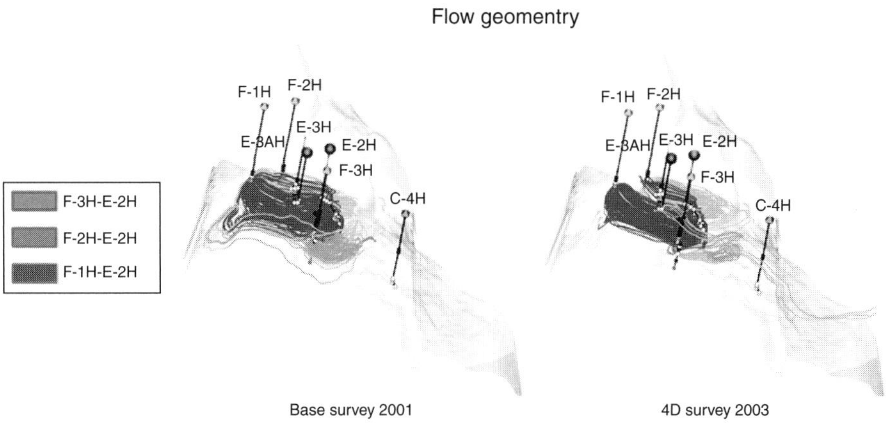

Figure 6.13 Well configuration and flow geometry during time lapse seismic surveys [from Rey et al. (2012)]. Reprinted with Permission. Copyright SEG. For the colour version, please refer to the plate section.

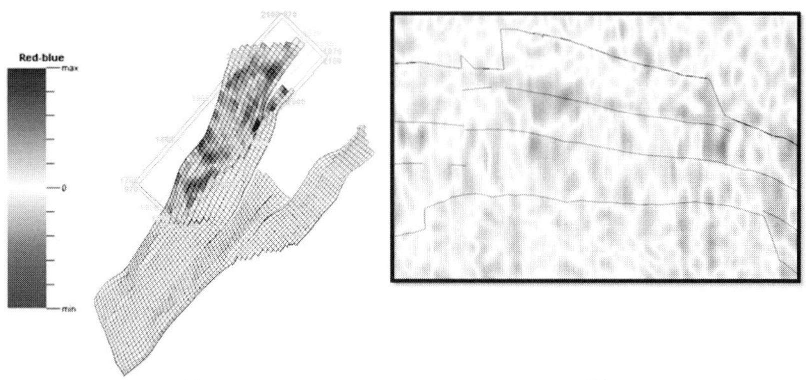

(a) (b)

Figure 6.14 (a) Time lapse seismic acoustic impedance changes displayed on the reservoir grid and (b) a transect along the section marked '1900' in (a) indicating hardening and softening of the rock because of reservoir production [from Rey et al. (2012)]. Reprinted with Permission. Copyright SEG. For the colour version, please refer to the plate section.

pair. This graphic is an essential tool for visualization of the effect of faults and barriers on the flow geometry. In addition to the production data, a set of 4D seismic surveys was acquired at the years 2001, 2003, and 2004. The seismic data were externally processed and provided for use in model calibration as post stack volumes of the reflection amplitude together with the corresponding horizons for the top and base of the reservoir.

The first step in our data calibration procedure is to invert the seismic volumes of reflection amplitude for changes in acoustic impedance. Figure 6.14 shows the results of the geophysical inversion and the upscaling of the seismic volume from the seismic grid resolution (25m × 25m) to the reservoir grid resolution (100m × 100m). Collocation of the seismic data in the reservoir grid allows us to perform a quantitative cell-by-cell evaluation of the match quality during the calibration procedure. Figure 6.14 also shows the extent of the seismic volume of the E-segment compared to the size of the entire reservoir. Considering the small time span between the consecutive surveys, it is reasonable to assume that the porosity does not vary as a result of subsidence of the reservoir, and that the temporal changes in acoustic impedance result only from changes in fluid phase saturations and/or reservoir pressure. In support of this assumption, the inverted seismic volume indicates the rise of the water-oil contact in the E-segment (Figure 6.14). The red color in the graphic, associated with a positive change in acoustic impedance, is most likely related to the incremental increase in rock stiffness corresponding to a water saturation increase through the production period.

Petroelastic model and sensitivity of acoustic impedance

The changes in seismic impedance due to reservoir pressure and fluid saturation changes are interpreted using a petroelastic model. In addition, we need to compute the sensitivity of the seismic response to the changes in reservoir properties. The trajectory-based method can be used to compute these sensitivities efficiently.

The elasticity of the reservoir rock and, therefore, its ability to propagate mechanical waves is determined by the rock matrix properties, the pore fluids and the reservoir pressure. For the range of seismic frequencies, Gassmann's equation (Gassmann, 1951b), discussed in Box 2.10 in Chapter 2, is an adequate representation of the elasticity of the bulk rock volume. The equation relates (1) the bulk modulus of the porous rock, also called the bulk modulus of the dry frame $K_d(\phi, p)$ which is a function of porosity and the lithostatic pressure, (2) the bulk modulus of the structural rock $K_g(\phi)$ which is an intrinsic property of the rock configuration and can be assumed to be constant under varying conditions from fluid phase changes in the porous space, and finally (3) the fluid bulk modulus $K_f(S_w, S_o, S_g, p, T)$ which is a function of the elastic properties of the complete mixture. Gassmann's equation is

$$K_s = K_d + \frac{\left(1 - K_d/K_g\right)^2}{\phi \left(K_f - K_g{}^{-1}\right)^{-1} + K_g{}^{-1}\left(1 - K_d/K_g\right)}, \tag{6.132}$$

where the composite fluid bulk modulus, K_f, is calculated from the properties of the fundamental components and using the harmonic or Reuss average of the properties weighted by the relative amount of each individual component. The Reuss average guarantees the lowest value of the modulus obtained by combining the individual elements (Mavko et al., 1998) and may be written as

$$\frac{1}{K_f} = \sum_{i=o,g,w} \frac{S_i}{K_i}. \tag{6.133}$$

For this application the relationship between the dry frame modulus, K_d, and the porosity of the rock was determined by laboratory observations made on cores and is expressed as the linear equations

$$\begin{aligned} K_d &= a - \varphi b \\ G_d &= c - \varphi d. \end{aligned} \tag{6.134}$$

The density, ρ, is computed as the weighted average of the densities of the individual components of the rock

$$\rho = (1 - \varphi)\rho_d + \phi \sum_{i=0,g,w} S_i \rho_i. \tag{6.135}$$

With the density and bulk modulus of the saturated rock, the compressional and shear velocities can be estimated under the assumptions of an isotropic media and of insensitivity of the shear modulus to the fluid inside the porous media. The compressional or p-wave velocity, V_p, is estimated as

$$V_p = \sqrt{\frac{K_s + 4/3G_d}{\rho}}. \tag{6.136}$$

The seismic impedance, Z, can then be written in terms of V_p as

$$Z = \rho V_p = \rho\sqrt{\frac{K_s + 4/3G_d}{\rho}}. \tag{6.137}$$

With the above petroelastic model we can compute the changes in the acoustic impedance as a function of pressure and saturation changes. For a simple two-phase (oil, water) system, Figure 6.15 shows an increase in acoustic impedance with increasing water saturation for a fixed pressure and a decrease of acoustic impedance with increasing pressure at a fixed saturation.

The sensitivity of acoustic impedance with respect to (absolute) permeability can be achieved from an expression of the differential of Equation (6.137). Under the assumption that temporal changes in acoustic impedance are a function of differences in the fluid saturation and the effective pore pressure of the rock, the differential of acoustic impedance is given by

$$S_Z = \frac{\delta Z}{\delta k} = \frac{\partial Z}{\partial S_w}\frac{\delta S_w}{\delta k} + \frac{\partial Z}{\partial S_g}\frac{\delta S_g}{\delta k} + \frac{\partial Z}{\partial p}\frac{\delta p}{\delta k}. \tag{6.138}$$

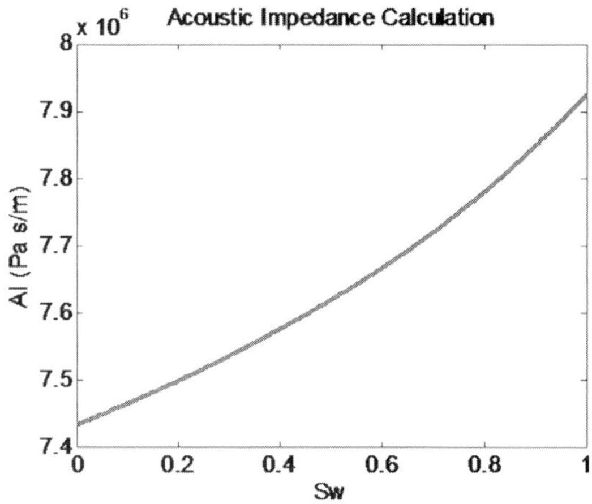

(a)

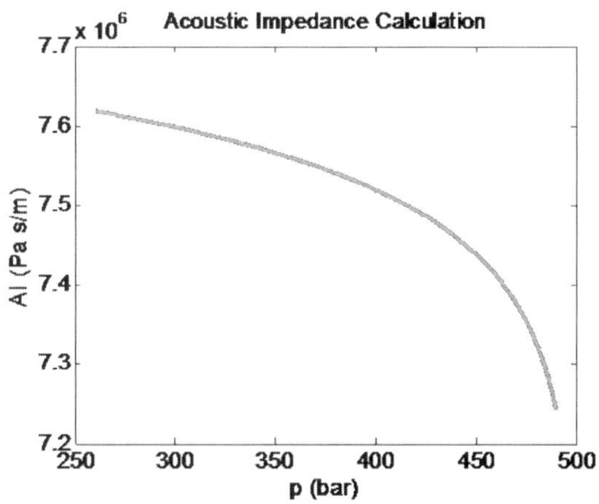

(b)

Figure 6.15 Acoustic impedance calculation sensitivity (a) with respect to water saturation changes under a fixed pressure (270 bar) and (b) with respect to pressure changes under a fixed saturation value (Sw = 0.5).

Here the partial derivatives of acoustic impedance, $\partial Z/\partial S_w$, $\partial Z/\partial S_g$, and $\partial Z/\partial p$ can be computed by numerical perturbation using the petroelastic models while the saturation sensitivity and pressure sensitivities can be computed using the trajectory-based methods. More details on these sensitivity computations can be found in Rey et al. (2012) and Watanabe et al. (2014).

Joint integration of seismic and multiphase flow data

The data integration is carried out using a deterministic approach in which a penalized misfit function is minimized. The misfit function quantifies the production and seismic data misfit together with regularization terms. The regularization terms consist of a norm constraint and a roughness constraint and are introduced to preserve the spatial continuity and geological realism in the updated model. In this manner, the model calibration is a balance between reducing the data misfit and minimizing the changes to the prior model, thereby maintaining geologic consistency. The penalized misfit function in this case is defined as

$$f(\delta \mathbf{k}) = \alpha_1 ||\delta \mathbf{d}_{seis} - \mathbf{G}_{seis}\delta \mathbf{k}|| + \alpha_2 ||\delta \mathbf{d}_{prod} - \mathbf{G}_{prod}\delta \mathbf{k}|| + \beta_1 ||\delta \mathbf{k}|| + \beta_2 ||\mathbf{L}\delta \mathbf{k}||. \tag{6.139}$$

In Equation (6.139), $\delta \mathbf{d}_{seis}$ is the vector of the misfit between the change in acoustic impedance derived after a geophysical seismic inversion and the change in simulated acoustic impedance. The sensitivity matrix $\mathbf{G}_{seis}$ contains the partial derivatives of the changes in acoustic impedance with respect to the reservoir parameters, that is, grid cell absolute permeability. The production information is incorporated to an extent controlled by the weighting factors α_1 and α_2. When the seismic information contains excess noise, or when scaling from the seismic to reservoir flow simulation grid is uncertain, then α_1 can be selected so that the seismic information is applied as a second smoothness constraint. The quantity $\delta \mathbf{k}$ is the vector of changes in the reservoir permeability field. The first penalty term is the norm constraint that penalizes deviation from the prior model. The second penalty term is the roughness constraint, where the operator $\mathbf{L}$ computes the second difference of each cell permeability, and has an effect analogous to the imposition of a prior variogram or covariance constraint. For a given step, the minimum of the objectve function is obtained using a least-squares solution of the augmented linear system of equations

$$\begin{pmatrix} \alpha_1 \mathbf{G}_{seis} \\ \alpha_2 \mathbf{G}_{prod} \\ \beta_1 \mathbf{I} \\ \beta_2 \mathbf{L} \end{pmatrix} \delta \mathbf{k} = \begin{pmatrix} \delta \mathbf{Z} \\ \delta \mathbf{d}_{prod} \\ \mathbf{0} \\ \mathbf{0} \end{pmatrix}. \tag{6.140}$$

where the scalars β_1 and β_2 determine the relative strengths of the norm and smoothness constraints. The weighting factors are typically determined empirically, requiring multiple trials, for a specific application. They are enforced only to the extent that implausible geologic features (i.e., artifacts) resulting from updates along streamline trajectories are avoided. Figure 6.16 shows the extent of the modifications in the acoustic impedance for selected layers. Even though the majority of the time lapse changes are related to the replacement of oil by water (i.e., positive changes), there are noticeable locations of an increase in acoustic impedance that are unrelated to water movement and that, we infer, are related to changes in pressure. These changes are located in areas that are initially fully

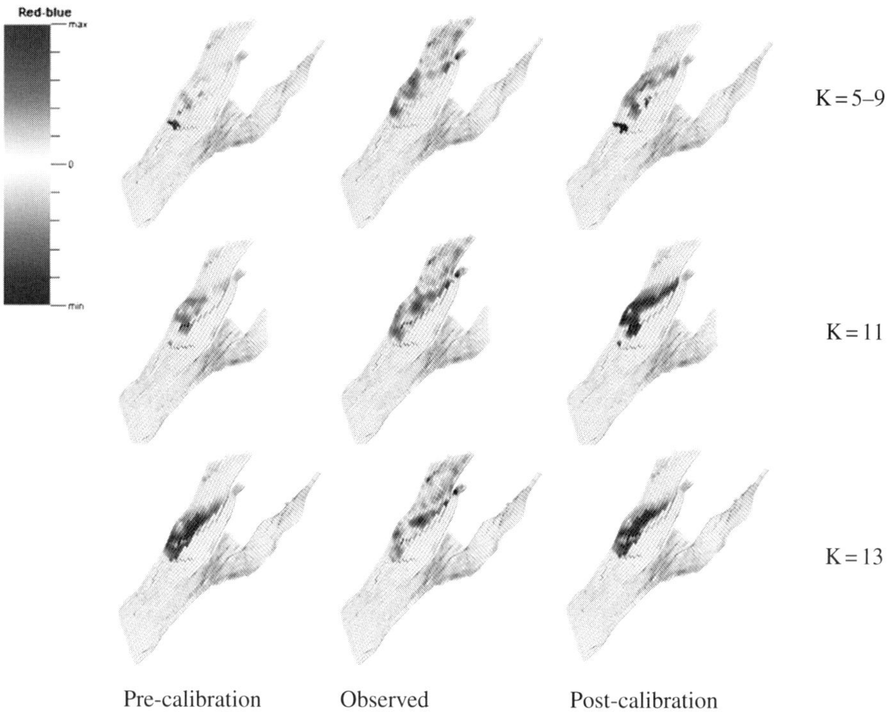

Figure 6.16 Acoustic impedence changes in selected layers before (pre-calibration) and after inversion (post-calibration) and their comparison with the observed values [from Rey et al. (2012)]. Reprinted with Permission. Copyright SEG. For the colour version, please refer to the plate section.

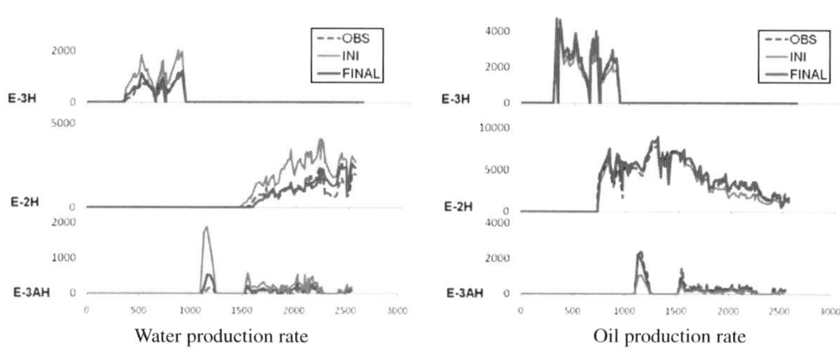

Figure 6.17 Comparison of the observed and calculated multiphase production response [from Rey et al. (2012)]. Reprinted with Permission. Copyright SEG. For the colour version, please refer to the plate section.

saturated with water; therefore, the decrease in bulk modulus is not related to changes in fluid saturations at those cells, but rather to changes in the average pressure. Rey et al. (2012) did not account for the pressure effects on the acoustic impedance calculations. This discrepancy was corrected in the subsequent work of Watanabe et al. (2014). They accounted for both pressure and saturation effects on the seismic response. Figure 6.17 shows the improvement in matches to the observed multiphase production data after joint calibration.

6.6 Summary

In this chapter we have discussed the application of asymptotic and trajectory-based methods to multiphase flow and transport in permeable media. As in the previous chapters, the asymptotic approach leads to an equation describing the phase or travel time of a propagating front, in this case the fluid saturation front. More importantly, the asymptotic approach is shown to naturally lead to a multiphase version of the streamline time of flight introduced in Chapter 5. The time of flight is used as a spatial coordinate to decouple the three-dimensional multiphase flow equation into one-dimensional equations along streamlines which are solved using analytic or numerical methods. The asymptotic method also allows for efficient computation of sensitivity of multiphase flow response to spatial variations in subsurface properties. The sensitivities can be used in conjunction with inverse modeling, leading to an efficient formalism for subsurface imaging using multiphase flow and transport data.

7

Coupled deformation and fluid flow

7.1 Introduction

With the continuing development of techniques for monitoring fluid flow using remote observations, there is an increasing need to model the associated deformation of the porous rock containing the fluid. Such modeling helps us understand displacements induced by fluid flow, as observed by geodetic methods and seismic time strains. Coupled modeling also provides insight into the deformation-induced flow associated with the passage of an elastic wave. Understanding this phenomena, and its relationship to the properties of a porous medium, are critical for effective time-lapse monitoring of fluid flow. Time-lapse monitoring provides a basis for the inverse problem: the determination of the properties of a porous medium using remote observations. In this chapter we develop solutions for coupled deformation and fluid flow. While one goal is the resulting efficiency associated with trajectory-based solutions for both forward and inverse modeling, the insight provided by the semi-analytic solutions is equally important to our understanding. The end result of our efforts will be similar to what we found in earlier chapters: an equation for the phase and for the trajectories. The primary difference will be the appearance of multiple phase velocities, revealing the existance of elastic-like fast waves, and slow diffusion-like modes of propagation.

In this chapter we consider two important situations in which the governing equations simplify. First, when a single fluid is present in a poroelastic medium, the non-linearity due to saturation-dependent properties disappears and the resulting equations are linear for small strain rates. Second, for the small but rapid displacements associated with the passage of an elastic wave through a poroelastic medium, we can linearize the equations about the existing saturation distribution.

7.2 Deformation in a porous body containing a single fluid

7.2.1 The governing equations

For fluid injection over a time period during which viscoelastic and plastic behavior are not significant, one may model the reservoir and the surrounding regions as a poroelastic medium. In this approximation, we assume that the porous skeleton behaves elastically while the injected material behaves as a Newtonian viscous fluid. We are free to adopt the constitutive models and the coupling terms discussed in Chapter 2. Thus, we can begin with the governing Equations (2.154) and (2.155). We can simplify these expressions by considering the case in which the body force terms $\mathbf{f}_s$ and $\mathbf{f}_f$ vanish and the convective terms are small enough to be neglected:

$$\varphi \rho_f \frac{\partial \dot{\mathbf{u}}_f}{\partial t} = \varphi \nabla \cdot \boldsymbol{\sigma}_f + \mathbf{d}_f, \tag{7.1}$$

$$(1 - \varphi)\, \rho_s \frac{\partial \dot{\mathbf{u}}_s}{\partial t} = (1 - \varphi)\, \nabla \cdot \boldsymbol{\sigma}_s - \mathbf{d}_f. \tag{7.2}$$

We denote all quantities associated with the fluid by the subscript f, and we have made use of the constraint $\mathbf{d}_f = -\mathbf{d}_s = \mathbf{d}$ on the interaction terms for the case of a single fluid [see Box 2.7]. We can now write everything in terms of the solid displacement $\mathbf{u}_s$ and the displacement of the fluid relative to that of the solid

$$\mathbf{w} = \varphi \left(\mathbf{u}_f - \mathbf{u}_s \right), \tag{7.3}$$

a common formulation in the field of poroelasticity. Note the weighting of the relative displacement by the porosity in the definition (7.3). If we add and subtract $\varphi \rho_f \partial \dot{\mathbf{u}}_s / \partial t$ from the left side of Equation (7.1), then we can write the system in terms of $\mathbf{u}_s$ and $\mathbf{w}$

$$\varphi \rho_f \frac{\partial \dot{\mathbf{u}}_s}{\partial t} + \varphi \rho_f \frac{\partial \dot{\mathbf{w}}}{\partial t} = \varphi \nabla \cdot \boldsymbol{\sigma}_f + \mathbf{d}_f \tag{7.4}$$

$$(1 - \varphi)\, \rho_s \frac{\partial \dot{\mathbf{u}}_s}{\partial t} = (1 - \varphi)\, \nabla \cdot \boldsymbol{\sigma}_s - \mathbf{d}_f. \tag{7.5}$$

We can remove the $\mathbf{d}_f$ term from Equation (7.5) by adding Equation (7.4) to it. The new system of equations that results is

$$\rho_f \frac{\partial \dot{\mathbf{u}}_s}{\partial t} + \rho_f \frac{\partial \dot{\mathbf{w}}}{\partial t} - \frac{\mathbf{d}_f}{\varphi} = \nabla \cdot \boldsymbol{\sigma}_f \tag{7.6}$$

$$\rho \frac{\partial \dot{\mathbf{u}}_s}{\partial t} + \varphi \rho_f, \frac{\partial \dot{\mathbf{w}}}{\partial t} = (1 - \varphi)\, \nabla \cdot \boldsymbol{\sigma}_s + \varphi \nabla \cdot \boldsymbol{\sigma}_f \tag{7.7}$$

where we have divided the first equation by φ, moved the drag force term $\mathbf{d}_f$ to the left-hand-side of the equation, and defined the composite density

$$\rho = (1 - \varphi)\, \rho_s + \varphi \rho_f. \tag{7.8}$$

In order to solve the system, we first need to write $\mathbf{d}_f$, $\boldsymbol{\sigma}_f$, and $\boldsymbol{\sigma}_s$ in terms of the velocities, $\dot{\mathbf{u}}_s$ and $\dot{\mathbf{w}}$, or the displacements.

An analysis of the interaction term $\mathbf{d}_f$

Regarding $\mathbf{d} = \mathbf{d}_f$, we are free to adopt a generalization of expression (2.156) from Chapter 2. Specifically, we can write $\mathbf{d}_f$ as

$$\mathbf{d}_f = \varphi \mu_f \mathcal{K}^{-1} \dot{\mathbf{w}}, \tag{7.9}$$

where $\mathcal{K}^{-1}$ is a convolutional operator whose Fourier transform is $1/k(\omega)$, the inverse of the dynamic permeability (Johnson et al., 1987). A time domain expression was derived by Plyushchchenkov and Turchaninov (2000) and incorporated into a numerical finite-difference code for modeling poroelastic propagation valid across all frequencies (Masson and Pride, 2010). In Box 2.8 we discuss a frequency-dependent expression for the dynamic permeability $1/k(\omega)$:

$$\frac{1}{k(\omega)} = \frac{1}{k_o}\left[\sqrt{1 - i\frac{\omega}{\omega_c}\Pi} - i\frac{\omega}{\omega_c}\right] \tag{7.10}$$

where Π is a pore geometry term

$$\Pi = 4\frac{\nu_\infty k_o}{\Lambda^2 \varphi}, \tag{7.11}$$

ω_c, which is the transition frequency signifying the crossover from viscous-dominated flow to flow dominated by inertial forces, given by

$$\omega_c = \frac{\mu_f \varphi}{\rho_f \nu_\infty k_o}, \tag{7.12}$$

where ν_∞ is the tortuosity at the high-frequency limit and k_o is the static permeability. The parameter Π also depends upon Λ, a variable that has an approximate interpretation as twice the ratio of pore volume to pore surface area (Johnson et al., 1987). For many porous media Π is approximately $1/2$ (Johnson et al., 1987). As noted in Box 2.8, this equation is the simplest model connecting the high- and low-frequency behavior. For most applications in the Earth Sciences, the transition or crossover frequency (ω_c) is much greater than the frequency of the propagating disturbance, so in most cases one may safely take $k(\omega) = k_o$. In fact, it appears that medium or mesoscale heterogeneity, such as patchy saturation of fluid or spatial variations in the material properties of the porous medium, may have a much more significant effect at the frequencies used in geophysical exploration (White, 1975; Johnson, 2001; Pride et al., 2004).

Substituting the expression (7.9) for $\mathbf{d}_f$ into Equation (7.6) and writing out all temporal derivatives explicitly gives

$$\rho_f \frac{\partial^2 \mathbf{u}_s}{\partial t^2} + \rho_f \frac{\partial^2 \mathbf{w}}{\partial t^2} - \mu_f \mathcal{K}^{-1} \frac{\partial \mathbf{w}}{\partial t} = \nabla \cdot \boldsymbol{\sigma}_f \tag{7.13}$$

$$\rho \frac{\partial^2 \mathbf{u}_s}{\partial t^2} + \varphi \rho_f \frac{\partial^2 \mathbf{w}}{\partial t^2} = (1 - \varphi) \nabla \cdot \boldsymbol{\sigma}_s + \varphi \nabla \cdot \boldsymbol{\sigma}_f \tag{7.14}$$

It makes physical sense that the interaction term $\mathbf{d}_f$ should involve a temporal convolution because the drag forces that it represents are due to viscous flow, a dissipative process, as well as inertial forcing. The effects of this 'memory kernel' on wave propagation in poroelastic media, along with an asymptotic solution, were discussed by Hanyga and Seredynska (1999).

We finish with a note on the relationship between the frequency-dependence of the dynamic permeability and the presence of derivatives of various orders in the time-domain expressions (7.13) and (7.14). First, in an effort to be mutually consistent, we apply the Fourier transform to Equations (7.13) and (7.14), bringing them both into the frequency domain

$$-\omega^2 \rho_f \mathbf{U}_s - \omega^2 \rho_k \mathbf{W} = \nabla \cdot \boldsymbol{\Sigma}_f \tag{7.15}$$

$$-\omega^2 \rho \mathbf{U}_s - \omega^2 \rho_f \mathbf{W} = (1 - \varphi) \nabla \cdot \boldsymbol{\Sigma}_s + \varphi \nabla \cdot \boldsymbol{\Sigma}_f, \tag{7.16}$$

where

$$\rho_k = \varphi \rho_f - \frac{\mu_f}{\omega k(\omega)} i. \tag{7.17}$$

And we denote the Fourier transform of the solid and fluid displacements by $\mathbf{U}_s$ and $\mathbf{W}$, respectively. Depending upon the nature of the dynamic permeability, $k(\omega)$ as a function of ω, Equations (7.15) and (7.16) can contain a variety of temporal derivatives upon transformation back into the time domain. In order to see this, consider the model of (Johnson et al., 1987), given by Equation (7.10). If we expand the quantity under the square root in a Taylor series in ω about zero, we obtain an infinite series in ω. Upon transforming this into the time domain, we have an operator containing an infinite series of derivatives with respect to time. The main point is that, depending upon the particular form of the dynamic permeability function, the governing equations can contain a variety of derivatives. Also note that the terms in the series may diminish quite rapidly if ω is much smaller then ω_c.

The fluid and solid stresses

The stresses and pressures may be written in terms of the spatial derivatives of the solid and fluid displacements as was done in Chapter 2. Note that one can argue, as in Box 2.11, that the coefficients in the expressions relating fluid and solid stresses to the fluid and solid displacements are frequency-dependent, particularly in the presence of variable fluid saturations (White, 1975; Johnson, 2001; Pride et al., 2004). Frequency-dependent coefficients will not stand in the way of our analysis. In fact, we will allow for a frequency-dependent permeability $k(\omega)$, such as that given by Equation (7.10). We do not invoke any explicit assumptions regarding the magnitude of ω. Rather, we shall assume that the microscopic and mesoscopic processes operate at length scales that are much smaller than the wavelength of the disturbance that we are modeling. In addition, we allow for a smoothly varying heterogeneity between boundaries and interfaces of a length scale that is much larger than the dominant wavelength of the disturbance. The large-scale heterogeneity will be reflected by the presence of spatially dependent coefficients in the governing equations.

Using the method of volume averaging, Pride et al. (1993) obtained explicit expressions for the the phase averaged stress-strain relations. We transform their results into the frequency domain and write them in terms of the poroelastic parameters K_u (undrained bulk modulus), C (Biot's coupling modulus), and M (the fluid storage coefficient) introduced in Chapter 2. If we also write the stress-strain relationship in terms of the dependent variables $\mathbf{U}_s$ and $\mathbf{W}$, and assume that the porosity is locally constant, we arrive at

$$(1-\varphi)\,\bar{\mathbf{\Sigma}}_s = [(K_u - \varphi C)\,\nabla \cdot \mathbf{U}_s + (C - \varphi M)\,\nabla \cdot \mathbf{W}]\,\mathbf{I} + \boldsymbol{\tau} \tag{7.18}$$

and

$$\bar{\mathbf{\Sigma}}_f = [C\nabla \cdot \mathbf{U}_s + M\nabla \cdot \mathbf{W}]\,\mathbf{I}, \tag{7.19}$$

where the overbar signifies the phase average of the transformed stress tensor, and $\boldsymbol{\tau}$ is the deviatoric stress tensor:

$$\boldsymbol{\tau} = G\left[\nabla\mathbf{U}_s + \nabla\mathbf{U}_s^T - \frac{2}{3}\nabla \cdot \mathbf{U}_s\mathbf{I}\right], \tag{7.20}$$

that part of the stress tensor representing the elastic shear stress. Equations (7.18) and (7.19) may be combined to give

$$(1-\varphi)\,\bar{\mathbf{\Sigma}}_s + \varphi\bar{\mathbf{\Sigma}}_f = [K_u\nabla \cdot \mathbf{U}_s + C\nabla \cdot \mathbf{W}]\,\mathbf{I} + \boldsymbol{\tau}. \tag{7.21}$$

The expressions (7.19) and (7.21) are compatible with the corresponding results in Chapter 2, Section 2.6.2, if we identify the phase averaged fluid stress tensor

$\bar{\boldsymbol{\Sigma}}_f$ with $\boldsymbol{\Sigma}_f$, the Fourier transform of $\boldsymbol{\sigma}_f$ in Chapter 2, and the Fourier transform of $\boldsymbol{\sigma}_s$ with

$$\boldsymbol{\Sigma}_s = (1 - \varphi)\,\bar{\boldsymbol{\Sigma}}_s + \varphi\bar{\boldsymbol{\Sigma}}_f. \tag{7.22}$$

The governing equations

Substituting the expressions (7.19) and (7.21) for the fluid and solid stresses in Equations (7.15) and (7.16) results in the final form of the governing equations

$$-\,\omega^2\rho_f\mathbf{U} - \omega^2\rho_k\mathbf{W} = \nabla\left(C\nabla\cdot\mathbf{U} + M\nabla\cdot\mathbf{W}\right) \tag{7.23}$$

$$-\omega^2\rho_s\mathbf{U} - \omega^2\rho_f\mathbf{W} = \nabla\left(K_u\nabla\cdot\mathbf{U} + C\nabla\cdot\mathbf{W}\right) + \nabla\cdot\boldsymbol{\tau}. \tag{7.24}$$

These equations spring from Biot's formulation of the mechanics of a poroelastic medium (Biot, 1956a,b, 1962b). A number of authors (Levy, 1979; Burridge and Keller, 1981; Pride et al., 1992; Pride and Berryman, 1998) have validated Biot's model of propagation in a poroelastic medium. The effects of the fluid enters these equations in several places. There are explicit coupling terms but there are also implicit couplings that enter through the frequency-dependent coefficients. The term ρ_k, given by Equation (7.17) is one example of such a coefficient. Additional frequency-dependent effects can enter through the other coefficients if we allow for processes such as small-scale patchy saturation and variations in the compressibility of the medium (White, 1975; Norris, 1993; Johnson, 2001; Pride et al., 2004) and squirt flow in microcracks (Mavko and Nur, 1975; O'Connell and Budiansky, 1977; Mavko and Nur, 1979; Dvorkin et al., 1995). Related to this is the heterogeneity of the porous material itself, due to variations in the properties of the grains, cementation, and other effects. The literature on this subject is extensive and the works by Berryman (1995) and Mavko et al. (1998) are helpful in understanding these effects. Broad overviews of the various complexities due to small-scale structure and flow are provided by the reviews of Pride (2005) and Müller et al. (2010).

One can solve the system of Equations (7.23) and (7.24) directly using numerical techniques such as finite differences (Masson et al., 2006; Masson and Pride, 2007, 2011), see Carcione et al. (2010) for a review. However, due to the very different nature of fluid flow and elastic deformation, there are processes that operate on very different time scales complicating numerical treatments. For example, as we shall see below, a poroelastic disturbance typically generates a fast wave that propagates elastically (see Figure 7.1) and a slow transient that may take minutes, hours, or even days to propagate. It can be difficult to bridge the gap between these time-scales using numerical approaches. That is one reason for turning to a semi-analytic treatment. Such methods also provide insight and flexibility, for example giving explicit expressions for the velocities, attenuation, and travel times of the various modes of propagation.

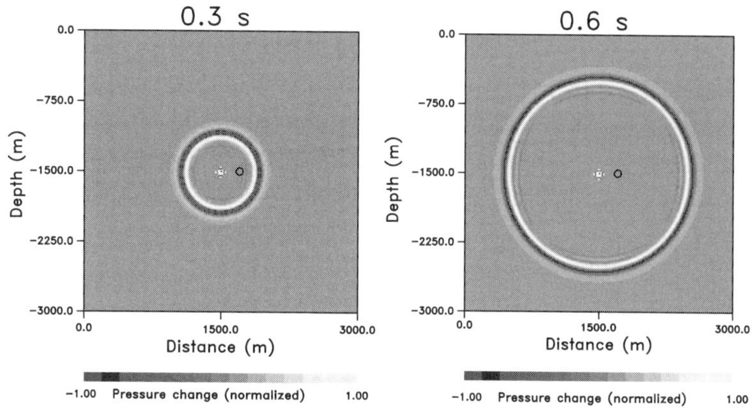

Figure 7.1 Fast elastic wave propagating outward from a fluid injection source at (1500 m, −1500 m).

7.2.2 A trajectory-based solution

We now turn our attention to the derivation of a trajectory-based, semi-analytic solution of Equations (7.23) and (7.24). By now the reader should be well versed in the asymptotic approach that we will adopt. However, as a refresher, we provide a brief motivation. The averaging techniques used to formulate the governing equations account for variations in properties at the mesoscopic length scale, of the order of a centimeter to a meter or so. We can represent this length scale by the variable m. In a typical experimental or field setting the mesoscopic scale is much smaller than the length scale of the propagating disturbance, which we denote by l. In addition to the small-scale variations, natural objects will also contain larger-scale spatial variations in properties as well as interfaces between regions with similar properties. We can treat the interfaces as boundaries and can introduce boundary conditions in order to match solutions in the various regions. It is the spatial variation in the properties of the medium, the heterogeneity, changing over a length scale of L, that we shall account for in developing an asymptotic solution. We assume that $L \gg l$, the heterogeneity is smoothly-varying with respect to the spatial variation the coupled front. It follows from this assumption that the ratio $\varepsilon = l/L$ is much smaller than one. We introduce the scaled spatial coordinates

$$\mathbf{X} = \varepsilon \mathbf{x}, \tag{7.25}$$

and consider the solution for $\mathbf{U}$ and $\mathbf{W}$ in the form of asymptotic power series in ε

$$\mathbf{U}(\mathbf{X}, \theta, \omega) = e^{i\theta} \sum_{l=0}^{\infty} \varepsilon^l \mathbf{U}_l(\mathbf{X}, \omega) \tag{7.26}$$

$$\mathbf{W}(\mathbf{X}, \theta, \omega) = e^{i\theta} \sum_{l=0}^{\infty} \varepsilon^l \mathbf{W}_l(\mathbf{X}, \omega) \tag{7.27}$$

where $\theta(\mathbf{x}, \omega)$ is a function, referred to as the phase, related to the travel time of the propagating disturbance and $\mathbf{U}_l$ and $\mathbf{W}_l$ are successive amplitude corrections. The partitioned or factored form, in which the phase occurs as exponential multiplier, is adopted because the governing equations are linear and we are working in the frequency domain. The functional forms (7.26) and (7.27) may be thought of as a plane wave expansion of a propagating disturbance. Because ε is assumed to be small, much less than 1, only the first few terms of these power series are likely to be significant.

As noted in Chapter 3, the trajectory-based approach belongs to the family of ray methods that have been used in many contexts. However, we do not restrict the technique to a particular class of equations, such as hyperbolic partial differential equations. Rather, we apply the approach to Equations (7.23) and (7.24) which may be of mixed character and contain solutions that display wave-like characteristics and solutions that are diffusive in nature (Pride, 2005). As shown in Chapter 4, if suitably generalized, one may derive an asymptotic solution to an equation governing a diffusive process. The power series representations (7.26) and (7.27) contain these generalizations, particularly with respect to the frequency-dependence of the terms in the expansion.

An asymptotic solution is obtained by writing the governing equations in terms of the coordinates $\mathbf{X}$, substituting in the asymptotic series solutions (7.26) and (7.27), and considering terms of successive order in ε. For smoothly-varying heterogeneity it is only necessary to consider the terms of low order in ε. We first need to rewrite Equations (7.23) and (7.24) in terms of $\mathbf{X}$, starting with the spatial derivatives. Using the chain rule, the derivatives may be re-written as

$$\frac{\partial \mathbf{U}}{\partial x_i} = \frac{\partial X_i}{\partial x_i}\frac{\partial \mathbf{U}}{\partial X_i} + \frac{\partial \theta}{\partial x_i}\frac{\partial \mathbf{U}}{\partial \theta}. \tag{7.28}$$

The differential operators, defined in terms of the partial derivatives with respect to the spatial coordinates, are likewise re-written as

$$\nabla \mathbf{U} = \varepsilon \overline{\nabla} \mathbf{U} + \nabla \theta \frac{\partial \mathbf{U}}{\partial \theta} \tag{7.29}$$

$$\nabla \cdot \mathbf{U} = \varepsilon \overline{\nabla} \cdot \mathbf{U} + \nabla \theta \cdot \frac{\partial \mathbf{U}}{\partial \theta} \tag{7.30}$$

where we have made use of the definition of $\mathbf{X}$, given by Equation (7.25), and $\overline{\nabla}$ denotes the gradient with respect to the components of the variable $\mathbf{X}$. Before converting all the differential operators in the governing equations, we can make a few simplifications. The first simplification is based upon the partitioned form of the series (7.26) and (7.27) with respect to the dependence on the phase θ, which leads us to conclude that

$$\frac{\partial \mathbf{U}}{\partial \theta} = i\mathbf{U} \tag{7.31}$$

$$\frac{\partial \mathbf{W}}{\partial \theta} = i\mathbf{W}. \tag{7.32}$$

Box 7.1 **Keeping the derivatives straight: dyadics**

One of the difficulties in treating the equations governing deformation and flow is the involved bookkeeping associated with the calculation of differential operators of vectors, particularly when second-order derivatives are present. The concept of a dyadic can be helpful in this regard, providing the systematics associated with matrix calculations as a tool for vector calculus. First we consider the representation of a vector, such as **U**, that can be written using classic vector notation:

$$\mathbf{U} = U_1\mathbf{i} + U_2\mathbf{j} + U_3\mathbf{k}.$$

A dyadic is similar in concept to the outer product of two vectors (Golub and Van Loan, 1989, p. 3). In the outer product elements such as **i** and **j** retain their identities rather then map into another element or into a scalar. The products, known as unit dyads, are written as **ij**. The order of the product matters and **ij** is not the same as **ji**. From the classic vector notation we can form a dyadic using simple algebra, making sure to keep track of the order of the products. For example, consider $\nabla\mathbf{U}$ which appeared earlier, we can define it as

$$\begin{aligned}
\nabla\mathbf{U} &= \left(\frac{\partial}{\partial x}\mathbf{i} + \frac{\partial}{\partial y}\mathbf{j} + \frac{\partial}{\partial z}\mathbf{k}\right)(U_1\mathbf{i} + U_2\mathbf{j} + U_3\mathbf{k}) \\
&= \frac{\partial U_1}{\partial x}\mathbf{ii} + \frac{\partial U_2}{\partial x}\mathbf{ij} + \frac{\partial U_3}{\partial x}\mathbf{ik} \\
&\quad + \frac{\partial U_1}{\partial y}\mathbf{ji} + \frac{\partial U_2}{\partial y}\mathbf{jj} + \frac{\partial U_3}{\partial y}\mathbf{jk} \\
&\quad + \frac{\partial U_1}{\partial z}\mathbf{ki} + \frac{\partial U_2}{\partial z}\mathbf{kj} + \frac{\partial U_3}{\partial z}\mathbf{kk}.
\end{aligned}$$

We can represent this quantity as a matrix if we let the first component denote the row of the matrix and we let the second component denote the column, leading to the array representation of $\nabla\mathbf{U}$, as given in Chapter 2. One can also formulate dyadics using a purely vector-matrix approach, and the idea of the outer product of two vectors:

$$\nabla\mathbf{U} = \begin{bmatrix} \partial/\partial x \\ \partial/\partial y \\ \partial/\partial z \end{bmatrix} \begin{bmatrix} U_1 & U_2 & U_3 \end{bmatrix}$$

$$= \begin{bmatrix} \partial U_1/\partial x & \partial U_2/\partial x & \partial U_3/\partial x \\ \partial U_1/\partial y & \partial U_2/\partial y & \partial U_3/\partial y \\ \partial U_1/\partial z & \partial U_2/\partial z & \partial U_3/\partial z \end{bmatrix}.$$

Since the same matrix arises, one can choose the approach best suited to the problem at hand.

The second simplification is notational, to denote the gradient of the phase by the vector $\mathbf{p} = \nabla\theta$. Thus, we can write the differential operators (7.29) and (7.30) as

$$\nabla\mathbf{U} = \varepsilon\overline{\nabla}\mathbf{U} + i\mathbf{p}\mathbf{U} \tag{7.33}$$

$$\nabla \cdot \mathbf{U} = \varepsilon\overline{\nabla} \cdot \mathbf{U} + i\mathbf{p} \cdot \mathbf{U} \tag{7.34}$$

and similarly for the other differential operators in the governing equations. Higher-order derivatives are obtained by the repeated application of the first-order operators. Equation (7.33) is actually a matrix of derivatives often referred to as a dyadic, a generalization of a vector. We made a passing reference to dyadic's in Chapter 2, when discussing the matrices of derivatives $\nabla\mathbf{v}$ and $\nabla\mathbf{u}$. However, confronted with calculations involving dyadics, it is now time to explore them in somewhat more depth, which we do in Box 7.1.

As an illustration of the computations involved, we shall consider the first of the governing Equations (7.23) explicitly. Substituting for the differential operators, using Equations (7.33) and (7.34) results in the equation

$$\begin{aligned} -\omega^2\rho_f\mathbf{U} - \omega^2\rho_k\mathbf{W} = {} & \varepsilon\overline{\nabla}\left[C\left(\varepsilon\overline{\nabla}\cdot\mathbf{U} + i\mathbf{p}\cdot\mathbf{U}\right) + M\left(\varepsilon\overline{\nabla}\cdot\mathbf{W} + i\mathbf{p}\cdot\mathbf{W}\right)\right] \\ & + i\mathbf{p}\left[C\left(\varepsilon\overline{\nabla}\cdot\mathbf{U} + i\mathbf{p}\cdot\mathbf{U}\right) + M\left(\varepsilon\overline{\nabla}\cdot\mathbf{W} + i\mathbf{p}\cdot\mathbf{W}\right)\right] \end{aligned}$$

Because we are assuming smoothly-varying properties between interfaces, ε is small and we shall neglect terms that must be of order ε^2 and greater. We are left with the expression

$$\begin{aligned} & \omega^2\rho_f\mathbf{U} + \omega^2\rho_k\mathbf{W} - C\mathbf{p}\mathbf{p}\cdot\mathbf{U} - M\mathbf{p}\mathbf{p}\cdot\mathbf{W} \\ & \quad + i\varepsilon\left[\overline{\nabla}\left(C\mathbf{p}\cdot\mathbf{U}\right) + C\mathbf{p}\overline{\nabla}\cdot\mathbf{U} + \overline{\nabla}\left(M\mathbf{p}\cdot\mathbf{W}\right) + M\mathbf{p}\overline{\nabla}\cdot\mathbf{W}\right] = 0. \end{aligned} \tag{7.35}$$

An application of the approach to the second governing Equation (7.24) results in a similar expression as indicated in the exercise below. Due to the presence of the deviatoric stress tensor $\boldsymbol{\tau}$, the second governing equation is more complicated. If we neglect terms that must be of order ε^2 or greater in the second governing equation, then we are left with

$$\omega^2 \rho_s \mathbf{U} + \omega^2 \rho_f \mathbf{W} - K_u \mathbf{pp} \cdot \mathbf{U} - C\mathbf{pp} \cdot \mathbf{W} - G\mathbf{p} \cdot \left[\mathbf{pU} + (\mathbf{pU})^T - \frac{2}{3}\mathbf{p} \cdot \mathbf{UI} \right]$$
$$+ i\varepsilon \left[\overline{\nabla} (K_u \mathbf{p} \cdot \mathbf{U}) + K_u \mathbf{p} \overline{\nabla} \cdot \mathbf{U} + \overline{\nabla} (C\mathbf{p} \cdot \mathbf{W}) + C\mathbf{p} \overline{\nabla} \cdot \mathbf{W} \right]$$
$$+ i\varepsilon \overline{\nabla} \cdot G \left[\mathbf{pU} + (\mathbf{pU})^T - \frac{2}{3}\mathbf{p} \cdot \mathbf{UI} \right]$$
$$+ i\varepsilon G\mathbf{p} \cdot \left[\overline{\nabla}\mathbf{U} + \left(\overline{\nabla}\mathbf{U} \right)^T - \frac{2}{3} \overline{\nabla} \cdot \mathbf{UI} \right] = 0. \tag{7.36}$$

The two governing Equations (7.35) and (7.36) appear rather formidable, even after neglecting terms of order ε^2 and greater. However, as we continue some simplifications will ensue. We first consider terms of the lowest order in ε, those of order $\varepsilon^0 \sim 1$. As we shall see, these terms will provide an equation for $p = |\mathbf{p}|$, allowing us to determine the phase $\theta(\mathbf{x}, \omega)$ in terms of the properties of the medium and the fluid. The kinematics of the propagating disturbance, the velocity and the travel time, as well as the attenuation, spring from these expressions.

Exercise 7.1. Derive asymptotic expression, given by Equation (7.36), for the the second governing Equation (7.24),

$$-\omega^2 \rho_s \mathbf{U} - \omega^2 \rho_f \mathbf{W} = \nabla (K_u \nabla \cdot \mathbf{U} + C \nabla \cdot \mathbf{W}) + \nabla \cdot \boldsymbol{\tau}$$

where $\boldsymbol{\tau}$ is given by

$$\boldsymbol{\tau} = G \left[\nabla\mathbf{U} + \nabla\mathbf{U}^T - \frac{2}{3} \nabla \cdot \mathbf{UI} \right].$$

7.2.3 Terms of order $\varepsilon^0 \sim 1$: an equation for the phase

Because $\varepsilon \ll 1$ the terms of lowest order in the asymptotic expansion are most important. Here we consider terms of zeroth-order in Equations (7.35) and (7.36). In order to do this we substitute the power series (7.26) and (7.27) into these equations and consider the zeroth-order terms. After factoring out $e^{i\theta}$ we can write the resulting expressions as a system of linear equations for the zeroth-order displacements $\mathbf{U}_0$ and $\mathbf{W}_0$.

$$\begin{bmatrix} \omega^2 \rho_f \mathbf{I} - C\mathbf{pp} \cdot \mathbf{I} & \omega^2 \rho_k \mathbf{I} - M\mathbf{pp} \cdot \mathbf{I} \\ \alpha \mathbf{I} - \beta \mathbf{pp} \cdot \mathbf{I} & \omega^2 \rho_f \mathbf{I} - C\mathbf{pp} \cdot \mathbf{I} \end{bmatrix} \begin{bmatrix} \mathbf{U}_0 \\ \mathbf{W}_0 \end{bmatrix} = \begin{bmatrix} \mathbf{0} \\ \mathbf{0} \end{bmatrix}, \tag{7.37}$$

where

$$\alpha = \omega^2 \rho - Gp^2, \tag{7.38}$$

$p^2 = \mathbf{p} \cdot \mathbf{p}$ and

$$\beta = K_u + \frac{1}{3} G. \tag{7.39}$$

A well-known theorem in linear algebra states that the system of Equations (7.37) has a non-trivial (non-zero) solution if the determinant of the coefficient matrix vanishes (Noble and Daniel, 1977, p. 203). Therefore, if Γ is the coefficient matrix of the system of Equations (7.37), then the requirement that there is a non-trivial solution is

$$\det \Gamma = \mathbf{0}. \tag{7.40}$$

This **characteristic equation** is a polynomial in the components of $\mathbf{p}$, with coefficients that depend upon the properties of the poroelastic medium and the fluid. Remember that $\mathbf{p} = \nabla \theta$, meaning that Equation (7.40) is actually a partial differential equation for $\theta(\mathbf{x}, \omega)$, a key function in our asymptotic solution. Furthermore, this function is directly related to the propagation velocity of disturbances within the poroelastic medium and hence the travel time of waves through the medium.

While it is possible to compute the determinant of the 6×6 matrix Γ directly, such an approach involves considerable algebra. Alternatively, one can use the fact that the determinant is given by the product of the eigenvalues of the coefficient matrix (Noble and Daniel, 1977, p. 264). Let us take a moment to recall the eigenvectors and eigenvalues associated with a matrix Γ. An eigenvector of a matrix, denoted by $\mathbf{e}$, is a vector that is not changed in direction when multiplied by the matrix Γ (Noble and Daniel, 1977, p. 257), rather it is rescaled by λ, the eigenvalue:

$$\Gamma \mathbf{e} = \lambda \mathbf{e}. \tag{7.41}$$

From this equation one can see how the vanishing of an eigenvalue of the matrix, for a non-zero eigenvector $\mathbf{e}$, is equivalent to Equation (7.37). We quickly run into a stumbling block if we pursue an approach based upon eigenvectors and eigenvalues. The eigenvalue problem associated with the coefficient matrix Γ leads to the same level of algebra as the computation of the determinant. In order to make headway we need to incorporate additional information in formulating our problem. We can find at least two sources of inspiration, one mathematical and one physical. Both considerations relate to the nature of the eigenvectors associated with the coefficient matrix in Equation (7.37).

Mathematically, we note that the coefficient matrix in Equation (7.37) partitions into a 2 by 2 block matrix, where each block is composed of a 3×3 sub-matrix.

Each sub-matrix is similar in structure: a scalar multiple of the identity matrix, $\mathbf{I}$, added to a scalar multiple of $\mathbf{pp} \cdot \mathbf{I}$, which is equivalent to the dyadic $\mathbf{pp}^T$. Any vector multiplied by the first factor will return a scalar multiple of itself while any vector multiplied by the second factor will return a scalar multiple of $\mathbf{p}$. This suggests that $\mathbf{p}$ is a potential eigenvector because multiplying it by both factors returns a scalar multiple of itself. In addition, any vector orthogonal to $\mathbf{p}$ will also return a scalar multiple of itself because it will simply eliminate the second factor, the dyadic $\mathbf{pp}^T$. This suggests that candidate eigenvectors are of the form

$$\mathbf{e}^l = \begin{bmatrix} y_1\mathbf{p} \\ y_2\mathbf{p} \end{bmatrix} \tag{7.42}$$

or

$$\mathbf{e}^\perp = \begin{bmatrix} s_1\mathbf{p}^\perp \\ s_2\mathbf{p}^\perp \end{bmatrix} \tag{7.43}$$

where $\mathbf{p}^\perp$ is a vector orthogonal to $\mathbf{p}$ and y_1, y_2, s_1, and s_2 are scalars. Note that the eigenvector $\mathbf{p}^\perp$, lying in the plane perpendicular to $\mathbf{p}$, is not unique. In fact $\mathbf{p}^\perp$ may lie anywhere in a two dimensional subspace spanned by two linear independent vectors perpendicular to $\mathbf{p}$. For now we shall ignore this ambiguity, as it will not influence the discussion that follows.

Physical considerations align quite nicely with the preceding mathematical analysis. In particular, in a homogeneous medium one can show that the vectors $\mathbf{U}_0$ and $\mathbf{W}_0$ must be in the same directions for a non-trivial solution to exist (Pride, 2005). Thus, the solution vectors are of the same general structure as (7.42) and (7.43). Using potentials, one can show that elastic waves in a homogeneous medium decouple into a longitudinal mode, in which the displacement is in the direction of propagation $\mathbf{p}$, and transverse modes with the displacement transverse to the direction of propagation, in a direction $\mathbf{p}^\perp$ (Aki and Richards, 1980b, p. 89). A similar decomposition into longitudinal and transverse modes is also meaningful for disturbances propagating in a homogeneous poroelastic medium containing a fluid (Pride, 2005). We now consider the two modes, longitudinal and transverse, in succession as they have rather different characteristics.

The longitudinal mode

Here we consider the longitudinal mode in which the solid and fluid displacements $\mathbf{U}_0$ and $\mathbf{W}_0$ are in the direction of propagation $\mathbf{p}$. For this mode the eigenvector is given by $\mathbf{e}^l$, defined in Equation (7.42) above, and for a vanishing eigenvalue, Equation (7.41) reduces to

$$\Gamma\mathbf{e}^l = \mathbf{0}. \tag{7.44}$$

For the specific form of $\mathbf{e}^l$, given by Equation (7.42), Equation (7.44) is equivalent to

$$\begin{bmatrix} (\omega^2\rho_f - Cp^2)\mathbf{I} & (\omega^2\rho_k - Mp^2)\mathbf{I} \\ (\alpha - \beta p^2)\mathbf{I} & (\omega^2\rho_f - Cp^2)\mathbf{I} \end{bmatrix} \begin{bmatrix} y_1\mathbf{p} \\ y_2\mathbf{p} \end{bmatrix} = \begin{bmatrix} \mathbf{0} \\ \mathbf{0} \end{bmatrix}. \tag{7.45}$$

As before, the condition that Equation (7.45) has a non-trivial solution is that the determinant of the coefficient matrix vanishes. So, we are still confronted with the calculation of the determinant of a 6×6 matrix. However, we can now take advantage of the structure of the coefficient matrix to reduce this computation to the determinant of a 2×2 matrix. Specifically, the coefficient matrix in Equation (7.45) is in the form of a tensor product of two matrices

$$\mathbf{M} \otimes \mathbf{I} = \begin{bmatrix} m_{11}\mathbf{I} & m_{12}\mathbf{I} \\ m_{21}\mathbf{I} & m_{22}\mathbf{I} \end{bmatrix}, \tag{7.46}$$

where m_{ij} are the elements of the matrix $\mathbf{M}$:

$$\mathbf{M} = \begin{bmatrix} (\omega^2\rho_f - Cp^2) & (\omega^2\rho_k - Mp^2) \\ (\alpha - \beta p^2) & (\omega^2\rho_f - Cp^2) \end{bmatrix}. \tag{7.47}$$

We can use following theorem for the determinant of a matrix composed of block submatrices (Silvester, 2000), in particular the determinant of the tensor product $\det(\mathbf{M} \otimes \mathbf{I})$:

$$\det(\mathbf{M} \otimes \mathbf{I}) = \det(\mathbf{M}) \det(\mathbf{I})^2 \tag{7.48}$$

where $\det(\mathbf{I})$ is squared because $\mathbf{M}$ is a 2×2 matrix. Because the determinant of $\mathbf{I}$ is just 1, we have that the vanishing of the determinant of the coefficient matrix in Equation (7.45) is equivalent to the condition:

$$\det \begin{bmatrix} (\omega^2\rho_f - Cp^2) & (\omega^2\rho_k - Mp^2) \\ (\omega^2\rho_s - Hp^2) & (\omega^2\rho_f - Cp^2) \end{bmatrix} = 0, \tag{7.49}$$

where we have made use of the definitions of α and β, Equations (7.38) and (7.39), and we have defined the undrained compressional modulus H as

$$H = K_u + \frac{4}{3}G. \tag{7.50}$$

The vanishing of the determinant, Equation (7.48), produces a quadratic equation for p^2

$$(p^2)^2 - \frac{\omega^2(\rho_k H + \rho_s M - 2\rho_f C)}{(HM - C^2)}p^2 + \frac{\omega^4(\rho_s\rho_k - {\rho_f}^2)}{(HM - C^2)} = 0 \tag{7.51}$$

with the two solutions indicating that there are two longitudinal modes. These two modes, known as the fast and slow waves, are a hallmark of poroelasticity. The fast longitudinal solution is the poroelastic equivalent of the elastic compressional wave, familiar to seismologists. As we shall see below, the slow wave is akin to a propagating pressure disturbance at sufficiently low frequencies, familiar to hydrologist and reservoir engineers. The two solutions are given by the explicit expression

$$p^2 = \frac{\gamma\omega^2}{2}\left[1 \pm \sqrt{1 - \frac{4}{\gamma^2}\frac{\left(\rho_s\rho_k - \rho_f{}^2\right)}{\left(HM - C^2\right)}}\right], \tag{7.52}$$

where the auxiliary parameter γ is

$$\gamma = \frac{\rho_k H + \rho_s M - 2\rho_f C}{HM - C^2}. \tag{7.53}$$

The solutions given by Equation (7.52) are generally complex, with non-zero imaginary components. In Figures 7.2 and 7.3 we plot the real and imaginary components for both the fast and slow wave solutions. Rather than plotting slowness, we plot its inverse, the speed of the disturbance. The spatial variations represent a simple example of a linear increase in wave speed with depth. For the fast longitudinal mode the imaginary component is usually small. The imaginary component increases if there is mesoscopic flow between unequilibrated patches of differing fluid saturations or regions with different compressibilities (White, 1975; Johnson, 2001; Pride et al., 2004).

Equation (7.52) is similar to an expression for the slowness of the longitudinal mode given by Pride (2005) for a plane wave in a homogeneous medium. However, the expression is valid for a medium with smoothly varying heterogeneity of arbitrary magnitude. Because $p^2 = \mathbf{p} \cdot \mathbf{p} = \nabla\theta \cdot \nabla\theta$, Equation (7.52) is a

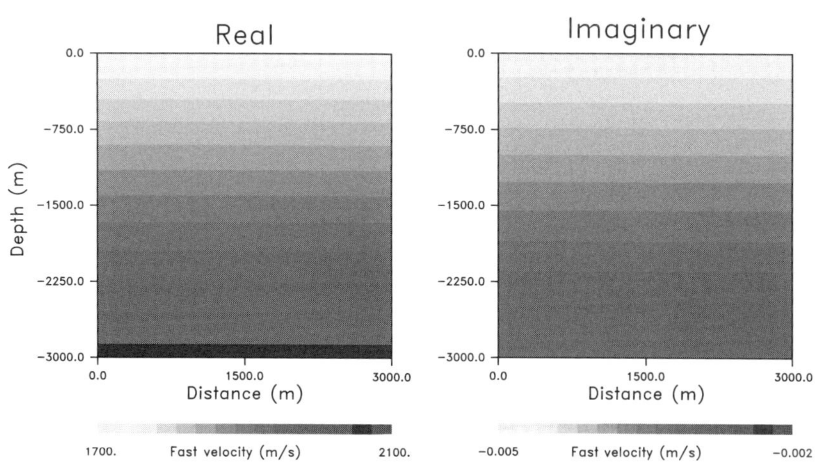

Figure 7.2 Real and imaginary components of the fast wave speed.

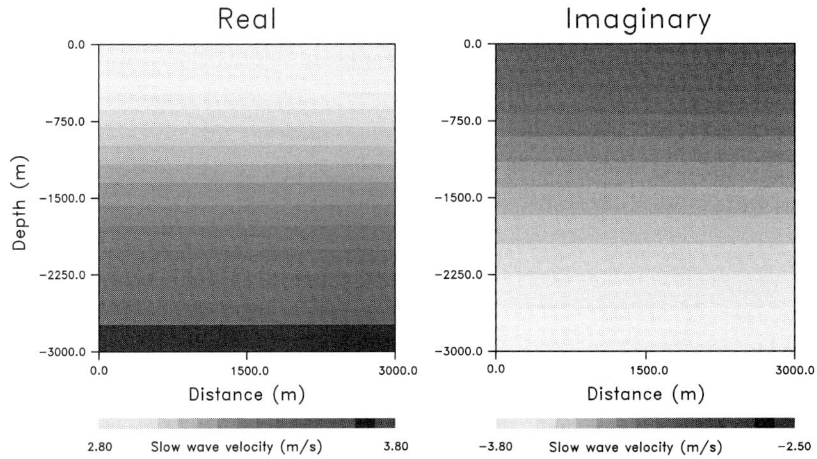

Figure 7.3 Real and imaginary components of the slow wave speed.

variant of the eikonal equation, a partial differential for the phase function $\theta(\mathbf{x}, \omega)$. The frequency-dependence is determined both explicitly, by the presence of the ω^2 term, and implicitly by the frequency-dependence of ρ_k which contains the dynamic permeability, $k(\omega)$, as well as ω, in its definition [see Equation (7.17)]. In fact, because we are not making any assumptions about the magnitude of the frequency, for a sufficiently smooth medium the results will hold even if the other moduli, such as H, M, and C, are frequency-dependent and complex.

The eikonal Equation (7.52), an example of a Hamilton–Jacobi equation, is our gateway to a trajectory-based solution. That is, we can use the ray equations from Box 3.1 in Chapter 3 to construct a solution that is defined along a trajectory through the model. Therefore, we begin with Equation (7.52) written in a form reminiscent of the results in Chapter 3,

$$F(\mathbf{x}, \mathbf{p}) = \mathbf{p} \cdot \mathbf{p} - \chi(\mathbf{x}, \omega) = 0, \tag{7.54}$$

where we have defined the function

$$\chi(\mathbf{x}, \omega) = \omega^2 \frac{\gamma}{2} \left[1 \pm \sqrt{1 - \zeta}\right] \tag{7.55}$$

which is just the right-hand-side of Equation (7.52). Note that χ can vary spatially and with frequency but does not depend upon $\mathbf{p}$. The variable ζ, contained within the square root in Equation (7.55), will have a strong influence on the longitudinal modes of propagation. An explicit expression for ζ is obtained by combining the variables under the square root in Equation (7.52) with the definition (7.53) of γ,

$$\zeta = \frac{4(\rho_s \rho_k - \rho_f{}^2)(HM - C^2)}{(\rho_k H + \rho_s M - 2\rho_f C)^2}. \tag{7.56}$$

Before we calculate the trajectories, let us make one observation regarding the parameters and coefficients in the equations, particularly those present in Equation (7.54). In the last few pages we have introduced a few quantities, such as γ and ζ, in order to reduce the complexity of some of the expressions. However, in all of the equations there are just three fundamental poroelastic constants, H, M, and C, and three densities, ρ_s, ρ_f, and ρ_k, that are the bedrock of the theory and each of the variables that we have introduced depends upon these quantities in some fashion.

Let us return to the trajectory calculation. Based upon the first two ray equations in Box 3.1 in Chapter 3, with $F(\mathbf{x}, \mathbf{p})$ given in Equation (7.54), we have

$$\frac{d\mathbf{x}}{ds} = 2\mathbf{p} \tag{7.57}$$

$$\frac{d\mathbf{p}}{ds} = \nabla\chi. \tag{7.58}$$

Because the function $F(\mathbf{x}, \mathbf{p})$ does not contain the dependent variable $\theta(\mathbf{x}, \omega)$ explicitly, these two coupled ordinary differential equations are sufficient to determine the trajectory $\mathbf{x}(s)$, given the frequency and the parameters describing both the poroelastic medium and the fluid. With a sufficient set of boundary conditions, such as a starting location for the trajectory and a take-off angle, or an initial value of $\mathbf{p}$ at $s = 0$, one can use numerical techniques for integrating the ordinary differential equations (Press et al., 1992) to determine $\mathbf{x}(s)$ and $\mathbf{p}(s)$. In Figures 7.4 and 7.5 we plot the ray paths from a source to five receivers for the fast and slow waves. The trajectories were computed using the ray Equations (7.57) and (7.58) for the complex velocity models plotted in Figures 7.2 and 7.3.

With the final ray equation from Box 3.1 we can derive an expression for the phase. We shall assume that the first two ray Equations (7.57) and (7.58) have been solved and that the trajectory is known. The final ray equation is

$$\begin{aligned}\frac{d\theta}{ds} &= \mathbf{p} \cdot \nabla_p F \\ &= 2\mathbf{p} \cdot \mathbf{p} \\ &= 2\chi\end{aligned} \tag{7.59}$$

if we make use of the functional form of $F(\mathbf{x}, \mathbf{p})$ in Equation (7.54). Using Equation (7.57) to substitute for $\mathbf{p}$ in the last equality of Equation (7.59) gives

$$\frac{dx}{ds} = 2\sqrt{\chi}. \tag{7.60}$$

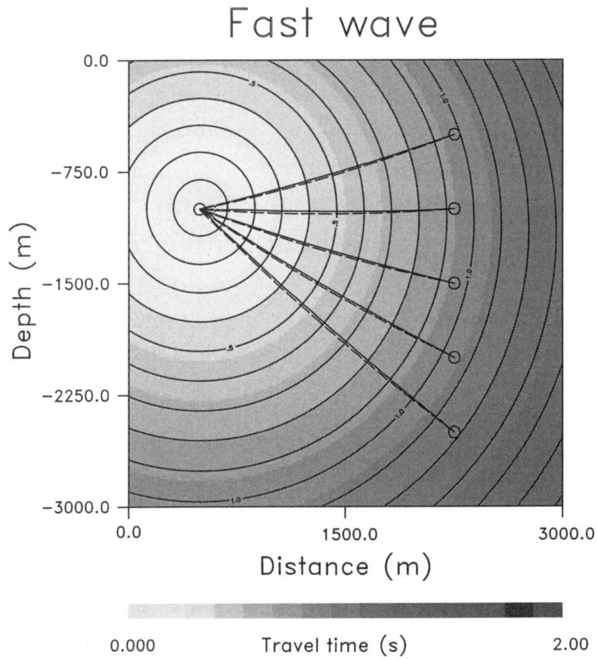

Figure 7.4 Trajectories associated with the fast wave calculated by solving the ray equations (solid) and by marching down the travel time gradient (dashed). The travel time contours were obtained by solving the eikonal equation.

Integrating Equation (7.59) from a known source location, $\mathbf{x}_s$, to a given observation point, $\mathbf{x}_o$, along the trajectory gives

$$\theta(\mathbf{x}_o, \omega) = \int_{\mathbf{x}_s}^{\mathbf{x}_o} 2\chi ds, \tag{7.61}$$

or, upon using Equation (7.60) to change the integration variable,

$$\theta(\mathbf{x}_o, \omega) = \int_{\mathbf{x}_s}^{\mathbf{x}_o} \sqrt{\chi} dx. \tag{7.62}$$

Equation (7.62) gives a closed-form, semi-analytic expression for the phase in terms of the parameters of the fluid and the porous medium at a particular frequency ω.

Instead of using the ray equations it is also possible to solve the Hamilton–Jacobi Equation (7.54) using a purely numerical approach such as finite differences (Sethian, 1999; Osher and Fedkiw, 2003). There are several advantages associated with these techniques, such as the fact that the solution is defined over the entire region of interest. One impediment to using the most efficient approaches (Sethian, 1999; Osher and Fedkiw, 2003) is the fact that the phase is generally a complex

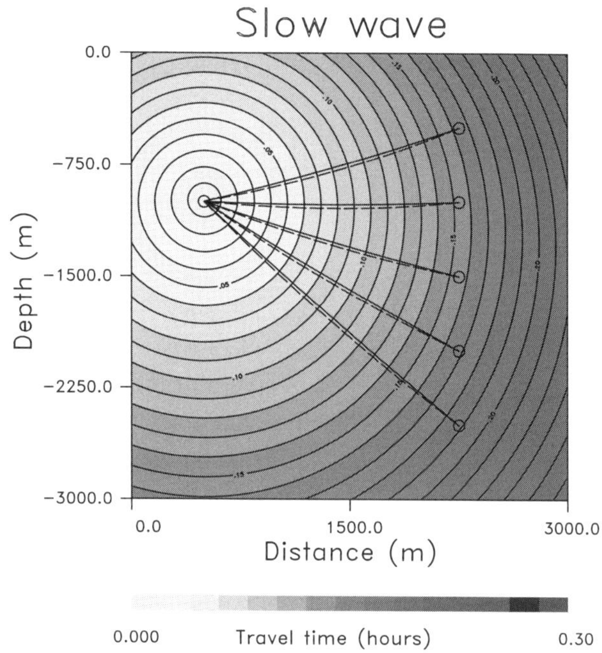

Figure 7.5 Trajectories associated with the slow wave calculated by solving the ray equations (solid) and by marching down the travel time gradient (dashed).

quantity and simple Fast Marching Methods cannot be applied directly. However, other approaches may be tried and it may be possible to extend techniques such as fast marching to partial differential equations involving a complex quantity. This is an area that would benefit from additional study. In Figures 7.4 and 7.5 we plot the travel time fields obtained using numerical solutions of the eikonal equations for the fast and slow waves. For these solutions we used the real components of the velocity models in Figure 7.2 and 7.3, neglecting the contributions of the imaginary parts. The trajectories obtained by marching down gradient of the travel time fields are also plotted in these figures.

The transverse mode

The displacement direction for the transverse mode, characterized by the vector $\mathbf{p}^{\perp}$ [see Equation (7.43)], is perpendicular to the propagation direction of the wavefront, given by $\mathbf{p}$. We have already noted that vectors perpendicular to $\mathbf{p}$ form a two-dimensional subspace lying in the plane of the wavefront. Therefore, a complete specification of the subspace requires two linear independent vectors. This requirement does not introduce any real difficulty because, having the vector $\mathbf{e}^{l}$ and another vector $\mathbf{e}^{\perp}$ perpendicular to it, it is relatively easy to calculate a third vector that is perpendicular to both of them using the vector cross product: $\mathbf{e}^{l} \times \mathbf{e}^{\perp}$.

Therefore, we shall only focus on determining a single eigenvector $\mathbf{e}^{\perp}$ associated with the transverse mode of propagation, satisfying

$$\Gamma \mathbf{e}^{\perp} = \mathbf{0}. \tag{7.63}$$

For $\mathbf{e}^{\perp}$ given above and the coefficient matrix Γ in Equation (7.37), this eigenvalue problem is equivalent to

$$\begin{bmatrix} \omega^2 \rho_f \mathbf{I} & \omega^2 \rho_k \mathbf{I} \\ \left(\omega^2 \rho_s - Gp^2\right) \mathbf{I} & \omega^2 \rho_f \mathbf{I} \end{bmatrix} \begin{bmatrix} s_1 \mathbf{p}^{\perp} \\ s_2 \mathbf{p}^{\perp} \end{bmatrix} = \begin{bmatrix} \mathbf{0} \\ \mathbf{0} \end{bmatrix} \tag{7.64}$$

because the terms containing $\mathbf{pp} \cdot \mathbf{I}$ vanish upon multiplication by $\mathbf{p}^{\perp}$. Again, there is a non-trivial solution to these equations if the determinant of the coefficient matrix vanishes. Using the same argument made in the preceding section, regarding the determinant of a tensor product of two matrices $\mathbf{M} \otimes \mathbf{I}$, one can show that the requirement that the determinant vanish reduces to

$$\det \begin{bmatrix} \omega^2 \rho_f & \omega^2 \rho_k \\ \left(\omega^2 \rho_s - Gp^2\right) & \omega^2 \rho_f \end{bmatrix} = 0, \tag{7.65}$$

a linear equation in p^2

$$\omega^4 \rho_f^2 - \left(\omega^2 \rho_s - Gp^2\right) \omega^2 \rho_k = 0 \tag{7.66}$$

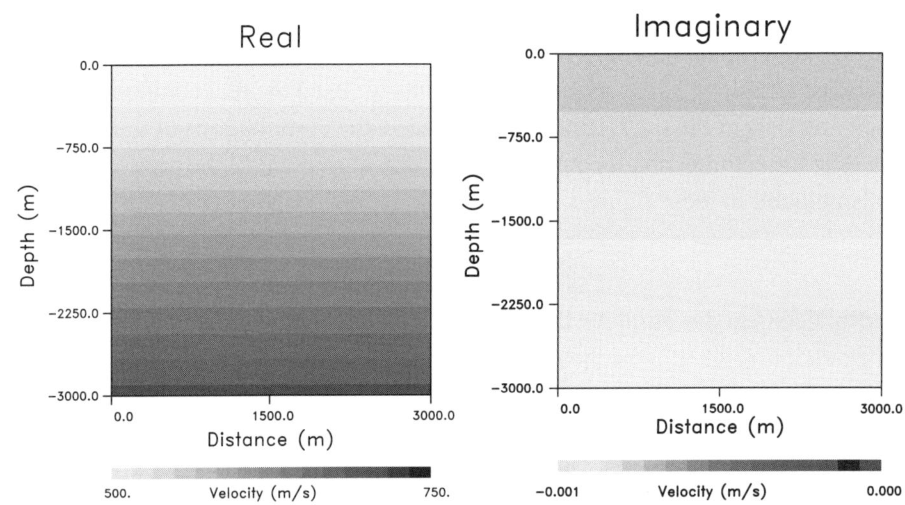

Figure 7.6 Real and imaginary velocities associated with the transverse mode of propagation in a porous solid.

with just one solution, given by

$$p^2 = \omega^2 \frac{\left[\rho_s - \left(\frac{\rho_f}{\rho_k}\right)\rho_f\right]}{G}. \tag{7.67}$$

In general, the solutions of Equation (7.67) are complex due to the presence of ρ_k. In Figure 7.5 we plot the real and imaginary components of the velocity of the transverse mode, which corresponds to the shear wave in an elastic medium. As with the fast longitudinal wave, the imaginary component is often small at seismic frequencies if there is no mesoscopic flow (Pride et al., 2004).

Equation (7.67) is an eikonal equation for the transverse mode and we may write it in the form of a function that depends upon $\mathbf{x}$ and $\mathbf{p}$

$$F(\mathbf{x}, \mathbf{p}) = \mathbf{p} \cdot \mathbf{p} - \omega^2 \frac{\left[\rho_s - \left(\frac{\rho_f}{\rho_k}\right)\rho_f\right]}{G} \tag{7.68}$$

as we did earlier. Note the implicit dependence upon $\mathbf{x}$ through the spatial variations of the medium properties G, ρ_s, and ρ_k. From this functional form springs the ray equations leading to the defining equations for the trajectories and to a semi-analytic expression for the phase. We will leave those calculations as an exercise for the

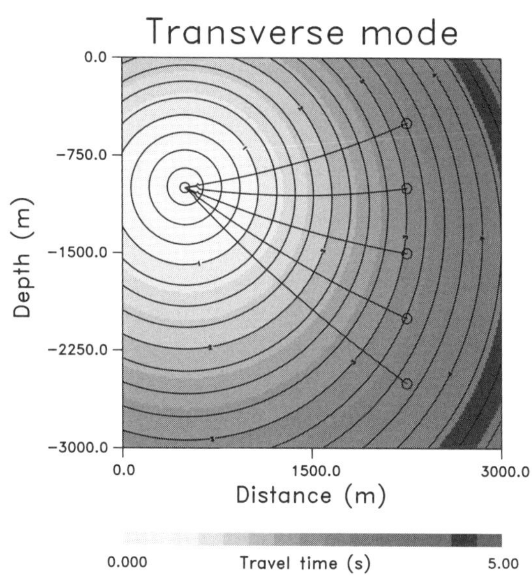

Figure 7.7 Travel time contours and trajectories associated with the transverse mode of propagation in the model plotted in Figure 7.6.

reader. In Figure 7.7 we plot the travel time contours obtained by solving the eikonal equation numerically with the imaginary component of the velocity set to zero.

Exercise 7.2. Starting from the form (7.68) use the ray equations, given in Box 3.1, to derive governing equations for the trajectory of the transverse mode and to find a semi-analytic expression for the phase.

Terms of order ε and Amplitudes

The first $[l = 0]$ terms in the series (7.26) and (7.27) form the zeroth-order approximation

$$\mathbf{U}(\mathbf{X}, \omega) = e^{i\theta}\mathbf{U}_0(\mathbf{X}, \omega) \tag{7.69}$$

$$\mathbf{W}(\mathbf{X}, \omega) = e^{i\theta}\mathbf{W}_0(\mathbf{X}, \omega). \tag{7.70}$$

For this level of approximation we need the functions $\theta(\mathbf{x}, \omega)$, $\mathbf{U}_0(\mathbf{x}, \omega)$, and $\mathbf{W}_0(\mathbf{x}, \omega)$. We have just seen how to find $\theta(\mathbf{x}, \omega)$ by considering terms of order $\varepsilon^0 \sim 1$. Equations defining the functions $\mathbf{U}_0(\mathbf{x}, \omega)$ and $\mathbf{W}_0(\mathbf{x}, \omega)$ follow if we consider terms of the next higher order in ε. We ask the reader to derive these equations in the exercise below. It is assumed that $\mathbf{U}_1$ and $\mathbf{W}_1$ retain their identity as longitudinal or transverse modes and thus satisfy the same zeroth-order condition (7.37) as $\mathbf{U}_0$ and $\mathbf{W}_0$. Therefore, the terms in the top line of Equation (7.35) vanish. As with the phase, the zeroth-order amplitude functions may also be identified with a specific mode of propagation. Hence, we can represent them in terms of scalar amplitude functions, $U_0(\mathbf{X}, \omega)$ and $W_0(\mathbf{X}, \omega)$, and the vector $\mathbf{p}$ or $\mathbf{p}^{\perp}$, depending upon the mode of propagation. Consider the longitudinal mode as an example. Here, we can write the zeroth-order amplitudes as

$$\mathbf{U}_0(\mathbf{X}, \omega) = U_0(\mathbf{X}, \omega)\mathbf{p} \tag{7.71}$$

$$\mathbf{W}_0(\mathbf{X}, \omega) = W_0(\mathbf{X}, \omega)\mathbf{p} \tag{7.72}$$

and substitute these representations into Equation (7.35). Finally, we can project Equation (7.35) onto the longitudinal vector $\mathbf{p}$ to produce a single linear equation for the two amplitude functions U_0 and W_0:

$$\mathbf{p} \cdot \overline{\nabla}\left(Cp^2 U_0\right) + Cp^2 \overline{\nabla} \cdot (U_0 \mathbf{p}) + \mathbf{p} \cdot \overline{\nabla}\left(Mp^2 W_0\right) + Mp^2 \overline{\nabla} \cdot (W_0 \mathbf{p}) = 0. \tag{7.73}$$

An additional constraint on U_0 and W_0 is obtained from the second governing Equation (7.36) using a similar analysis. Because these equations are linear they may be solved using numerical methods, either directly or by converting the equations to a system of ordinary differential equations. Note that the equations require that $\mathbf{p}$, and hence θ, be known, either throughout the region of interest, or along the trajectory connecting the observation point and the source point.

Exercise 7.3. Find the terms of order ε, the next level of approximation in the asymptotic expansion. Consider the first governing Equation (7.35) if we substitute the series expansions (7.26) and (7.27) and retain terms of order ε.

$$\omega^2\rho_f\mathbf{U}_1 + \omega^2\rho_k\mathbf{W}_1 - C\mathbf{pp}\cdot\mathbf{U}_1 - M\mathbf{pp}\cdot\mathbf{W}_1 \\ + i\left[\overline{\nabla}\left(C\mathbf{p}\cdot\mathbf{U}_0\right) + C\mathbf{p}\overline{\nabla}\cdot\mathbf{U}_0 + \overline{\nabla}\left(M\mathbf{p}\cdot\mathbf{W}_0\right) + M\mathbf{p}\overline{\nabla}\cdot\mathbf{W}_0\right] = 0.$$

7.2.4 The nature of the slow and fast longitudinal modes at low frequencies

To gain some insight into the nature of the two longitudinal modes of propagation, the fast and slow waves, we can consider the semi-analytic expressions in the limit of low frequencies. In particular, let us examine the zeroth-order expression

$$\mathbf{U}(\mathbf{X},\omega,\theta) = e^{i\theta(\mathbf{x},\omega)}\mathbf{U}_0(\mathbf{X},\omega). \tag{7.74}$$

To some extent the limiting process will constrain the heterogeneity to be increasingly smooth in order that ε remain a small number. In the extreme case we may require the medium to be homogeneous. This is not really an issue because we are simply trying to gain some insight into the physical nature of the fast and slow modes. Equation (7.62) provides an explicit expression for the phase in terms of $\chi(\mathbf{x},\omega)$. Using the definition of $\chi(\mathbf{x},\omega)$ provided by Equation (7.55) we can write the phase function as an integral over the trajectory $\mathbf{x}(s)$,

$$\theta(\mathbf{x},\omega) = \omega\int_{\mathbf{x}}\sqrt{\frac{\gamma}{2}\left[1\pm\sqrt{1-\zeta}\right]}dx. \tag{7.75}$$

The integral representation $\chi(\mathbf{x},\omega)$ contains an explicit ω frequency dependence as well as an implicit dependence through the two parameters γ and ζ given by Equations (7.53) and (7.56), respectively. If the presence of ω in the expressions for γ and ζ is still not evident, it is because it is hidden within

$$\rho_k = \rho_f - \frac{\mu_f}{\omega k(\omega)}i. \tag{7.76}$$

It is this quantity that will provide the starting point for our low-frequency exploration. The frequency dependence of $k(\omega)$ could present a complication in general. However, because we are interested in the low-frequency limit, we can use the results in Box 2.8 to note that $k(\omega)$ approaches a constant k_o as the frequency approaches small values. As noted by Pride (2005), for most applications in the Earth sciences we are operating in the 'low-frequency' regime. By way of clarification, in the limit we will be taking the frequency to be a small number, denoted by ϵ, but not down to zero. Thus, we will have

$$\lim_{\omega \to \epsilon} \rho_k \approx -\frac{\mu_f}{\omega k_o} i, \tag{7.77}$$

where we have retained ω in the expression because we are interested in the frequency-dependence of the solution. Substituting this into the expression for ζ and taking the limit, we find a linear dependence on ω at low frequencies:

$$\lim_{\omega \to \epsilon} \zeta \approx -4 \frac{\rho_s k_o \left(HM - C^2\right)}{\mu_f H^2} i\omega. \tag{7.78}$$

The final limit that we shall need is

$$\lim_{\omega \to \epsilon} \gamma \approx \frac{\mu_f H}{k_o \left(HM - C^2\right)} \frac{i}{\omega} \tag{7.79}$$

Because of the linear frequency dependence of ζ in the limit of low frequencies, we can expand the square root $\sqrt{1 - \zeta}$ in a Taylor's series, retaining only the first two terms:

$$\chi(\mathbf{x}, \omega) \approx \frac{\gamma \omega^2}{2} \left[1 \pm \left(1 - \frac{1}{2} \zeta \right) \right]. \tag{7.80}$$

We have laid the groundwork and can now consider each of the signs in turn in order to examine the slow and fast wave solutions associated with a propagating disturbance in a poroelastic medium.

The slow wave

Consider the solution associated with the positive sign in the expression (7.55) for $\chi(\mathbf{x}, \omega)$. For real values of $\sqrt{1 - \zeta}$, this sign will produce the greatest value for the slowness magnitude $p = \mathbf{p} \cdot \mathbf{p}$ [see Equation (7.54)]. Because p is the reciprocal of the wave speed, this disturbance will travel slower than the other solution. For this reason this particular longitudinal mode is known as the slow wave. Let us consider the zeroth-order solution in the limit of low frequency in order to derive a semi-analytic solution and to gain some insight into the character of the slow wave. Taking the positive sign in (7.80), and substituting in the limiting expressions for ζ and γ produces

$$\chi_{slow}(\mathbf{x}, \omega) \approx \frac{i\omega\mu_f}{k_o} \frac{H}{HM - C^2} - 4\frac{\rho_s \omega^2}{H}$$

which, is dominated by the first term for small values of ω. Returning to the integral expression (7.62), the approximate low-frequency expression for the phase of the slow wave is given by

$$\theta_{slow}(\mathbf{x}, \omega) \approx \sqrt{i\omega}\tau(\mathbf{x}) \tag{7.81}$$

where we have separated the frequency-dependence and the spatial depencence and gathered all terms that depend upon $\mathbf{x}$ into a single function defined by an integral (7.62) along the trajectory

$$\tau(\mathbf{x}) = \int_{\mathbf{x}} \sqrt{\frac{\mu_f}{k_o} \frac{H}{\left(HM - C^2\right)}} dx. \tag{7.82}$$

Given the explicit form for $\theta_{slow}(\mathbf{x}, \omega)$, we can return to the zeroth-order expression for $\mathbf{U}(\mathbf{X}, \omega, \theta)$, given by Equation (7.74). The zeroth-order expression for the slow wave takes the form

$$\mathbf{U}_{slow}(\mathbf{X}, \omega) = e^{\sqrt{-i\omega}\tau(\mathbf{x})} \mathbf{U}_0(\mathbf{X}, \omega)$$

where we have made use of the fact that i is $\sqrt{-1}$. Taking the inverse Fourier transform to obtain a time domain expression gives (Virieux et al., 1994)

$$\mathbf{u}_{slow}(\mathbf{x}, t) = \frac{\tau}{2\sqrt{\pi t^3}} e^{-\tau^2/4t} H(t) * \mathbf{u}_0(\mathbf{X}, t) \tag{7.83}$$

where $H(t)$ is the Heaviside or step-function. We have already encountered a solution of this form in Chapter 4. This is a diffusive wave that decays rapidly away from the source. The quantity $\tau(\mathbf{x})$ is related to the propagation time of the disturbance but requires some interpretation. In essence, for an impulsive source $\tau(\mathbf{x})$ can be related to the arrival time of the peak displacement. As shown in Chapter 4, for an impulsive source the condition for the peak of the displacement specified by Equation (7.83) is the vanishing of

$$\frac{\partial \mathbf{u}_{slow}}{\partial t} = \mathbf{u}_0(\mathbf{x}) e^{-\tau^2(\mathbf{x})/4t} \left[-\frac{3}{2\sqrt{t^5}} + \frac{\tau^2(\mathbf{x})}{4\sqrt{t^7}} \right]. \tag{7.84}$$

For finite values of the phase function $\tau(\mathbf{x})$ and non-zero times, the exponential term is never zero. Hence, the condition for an extremum is given by the vanishing of the quantity within the square brackets in Equation (7.84), when either t approaches infinity or when

$$6t = \tau^2(\mathbf{x}).$$

Thus, we may interpret the phase function $\tau(\mathbf{x})$ in terms of the time at which the peak displacement is observed, denoted by T_{max}. We can use this fact to produce a relationship between an observable quantity, the peak deformation associated with the passing of the slow wave, and the properties of the medium, including the permeability

$$\sqrt{T_{max}} = \int_{\mathbf{x}} \sqrt{\frac{\mu_f}{k_o} \frac{H}{\left(HM - C^2\right)}} dx. \tag{7.85}$$

As we will illustrate in the Application section, Equation (7.85) can serve as the basis for a form of diffusive wave travel time tomography, similar to that described in Chapter 4.

The fast wave

Now consider the characteristics of the fast wave in a poroelastic medium. From the preceding example, we see that it travels at speeds akin to that of an elastic or seismic wave. If we take the minus sign, expression (7.80) for χ reduces to

$$\chi_{fast}(\mathbf{x}, \omega) = \frac{\gamma\zeta\omega^2}{4}$$

which, in the limit of low-frequency becomes

$$\chi_{fast}(\mathbf{x}, \omega) \approx \omega^2 \frac{\rho_s}{H}.$$

The expression for the phase of the fast wave is the integral (7.62) of the form

$$\theta_{fast}(\mathbf{x}, \omega) \approx \omega \int_{\mathbf{x}} \sqrt{\frac{\rho_s}{H}} dx = \omega\tau(\mathbf{x}), \tag{7.86}$$

where the function $\tau(\mathbf{x})$ contains all of the terms that are functions of $\mathbf{x}$,

$$\tau(\mathbf{x}) = \int_{\mathbf{x}} \sqrt{\frac{\rho_s}{H}} dx. \tag{7.87}$$

Note, in this approximation the phase of the fast wave only depends upon the density of the solid ρ_s and H and not upon the flow properties of the medium. The zeroth-order expression for the fast wave is then

$$\mathbf{U}_{fast}(\mathbf{X}, \omega) = e^{i\omega\tau(\mathbf{x})}\mathbf{U}_0(\mathbf{X}, \omega)$$

which, upon taking the inverse Fourier transform, provides the time-domain expression

$$\mathbf{u}_{fast}(\mathbf{x}, t) = \delta\left[t - \tau(\mathbf{x})\right] * \mathbf{u}_0(\mathbf{x}, t) = \mathbf{u}_0\left(\mathbf{x}, t - \tau(\mathbf{x})\right) \tag{7.88}$$

where δ signifies the 'delta' or impulse function. In this case $\tau(\mathbf{x})$ signifies the travel time along the trajectory $\mathbf{x}(s)$ from the source to the observation point. Thus, the fast wave propagates as an elastic wave and only attenuates with distance due to geometrical spreading given by the transport equation and contained in $\mathbf{U}_0(\mathbf{X}, \omega)$.

7.3 A porous body containing three fluids

There are important situations in which a porous medium contains two or more fluids. Water, gas, and hydrocarbons may saturate a petroleum reservoir in spatially varying fractions with significant economic consequences. Phase changes, such as those in a geothermal reservoir, will likewise produce variable water, steam, and gas saturations at depth. In this section we discuss the extension of the previous analysis to the case in which up to three fluids occupy the pore space. We shall not be using this in the application concluding this chapter, so the casual reader may skip this section and proceed directly to the application.

As in the case of a single phase, the presence of the fluids will influence the response of a porous body to imposed stresses. Furthermore, the porous elastic solid is still characterized by the porosity φ. However, the pores may now be filled with some combination of three fluids, and we will denote the saturation (fraction) of the n-th fluid by S_n. Because the pore is taken to be fluid-filled, the saturations sum to unity

$$\sum_{n=1}^{3} S_i = S_1 + S_2 + S_3 = 1. \tag{7.89}$$

It will be convenient to denote the fraction of a given unit volume of material for the solid

$$\alpha_s = (1 - \varphi) \tag{7.90}$$

and for the three fluids

$$\alpha_n = \varphi S_n. \tag{7.91}$$

As noted in Chapters 2 and 6, the presence of multiple fluids introduces complications on several levels. The chief difficulty is that it renders the problem non-linear, because the evolution of the saturation is severely impacted by the current saturated state. That is, due to the ability of one fluid to block the flow of another, the fractional flow of the fluid will depend upon the existing saturations. For the case considered here, the passage of an elastic wave, we linearize the problem by assuming that the saturation changes induced by the wave are small. Because the formulation is a direct extension of the derivation in the previous section, and the results are discussed in detail in Vasco (2013), our discussion will be very brief. We will merely present the critical steps and indicate how the single fluid approach from the previous section generalizes when several fluids are involved.

Following the approach taken in Section 7.2, suitably generalized for the case of three fluids, one can produce the four governing equations where the first equation is indexed by $n = 1, 2, 3$ and actually represents three equations:

$$\nu_n \mathbf{U} + \Gamma_n \mathbf{W}_n = \nabla \left[C_{ns} \nabla \cdot \mathbf{U} + \sum_{j=1}^{3} M_{nj} \nabla \cdot \mathbf{W}_n \right] \tag{7.92}$$

$$\nu_s \mathbf{U} + \sum_{j=1}^{3} \xi_j \mathbf{W}_j = \nabla \left[K_u \nabla \cdot \mathbf{U} + \sum_{j=1}^{3} C_{sj} \nabla \cdot \mathbf{W}_j \right] + \nabla \cdot \boldsymbol{\tau}, \tag{7.93}$$

the analogues of Equations (7.23) and (7.24). In this derivation we have linearized the equations by assuming that the saturation variations due to the passage of the wave are small. In addition, we have transformed the equations into the frequency domain so dependent variables $\mathbf{U}(\mathbf{x}, \omega)$ and $\mathbf{W}_n(\mathbf{x}, \omega)$ for $n = 1, 2, 3$ are the displacements of the solid and fluids as a function of spatial location and frequency. To keep the equations in a compact form, we have defined the coefficients

$$\nu_s = \alpha_s \rho_s \omega^2, \tag{7.94}$$

$$\nu_n = \alpha_n \rho_n \omega^2, \tag{7.95}$$

$$\xi_j = \alpha_j \rho_j \frac{\mu_j}{k(\omega)} \omega,$$

$$\Gamma_n = \alpha_n \rho_n \left[\omega - \frac{\mu_n}{k(\omega)} \right] \omega,$$

where $k(\omega)$ is the dynamic permeability, discussed in Box 2.8.

The system of linear partial differential Equations (7.92) and (7.93) can be solved using a numerical method such as finite-differences or the spectral-element method (Carcione et al., 2010; Morency et al., 2011). However, as in the single-fluid case, due to the presence of slow diffusive modes and fast, elastic-like modes, numerical codes can require considerable computational resources. As an alternative, we can consider the asymptotic, trajectory-based approach discussed in Section 7.2.2. It is possible to follow an entirely analogous procedure for a porous medium containing three fluids, from the governing equations, through the asymptotic representation of a solution, to the equations that result from retaining only zeroth-order terms in the representation (Vasco, 2013). The result is a generalization of the system of Equation (7.37) for a single fluid, to the zeroth-order equations for a porous medium containing three fluids,

$$\boldsymbol{\Omega} \mathbf{V}_o = \mathbf{0} \tag{7.96}$$

where $\boldsymbol{\Omega}$ is the 12×12 coefficient matrix

$$\boldsymbol{\Omega} = \begin{bmatrix} \alpha \mathbf{I} - \beta \mathbf{p}\mathbf{p}^T & \xi_1 \mathbf{I} - C_{s1} \mathbf{p}\mathbf{p}^T & \xi_2 \mathbf{I} - C_{s2} \mathbf{p}\mathbf{p}^T & \xi_3 \mathbf{I} - C_{s3} \mathbf{p}\mathbf{p}^T \\ \nu_1 \mathbf{I} - C_{1s} \mathbf{p}\mathbf{p}^T & \Gamma_1 \mathbf{I} - M_{11} \mathbf{p}\mathbf{p}^T & -M_{12} \mathbf{p}\mathbf{p}^T & -M_{13} \mathbf{p}\mathbf{p}^T \\ \nu_2 \mathbf{I} - C_{2s} \mathbf{p}\mathbf{p}^T & -M_{21} \mathbf{p}\mathbf{p}^T & \Gamma_2 \mathbf{I} - M_{22} \mathbf{p}\mathbf{p}^T & -M_{23} \mathbf{p}\mathbf{p}^T \\ \nu_3 \mathbf{I} - C_{3s} \mathbf{p}\mathbf{p}^T & -M_{31} \mathbf{p}\mathbf{p}^T & -M_{32} \mathbf{p}\mathbf{p}^T & \Gamma_3 \mathbf{I} - M_{33} \mathbf{p}\mathbf{p}^T \end{bmatrix}, \tag{7.97}$$

p is the phase gradient $\nabla\theta$, we have defined the coefficients

$$\alpha = \nu_s - Gp^2,$$

$$\beta = K_u + \frac{1}{3}G,$$

and $\mathbf{V}_o$ is the extended vector $(\mathbf{U}^0, \mathbf{W}_1^0, \mathbf{W}_2^0, \mathbf{W}_3^0)^T$ containing the zeroth-order solid and fluid displacement amplitudes. The system of linear Equations (7.96) has a non-trivial (non-zero) solution $\mathbf{V}_o$ if, and only if, the determinant of the coefficient matrix $\boldsymbol{\Omega}$ vanishes [see the discussion in Section 7.2.3]. This condition produces a polynomial equation in the components of **p**. Because $\mathbf{p} = \nabla\theta$, the polynomial equation is actually a non-linear partial differential equation for the phase, similar to the eikonal equation found in the theory of high-frequency elastic wave propagation.

Note that the structure of the coefficient matrix (7.97) is similar to that of the matrix (7.37) for the single phase case with block elements composed of the identity matrix **I** and the dyadics $\mathbf{pp}^T$. This suggests that a vector $\mathbf{e}^l$ composed of multiples of **p**

$$\mathbf{e}^l = \begin{bmatrix} y_1\mathbf{p} \\ y_2\mathbf{p} \\ y_3\mathbf{p} \\ y_4\mathbf{p} \end{bmatrix}$$

is a likely eigenvector for the matrix that allows us to simplify the calculation of the determinant of $\boldsymbol{\Omega}$. This vector is associated with the longitudinal mode of propagation, discussed in detail in the previous section. A vector $\mathbf{e}^\perp$ composed of multiples of $\mathbf{p}^\perp$, a vector perpendicular to **p**,

$$\mathbf{e}^\perp = \begin{bmatrix} s_1\mathbf{p}^\perp \\ s_2\mathbf{p}^\perp \\ s_3\mathbf{p}^\perp \\ s_4\mathbf{p}^\perp \end{bmatrix}$$

is another potential eigenvector, as in Section 7.2.3. The solid and fluid displacements associated with the vector $\mathbf{e}^\perp$ are transverse (perpendicular) to the direction of propagation, similar to an elastic shear wave. Let us consider each of these modes in turn.

The longitudinal mode

Substituting the vector $\mathbf{e}^l$ into the eigenvalue problem, which for a vanishing eigenvalue is equivalent to Equation (7.96), results in a homogeneous linear system of equations. The coefficient matrix of this system is highly structured, in fact it is in the form of a tensor product

$$\boldsymbol{\Omega} = \mathbf{M} \otimes \mathbf{I},$$

[see Equation (7.46)], where

$$\mathbf{M} = \begin{bmatrix} \nu_s - Hs & \xi_1 - C_{s1}s & \xi_2 - C_{s2}s & \xi_3 - C_{s3}s \\ \nu_1 - C_{1s}s & \Gamma_1 - M_{11}s & -M_{12}s & -M_{13}s \\ \nu_2 - C_{2s}s & -M_{21}s & \Gamma_2 - M_{22}s & -M_{23}s \\ \nu_3 - C_{3s}s & -M_{31}s & -M_{32}s & \Gamma_3 - M_{33}s \end{bmatrix}, \tag{7.98}$$

$s = p^2$, and $H = K_u + 4/3G$. For matrices composed of identical block sub-matrices, the identity matrix in this case, one can show that

$$\det(\Omega) = \det(\mathbf{M} \otimes \mathbf{I}) = \det(\mathbf{M}) \det(\mathbf{I})^4 = \det(\mathbf{M})$$

(Silvester, 2000). Hence, the condition that the determinant of Ω vanishes is equivalent to the vanishing of the determinant of **M**. The determinant of **M** is a quartic polynomial in $s = p^2$

$$Q_4 s^4 - Q_3 s^3 + Q_2 s^2 - Q_1 s + Q_0 = 0 \tag{7.99}$$

that has four complex roots in general. The coefficients Q_0, Q_1, Q_2, Q_3, and Q_4 are given in Vasco (2013) in terms of the medium and fluid parameters. The quartic equation can be solved either numerically or analytically. An explicit solution can be derived in terms of the solutions of a related cubic equation, the resolvent cubic (Faucette, 1996; Nickalls, 2009). Using a numerical approach, we computed the roots of the quartic (7.99), for a disturbance with a center frequency of 100 Hz propagating in a porous medium containing oil, water, and gas. In Figure 7.8 we plot the real component of the phase velocity, given by

$$c = \frac{\omega}{p}$$

as a function of water and gas saturation. The asymptotic representation consists of series of the same form as (7.26) and (7.27). From the zeroth-order terms of the series, $\mathbf{U}(\mathbf{x}, \theta, \omega) = e^{i\theta}\mathbf{U}_0(\mathbf{X}, \omega)$ and $\mathbf{W}(\mathbf{x}, \theta, \omega) = e^{i\theta}\mathbf{W}_0(\mathbf{X}, \omega)$, it is evident that the imaginary part of the phase $\theta(\mathbf{x}, \omega)$ gives rise to attenuation of the propagating disturbance. The attenuation coefficient is the imaginary component of p. In Figure 7.9 we plot the attenuation coefficient as a function of the water and gas

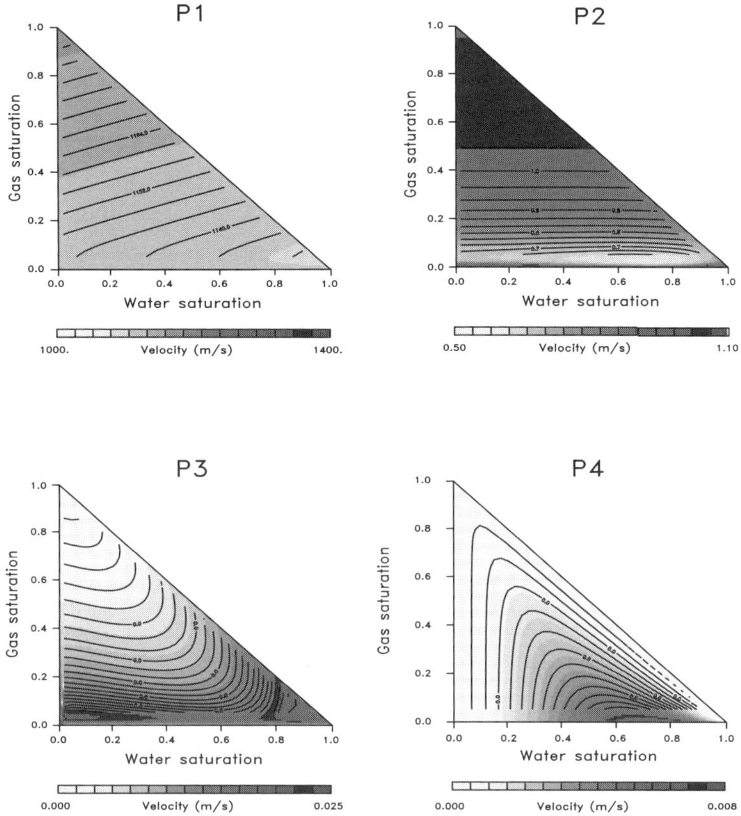

Figure 7.8 Velocities resulting from the four solutions of the quartic Equation (7.99). The velocities correspond to the four longitudinal waves in a porous medium containing three fluids. The velocities are plotted as functions of the water and gas saturation. The oil saturation is given by $1 - S_w - S_g$ and is zero along the diagonal of the triangle and 1 at the origin.

saturations. The fast wave (P1) experiences little attenuation and attenuation generally increases from P2 through P3 to P4.

The transverse mode

For this mode of propagation the displacements are perpendicular to the direction of propagation $\mathbf{p}$. The eigenvector is $\mathbf{e}^{\perp}$ and the system of equations associated with the vanishing eigenvalue is

$$\begin{bmatrix} \alpha\mathbf{I} & \xi_1\mathbf{I} & \xi_2\mathbf{I} & \xi_3\mathbf{I} \\ \nu_1\mathbf{I} & \Gamma_1\mathbf{I} & \mathbf{0} & \mathbf{0} \\ \nu_2\mathbf{I} & \mathbf{0} & \Gamma_2\mathbf{I} & \mathbf{0} \\ \nu_3\mathbf{I} & \mathbf{0} & \mathbf{0} & \Gamma_3\mathbf{I} \end{bmatrix} \begin{bmatrix} t_s\mathbf{p}^{\perp} \\ t_1\mathbf{p}^{\perp} \\ t_2\mathbf{p}^{\perp} \\ t_3\mathbf{p}^{\perp} \end{bmatrix} = \begin{bmatrix} \mathbf{0} \\ \mathbf{0} \\ \mathbf{0} \\ \mathbf{0} \end{bmatrix}. \quad (7.100)$$

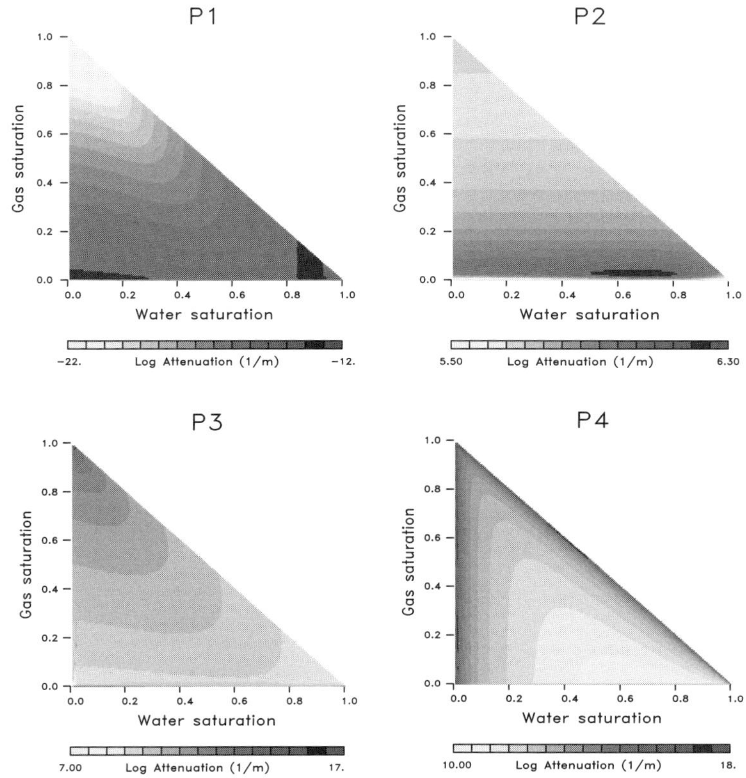

Figure 7.9 Attenuation resulting from the four solutions of the quartic Equation (7.99). The attenuation coefficients are plotted as functions of the water and gas saturation.

This equation has a non-trivial solution if the determinant of the coefficient matrix vanishes. Due to the structure of the matrix, it is a tensor product $\mathbf{M} \otimes \mathbf{I}$, this condition is equivalent to

$$\det \begin{bmatrix} \nu_s - G_m p^2 & \xi_1 & \xi_2 & \xi_3 \\ \nu_1 & \Gamma_1 & 0 & 0 \\ \nu_2 & 0 & \Gamma_2 & 0 \\ \nu_3 & 0 & 0 & \Gamma_3 \end{bmatrix} = 0 \tag{7.101}$$

where we have used the fact that $\alpha = \nu_s - Gp^2$. The vanishing of the determinant in Equation (7.101) produces a quadratic equation for p with no linear term. In Figure 7.10 we plot the real velocity of the transverse mode as a function of water and gas saturation. In addition, plot the attenuation, obtained from the imaginary component of the velocity. The attenuation coefficient is generally quite small, signifying minor attenuation.

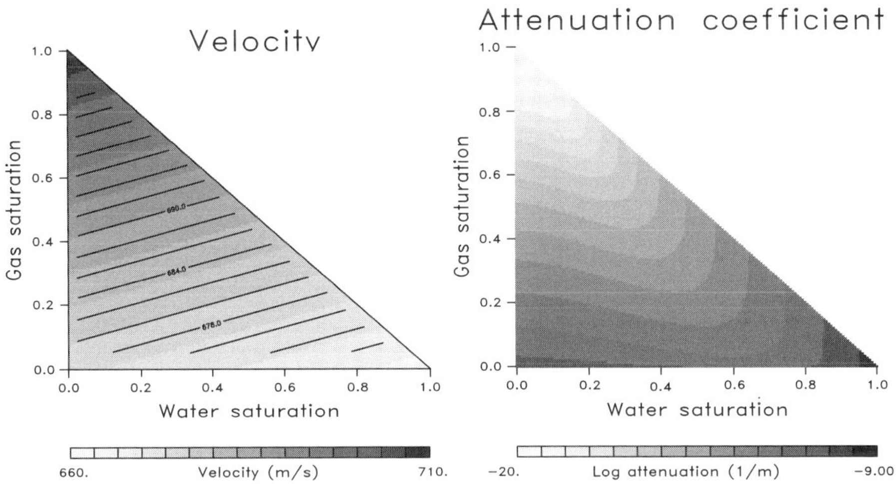

Figure 7.10 Velocity and attenuation coefficient associated with the transverse mode of propagation.

7.4 Application

7.4.1 Diffusive onset time tomography

We can make use of the insight provided by the analysis in Section 7.2.4 to develop a method for imaging flow properties. First, we have seen that the arrival time of the fast wave is not directly sensitive to permeability, at least in the low-frequency approximation. On the other hand, the velocity, and hence the travel time, of the slow wave is influenced by the permeability, as indicated by Equation (7.85). While the slow wave is diffusive in nature, it is certainly possible to propagate transient pressure disturbances for some distance from injection or production wells. That is, with sufficient mass fluxes, one can transmit observable fluid pressure changes and associated deformation, possibly for long distances if the permeability is sufficiently large. The resulting pressure changes may be detectable using pressure transducers in boreholes or using geophysical imaging techniques. In fact, such propagation is used for transient pressure testing, such as interference tests, in reservoir engineering and hydrology. Note that the fast wave is still extremely useful in detecting saturation and fluid pressure related changes. Indeed, that is the rationale behind the time-lapse imaging of reservoir processes, a very important aspect of geophysical monitoring.

Because its travel time depends upon k_o, it should be possible to use its arrival times of the slow wave to say something about the flow characteristics of a porous medium. In fact, we can take advantage of the long propagation times, using geophysical methods to capture the movement of slow transients in some detail.

Wide-spread seismic monitoring of such transients will require improvements in current practice because pressure changes can propagate through a reservoir in hours, days, or months, depending upon the flow characteristics. Therefore, routine seismic surveys will typically alias the fluid pressure signal. New techniques, such as permanent arrays, and continuous active seismic sources (Daley et al., 2011; Ajo-Franklin et al., 2012) are well suited to track fluid pressure-related changes. Geodetic methods, such as tiltmeters and Interferometric Synthetic Aperture Radar (InSAR) can detect fluid-related changes with a temporal resolution of minutes, weeks, and months. Thus, in favorable cases geodetic methods can be used to monitor flow and estimate flow properties (Vasco et al., 2001; Vasco, 2004b; Vasco et al., 2008a; Rucci et al., 2010).

Our basic approach is a straight-forward adaptation of Equation (7.85), akin to the crosswell pressure imaging technique presented in Chapter 4. However, the method is closer in spirit to the x-ray imaging of brine injection discussed in Chapter 5. That is, rather than use crossing rays and varous source-receiver combinations to provide spatial resolution characteristic of crosswell imaging, we use the temporal sampling to resolve variations in properties along trajectories. The approach will become clearer as we illustrate it below.

7.4.2 Geological storage of carbon dioxide at In Salah, Algeria

At a site near In Salah, Algeria excess carbon dioxide extracted from raw natural gas was reinjected back into a sandstone formation for geological storage. The roughly 20m thick reservoir was overlain by approximately one kilometer of shale and an additional kilometer of interbedded sand and shale. In an effort to monitor the fate of the injected carbon dioxide, Interferometric Synthetic Aperture Radar (InSAR) was used to observe any surface deformation associated with the storage effort (Vasco et al., 2008a, 2010). In this approach electromagnetic radiation from an orbiting satellite is reflected off the surface of the Earth and the complex return is recorded and processed (Ferretti, 2014). Differences in the processed signals from repeated satellite passes are used to infer changes in range, essentially displacements along a line extending from the reflection point to a reference point in space. Given the altitude of the satellite orbit the range change is most sensitive to the vertical displacement. The reference point is related to the average position of the satellite source. In Figure 7.11 we plot four images of cumulative range change at various times since the injection of carbon dioxide began. In all there were 28 snapshots of range change covering a period of 950 days, providing roughly monthly coverage, although there could be gaps of several months in some instances.

Due to the presence of a thick sequence of impermeable shale in the overburden at In Salah, it was thought that the flow of carbon dioxide would be contained

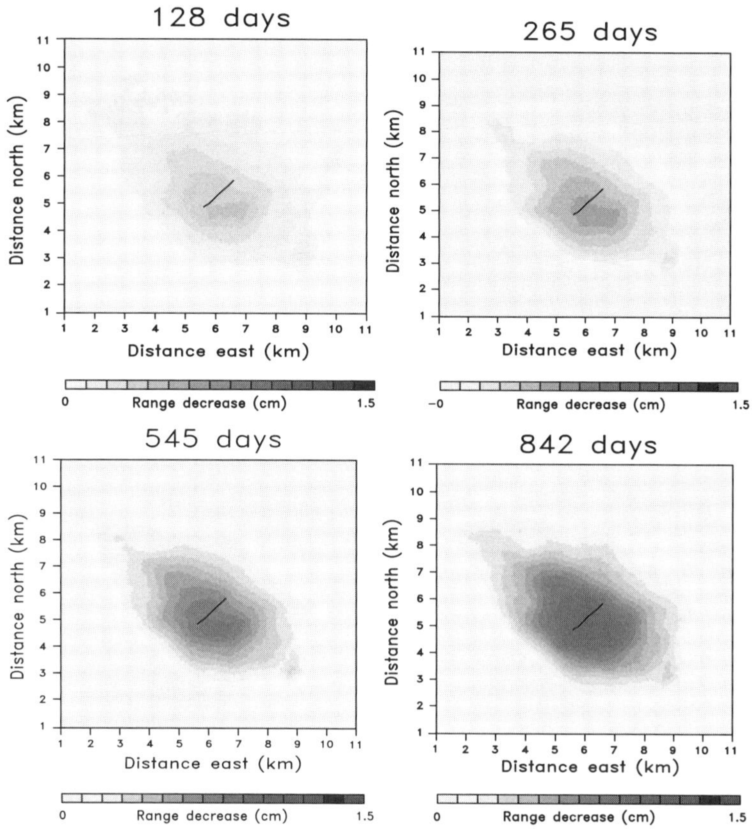

Figure 7.11 Four snapshots of cumulative range change due to surface deformation caused by the injection of carbon dioxide at depth. The horizontal section of the injection well within the reservoir interval is indicated by the black line.

within the reservoir. Therefore, our model assumes that the injected mass produces pressure and volume changes within the reservoir, leading to deformation at the surface. The rapid onset of injection, followed by injection at a rather steady rate, led to a propagating coupled pressure and deformation front. The deformation propagates elastically to the surface where it is observed as range change. We first use the range change sequence to infer volume change within the 20m thick reservoir. This is a linear inverse problem, assuming that the overburden behaves elastically over the roughly one month interval between satellite passes. The approach has been described in detail elsewhere (Vasco, 2004b; Vasco et al., 2008a, 2010; Rucci et al., 2010). The reservoir interval underlying the region plotted in Figure 7.11 was sub-divided into a 33 by 33 grid of blocks. We inverted for the sequence of volume change associated with each grid block, obtaining a history of reservoir volume change. Based upon a poroelastic model for the reservoir we mapped the

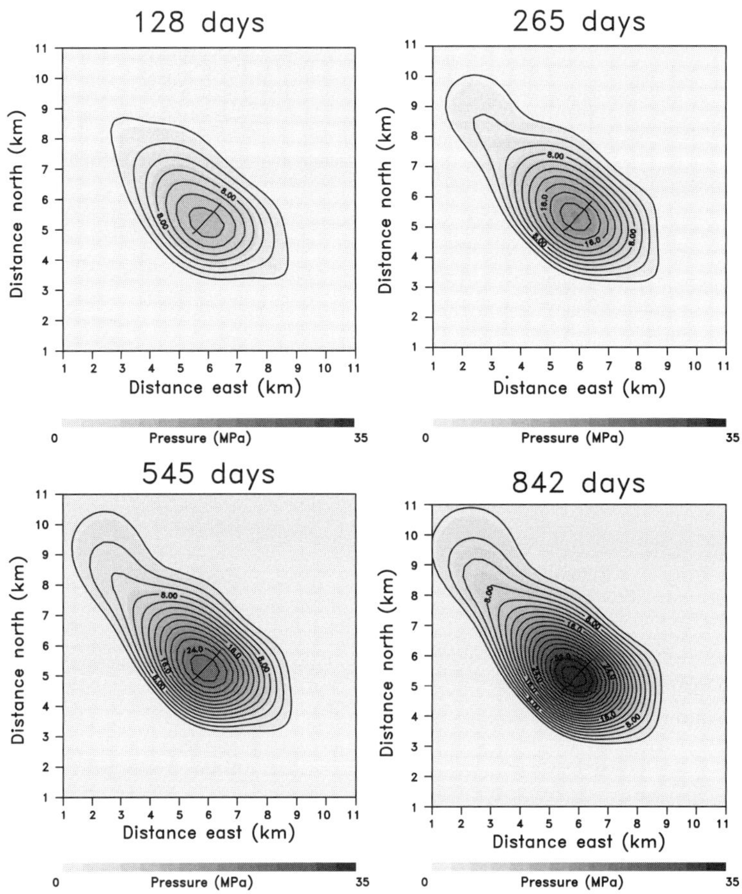

Figure 7.12 Inferred pressure evolution within the reservoir, obtained from an inversion of the range changes plotted in Figure 7.11.

estimated grid block volume changes into reservoir pressure changes since the start of injection, plotted in Figure 7.12. The calculated reservoir volume changes provide a time series within each grid block may be used to determine the arrival time of the pressure disturbance at that location. That is, using the formulation of Section 7.2, in particular Equations (7.83), (7.84), and (7.85), we estimate the arrival time T_{max}. Note that we must modify this approach to account for the differences in the time variation of the sources. At In Salah, the injection starts suddenly and is then maintained at a high rate, essentially for the entire time interval under consideration. The step-function rate is the integral of the impulsive or delta-function source that we considered in the derivation above. We can use the results from the previous section if we take the derivative of the time series for each grid block. Specifically, we construct a representation of the function, using cubic splines for

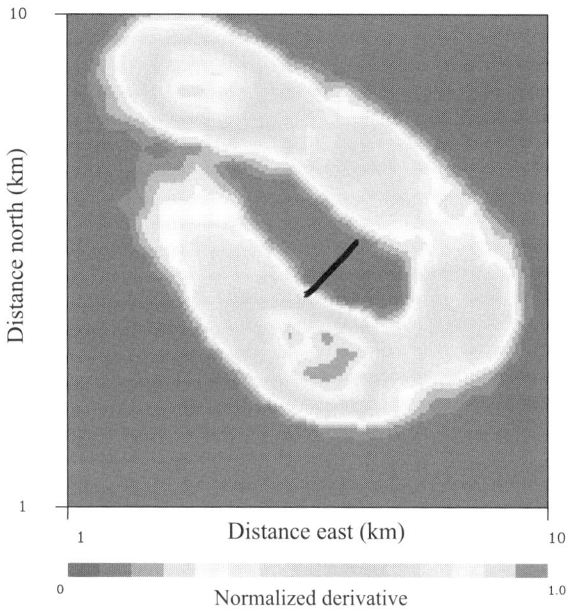

Figure 7.13 Normalized time derivative of reservoir volume change after approximately 300 days of injection. The intersection of the horizontal well and reservoir is indicated by the solid black line. For the colour version, please refer to the plate section.

example, and use the representation to differentiate the time series. An example of the spatial distribution of the normalized derivative within the reservoir, about 300 days after the start of injection, is shown in Figure 7.13. Note the resemblance to the propagating pressure fronts plotted in Figure 1.4 in Chapter 1. The arrival time T_{max} corresponds to the time at which the derivative attains its peak value and is plotted in Figure 7.14.

While we have converted the volume change to pressure change, such a conversion is unnecessary and introduces potential errors due to variations in the mechanical properties within the reservoir (Rucci et al., 2010). For example, parts of the reservoir that are highly compressible might be assumed to have large pressure changes if the same rock properties are used throughout the model. One of the advantages of the arrival time T_{max} is that its estimated value is less sensitive to such variations in mechanical properties than are the magnitudes of pressure changes. If the reservoir behaves linearly over the monthly time intervals, then the time at which grid block volume and the peak pressure are changing the most rapidly should roughly coincide. This idea has been abstracted to the notion of an onset time (Vasco et al., 2014, 2015). We have already encountered onset times in the application presented in Chapter 5 involving a brine injection experiment. An onset time is associated with time-lapse monitoring and is the time at which an observed quantity,

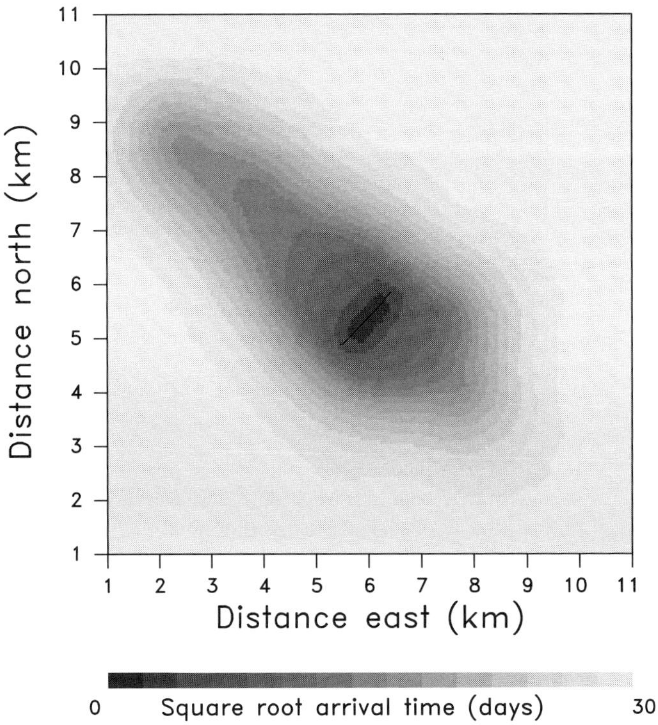

Figure 7.14 Square root of the arrival time at each grid block.

such as a seismic reflection amplitude, begins to change from its background value. When monitoring experiments or field activities involving fluid flow, onset times are frequently related to saturation or pressure changes. In such cases onset times are typically influenced by flow properties but are not as sensitive to rock physics models as are the magnitudes of the time-lapse changes (Vasco et al., 2014, 2015).

Returning to the constraint (7.85) provided by T_{max}, we consider the discrete representation of the reservoir as grid blocks. Because the properties are assumed to be constant in each grid block, the trajectory is a linear segment within the cell. We can write the integral as a sum over the linear segments in each grid block

$$\sqrt{T_k} = \sum_{i \in X_k} \sqrt{\frac{\mu_i}{k_i} \frac{H_i}{\left(H_i M_i - C_i^2\right)}} \Delta x_{ik}, \tag{7.102}$$

for the $k-th$ peak travel time and the sum is over all the grid blocks intersected by the trajectory X_k. Here Δx_{ik} denotes the length of the $k-th$ ray segment within the $i-th$ grid block. It is not possible to resolve all of model parameters in the sum (7.102). In fact one can only resolve an effective property related to the diffusivity of the transient disturbance, denoted by κ_i,

$$\kappa_i = \sqrt{\frac{\mu_i}{k_i} \frac{H_i}{\left(H_i M_i - C_i^2\right)}} \tag{7.103}$$

allowing us to write the constraint along the trajectory as

$$\sqrt{T_k} = \sum_{i \in X_k} \kappa_i \Delta x_i. \tag{7.104}$$

Before we advance any further in our discussion of the solution we need to make a few observations regarding the construction of the trajectories. First, we are able to construct the spatial variation of the phase within the model from the time-lapse estimates of the volume change in the reservoir and from the fact that

$$\tau(\mathbf{x}) = \sqrt{6 T_{max}(\mathbf{x})}. \tag{7.105}$$

Second, knowing the spatial variation of the phase allows us to determine the trajectory $\mathbf{x}(s)$ using the ray Equation (7.57) and the fact that $\mathbf{p} = \nabla\theta$. We can obtain the trajectory simply by marching down the phase gradient from the point at which we have a travel time estimate back to the well. In Figure 7.15 we plot the

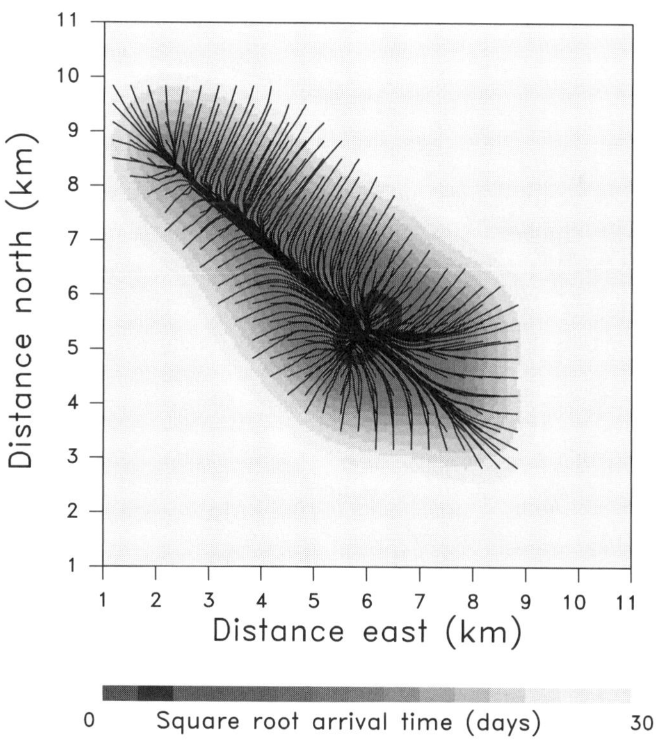

Figure 7.15 Trajectories signifying propagation from the injection well to grid blocks in the model with resolvable volume changes.

trajectories associated with the arrival time field shown in Figure 7.14. Because we are using 28 snapshots of range change to calculate the arrival time field plotted in Figure 7.15, we can resolve propagation to various distances from the well, allowing us to resolve variations in travel times along the trajectories. This aspect of our data provides spatial resolution without the need for crossing rays, a unique characteristic of this approach. Finally, we note that our ability to calculate the trajectories from the estimated arrival times allows us to effectively linearize the inverse problem. This a very useful feature because the large spatial variations in properties such as permeability can lead to significant non-linearity and severe ray bending. Furthermore, the formalism for the solution and assessment of an inverse problem, such as the calculation of resolution and covariance matrices described in Chapters 4 and 5, is in a much more satisfactory state when the problem is linear (Menke, 1989; Parker, 1994; Aster et al., 2013).

Given a set of arrival times and trajectories, such as those shown in Figure 7.15 respectively, we can construct a linear system of equations for the

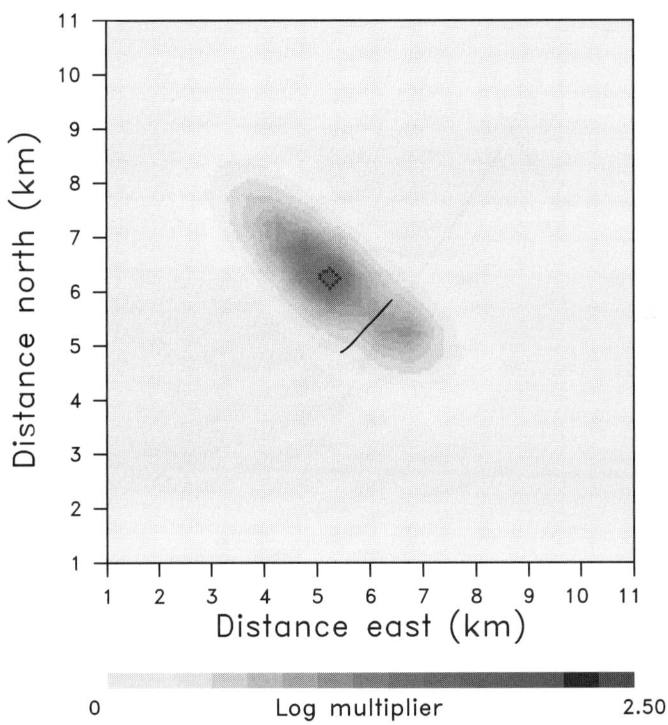

Figure 7.16 The logarithm of the multipliers of $1/\kappa_0$ where κ_0 is the background value of the effective parameter given by Equation (7.103). Higher values of the multiplier correlate with higher permeability.

effective diffusivity in each grid block. We write the set of linear equations in matrix-vector form

$$\Upsilon = \mathbf{M}\kappa \tag{7.106}$$

where Υ is a vector with elements $\sqrt{T_k}$, the coefficients of matrix $\mathbf{M}$ are the trajectory lengths of the path in each grid block, and κ is a vector containing the effective parameters κ_i for all grid blocks. If the path does not traverse a particular grid block then the corresponding entry in $\mathbf{M}$ is zero. We have already described techniques for solving such sparse linear systems in Chapters 4 and 5, so we refer the reader to those discussions for more details.

The results of a constrained solution of Equation (7.106) for κ are plotted in Figure 7.16. The results suggest a linear feature that is compatible with a high permeability zone. It was noted in Vasco et al. (2008a) that the feature correlates with a fault inferred from a depression in seismic time horizons from an earlier 1997

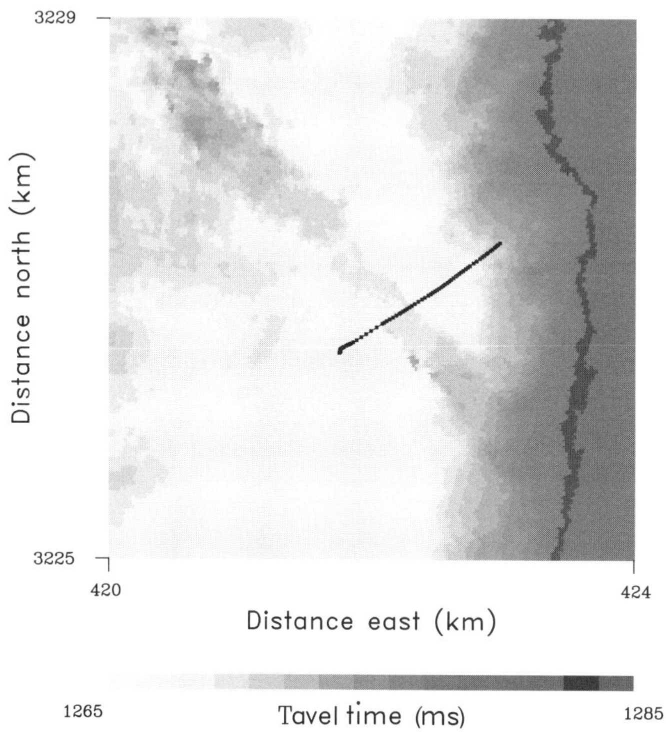

Figure 7.17 Seismic time horizon for a reflector near the base of the shale overburden at In Salah. Topography of the horizon indicates a depression, likely due to velocity push-down induced by the accumulation of injected carbon dioxide within a fault/fracture zone. The line of black squares indicates the location of injection well KB-502 within the reservoir layer.

survey. This interpretation was supported by further analysis of multicomponent InSAR data (Rucci et al., 2013). Furthermore, a fault/fracture zone of this nature and orientation was also shown to fit a double-lobbed pattern of range change associated with the injection of carbon dioxide at a nearby well, KB-502 (Vasco et al., 2010). Confirmation of the role of these large-scale, linear fault/fracture zones was provided by observations from a follow-up seismic survey acquired in 2009 (Gibson-Poole and Raikes, 2010; Zhang et al., 2015). A depression in a seismic time horizon just above the reservoir, likely due to the velocity deviations induced by the injection of carbon dioxide, images a large, linear feature intersecting the injection well KB-502 (Figure 7.17). The orientation and location of this linear feature is well correlated with with the observed surface deformation (Gibson-Poole and Raikes, 2010).

7.5 Summary

Accurate and efficient modeling of coupled deformation and flow is becoming increasingly important in reservoir monitoring and characterization. The complicated nature and the wide range of temporal and spatial scales of flow-related deformation can prove challenging for purely numerical techniques. We have seen that an asymptotic, or trajectory-based method, produces semi-analytic expressions for the velocity, attenuation, and path of a propagating disturbance. Thus, the approach provides insight into the factors controlling the observable features of a transient disturbance. As we have seen in the example, we can use this knowledge to formulate and solve the inverse problem, allowing us to infer the spatial variation of properties determining flow.

8

Appendix: a guide to the accompanying software

The trajectory-based approach for modeling fluid flow is intrinsically geometrical. As such, it can help to develop some intuition and provide visual aid in understanding flow and transport. Here we describe two software packages Fronts3D and Trace3D for modeling pressure front propagation and streamline simulation, respectively, provided to the reader with the intent of fostering such insight. The software, example data files, and user's manual can be downloaded from the weblink to the online resources at:

www.cambridge.org/vasco

There is also a graphical user interface for PC users, allowing for a menu-driven experience as shown in Figure 8.1.

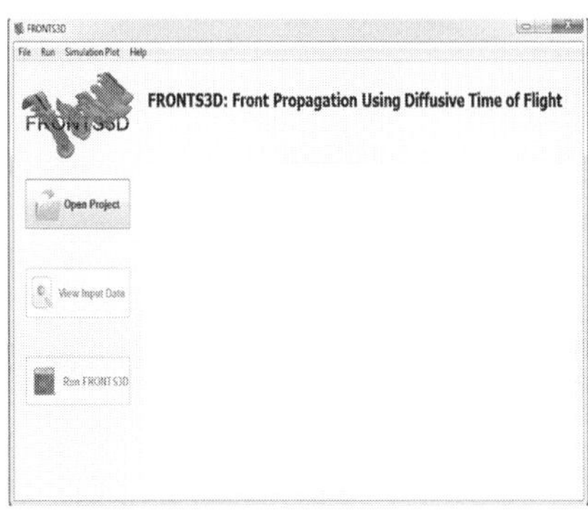

Figure 8.1 The Fronts3D graphical user interface.

8.1 Fronts3D: computing pressure propagation by Fast Marching

Fronts3D provides the reader with a hands-on introduction to the concepts discussed in Chapter 4. In particular, the routines allow for the simulation of diffusive pressure front propagation using a fast eikonal solver, the Fast Marching Method (Sethian, 1999). The numerical algorithm outlined in Section 4.3.1, and implemented in Fronts3D, allows the reader to determine the spatial variation of the diffusive phase within the porous medium. We illustrate the application of the Fronts3D software by computing the pressure distribution for a single producing well in a heterogeneous permeability field [see Exercise 4.2]. The permeability distribution, along with the well location, is shown in Figure 8.2(a). The solution of the non-linear partial differential equation for the phase $\sigma(\mathbf{x})$,

$$\nabla\sigma \cdot \nabla\sigma - \frac{1}{\kappa} = 0,$$

shown in Figure 8.2(b), may be found using the Fast Marching method, as discussed in Section 4.3.1. This algorithm is implemented in the Fronts3D software package. The kinematics of the pressure diffusion within the porous medium may be used to visualize the drainage volume as a function of time, using the relationship (4.50)

$$\sigma(\mathbf{x}) = \sqrt{T_{max}},$$

between the phase and T_{max}, the arrival time of the peak pressure generated by an impulsive source [also see Exercise 4.1]. The time evolution of the drainage volume is shown in Figure 8.3.

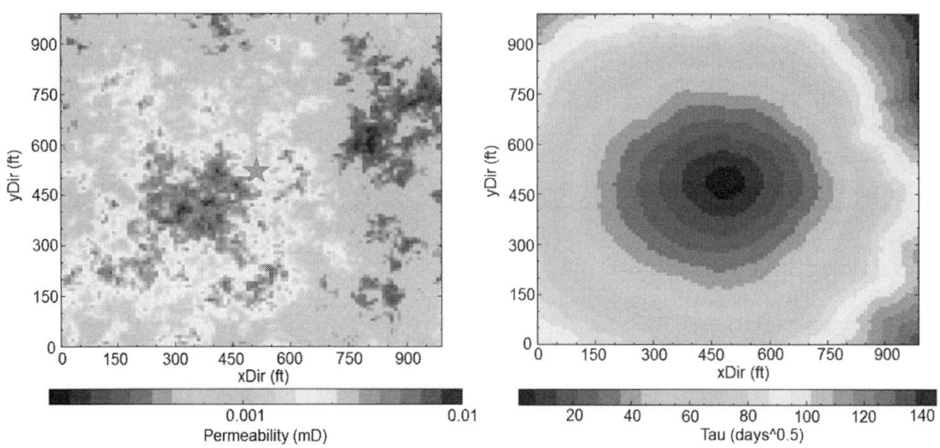

(a) Permeability distribution and well location. (b) Diffusive time of flight.

Figure 8.2 Pressure propagation in a heterogeneous permeability field with a central producing well.

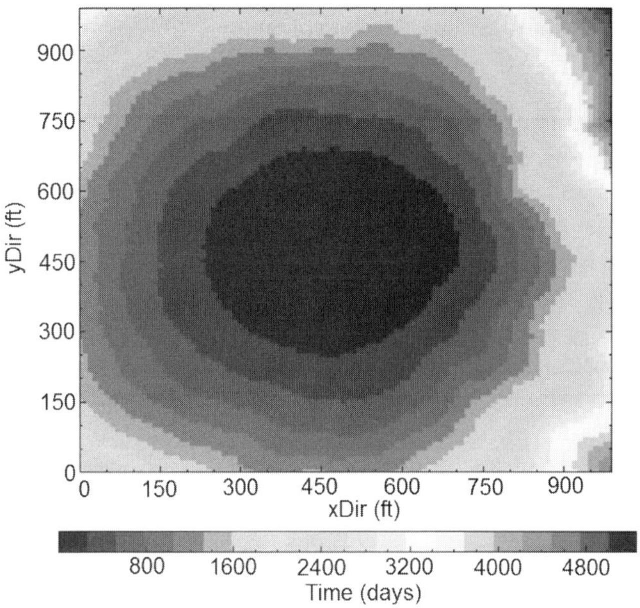

Figure 8.3 Well drainage volume.

The phase provides a coordinate for the reduction of the diffusion equation in three space and one time dimensions to an equation in a single spatial and a single time dimension:

$$\frac{1}{w(\sigma)} \frac{\partial}{\partial \sigma} \left[w(\sigma) \frac{\partial p}{\partial \sigma} \right] = \frac{\partial p}{\partial t},$$

as discussed in Section 4.3.2 and Box 4.2 in Chapter 4 [also see Exercise 4.8]. This reduced equation is solved numerically in Fronts3D to obtain the pressure distribution within the reservoir. The reduction leads to substantial savings in computation time compared to a full three dimensional numerical simulation. The pressure distribution calculated by Fronts3D, corresponding to 3600 days of injection into the model shown in Figure 8.2, is displayed in Figure 8.4.

8.2 Trace3D: software for trajectory-based modeling and inversion

Trace3D is a three dimensional trajectory-based software package for both forward and inverse modeling of flow and transport in porous media. It illustrates the concepts related to trajectory/streamline-based simulation and inversion introduced in Chapters 5 and 6. In the routines contained in Trace3D the time of flight is used as a spatial coordinate to decouple the three dimensional transport equation into a set of one dimensional equations along the trajectories. The one dimensional equations

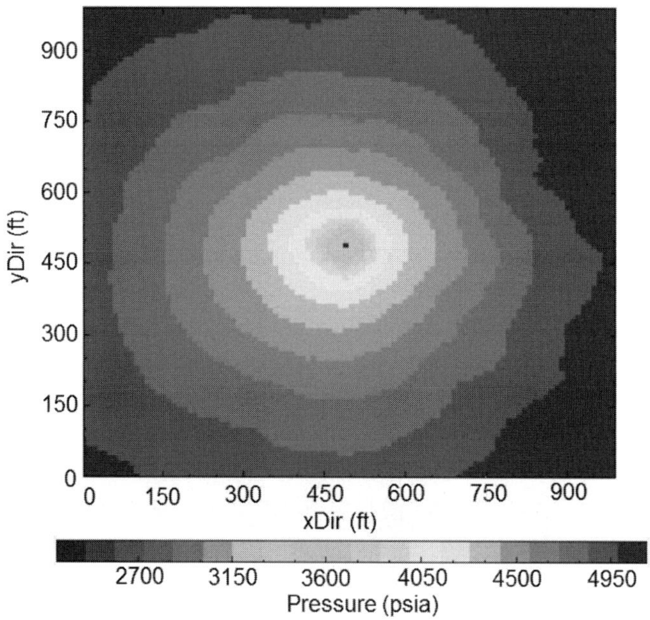

Figure 8.4 Pressure distribution at 3600 days.

are solved using analytic and numerical methods, leading to substantial savings in computation time compared to a full three dimensional flow simulation. The inversion module utilizes streamline-based semi-analytic sensitivities to update the spatial permeability distribution. In the inversion module one may match fluid front arrival times, fluid front amplitudes, and the flowing bottom-hole pressure.

8.2.1 Forward modeling

We illustrate the use of the Trace3D software by simulating water injection into a five-spot pattern with a central injector and four neighboring producers [see Exercises 6.1 and 6.2]. The heterogeneous permeability field for this example is shown in Figure 8.5. The pressure distribution is computed by solving the scalar partial differential equation (6.13),

$$\varphi \bar{c} \frac{\partial \bar{p}}{\partial t} - \frac{1}{\rho_w} \nabla \cdot [\rho_w \lambda_w \nabla \bar{p}] - \frac{1}{\rho_n} \nabla \cdot [\rho_n \lambda_n \nabla \bar{p}] = \hat{\bar{Q}}$$

using a finite-difference algorithm, where $\bar{p}$ is the average fluid pressure. From the pressure field, $\bar{p}$, we can compute the velocity field, $\mathbf{q}$ using Darcy's law, Equation (2.90)

$$\mathbf{q} = -\frac{k}{\mu}\nabla\bar{p},$$

where k is the absolute permeability and μ is the fluid viscosity. The velocity field can then be used for generating the trajectories shown in Figure 8.6. Note that for this multiphase case, the streamline trajectories are constructed using the total phase velocity $\mathbf{q}_t = \mathbf{q}_w + \mathbf{q}_n$, where $\mathbf{q}_w$ is the wetting phase and $\mathbf{q}_n$ is the non-wetting phase. The time of flight is obtained by integrating the total velocity along the trajectories, as given in Equation (6.53),

$$\tau(s) = \int_{\mathbf{x}(0)}^{\mathbf{x}(s)} \frac{\varphi}{|\mathbf{q}_t|} d\mathbf{x}$$

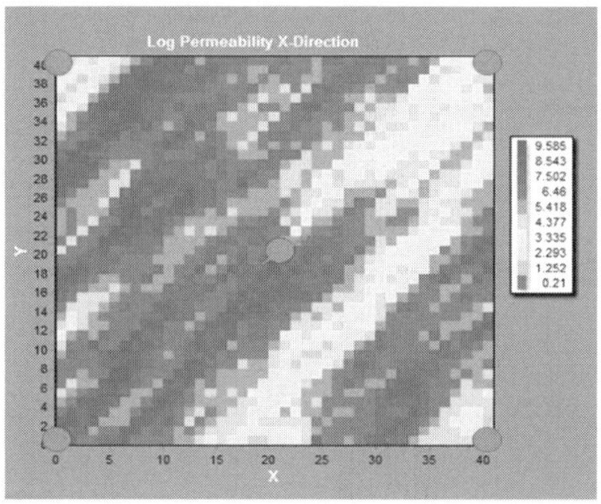

Figure 8.5 Permeability distribution and well configuration for the synthetic test illustrating Trace3D.

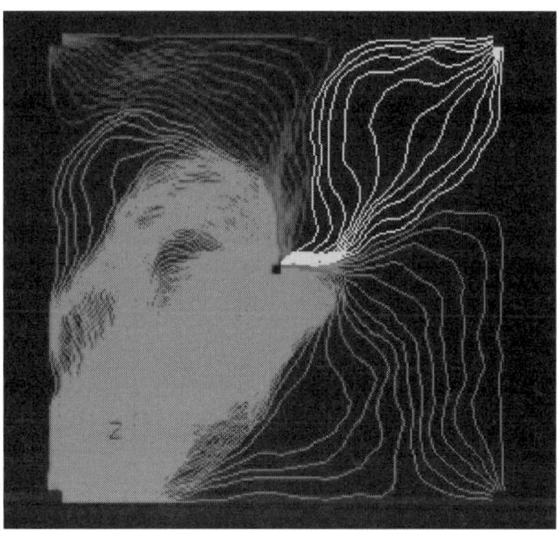

Figure 8.6 Streamline trajectories.

where the integral is taken along the trajectory $\mathbf{x}(s)$ from the source to the observation point at a distance s along the trajectory. The time of flight can be computed by starting either from an injector or starting from a producer [see Section 5.5.1]. The time of flight contours from an injector display the area, or a volume in three dimensions, swept by a neutral tracer. The time of flight from the producers displays the regions drained by the respective producing wells. The streamline bundles provide a convenient mechanism to partition the reservoir using injection or producing wells as shown in Chapter 5, Section 5.5.1. Finally, as in Fronts3D, we can use the time of flight as a spatial coordinate transform the four dimensional saturation equation (three space variables and one time variable) into a two dimensional differential equation (one space variable, τ, and one time variable)

$$\frac{\partial S}{\partial t} + \frac{df_w}{dS}\frac{\partial S}{\partial \tau} = 0.$$

In Trace3D, this equation is solved using stable and efficient numerical methods, as detailed in Chapter 6, Section 6.4.2, to compute the saturation histories at the individual wells. Note that this equation must be formulated and solved along each of the streamlines.

8.2.2 Inverse modeling

The Trace3D package has routines for solving the inverse problem, where one estimates the spatial variation of permeability based upon observations of flow and transport. In this case the observations are water-cut responses from 8 wells surrounding a single water injector [see Exercise 6.10]. These observations include the water breakthrough times, as well as the full watercut (water fraction) histories, at the 8 adjacent wells. The reference model, the heterogeneous permeability field to be reconstructed, is shown in Figure 8.7(a). The water-cut observations at the well

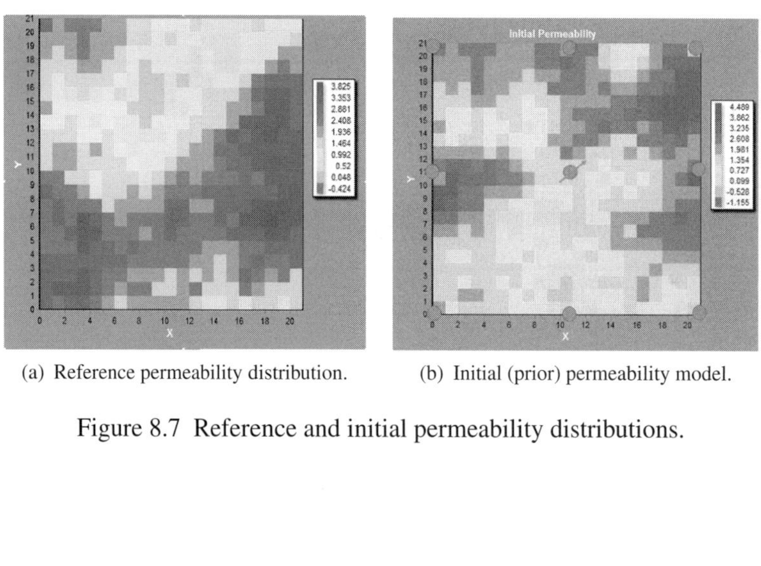

(a) Reference permeability distribution. (b) Initial (prior) permeability model.

Figure 8.7 Reference and initial permeability distributions.

have been generated using the reference permeability distribution and the forward modeling capability of Trace3D.

The first step involves matching of the water breakthrough or arrival times at the observation wells using streamline-derived analytical sensitivities, Equation (6.120–6.122). For example, the sensitivity of the water front arrival time to changes in permeability at $\mathbf{x}$ is given by

$$\frac{\partial T}{\partial k(\mathbf{x})} = -\frac{1}{k_o(\mathbf{x})|\mathbf{v}_o(\mathbf{x})|}$$

where the subscript o indicates that the quantities correspond to the background model values. This is a semi-analytic quantity that may be constructed from the background model and a single reservoir simulation. The deviation in an arrival time due to permeability variations along a flow path $\mathbf{x}_o(s)$ from the injector to the receiver is given by

$$\delta T = \int_{\mathbf{x}_o(s)} \frac{1}{|\mathbf{v}_o|} \frac{\delta k}{k_o} ds.$$

The reader may produce this expression by following the derivation (6.116)–(6.118) for porosity, as requested in Exercise 6.9. The arrival times from the reference model are compared with those of the initial reservoir model in Figure 8.8 and, as expected, show large discrepancies from the synthetic arrival times of the reference model (the 'data').

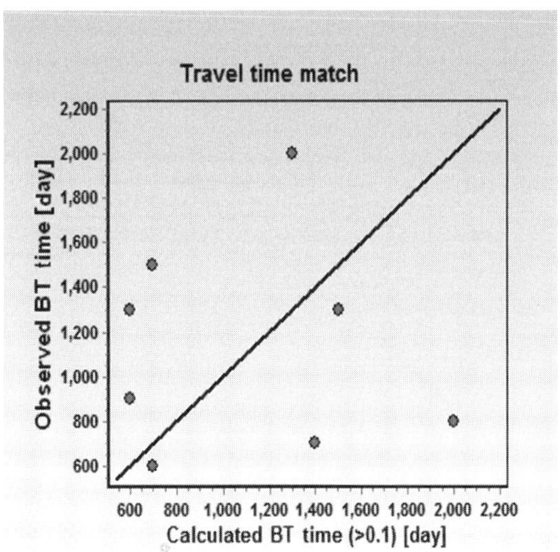

Figure 8.8 Arrival time match based upon the initial model.

The arrival time inversion implemented in Trace3D is an iterative, linearized updating of the model described in Section 5.5.3 of Chapter 5. It follows from the discretization of the preceding integral, producing a linear constraint on the model perturbations along the trajectory. A collection of such constraints, in this case 8 travel times, but often many more in actual field settings, produces a linear system of equations relating the travel time perturbations $\delta\mathbf{T}$ to perturbations in permeabilities $\delta\mathbf{k}$,

$$\delta\mathbf{T} = \mathbf{M}\delta\mathbf{k}.$$

The coefficients of the matrix $\mathbf{M}$ result from the discretization of the preceding integral. As discussed in Chapter 5, Section 5.5.3, penalty terms are added to stabilize the solution of the linear system of equations and a regularized least-squares approach is adopted producing the linear system (5.85) of normal equations

$$\left[\mathbf{M}^t\mathbf{M} + w_n\mathbf{I} + w_r\mathbf{L}^t\mathbf{L}\right]\delta\mathbf{k} = \mathbf{M}^t\delta\mathbf{T}.$$

This sparse linear system is solved for $\delta\mathbf{k}$ using an iterative solver. The perturbation $\delta\mathbf{k}$, is added to the initial or starting permeability model, plotted in Figure 8.7(b). The iterative updating continues until the error reduction levels off or the observations are fit within their estimated uncertainties. The permeability changes due to the iterative updating are shown in Figure 8.10(a). After the inversion, the arrival time match for the updated model is shown in Figure 8.9. The arrival time

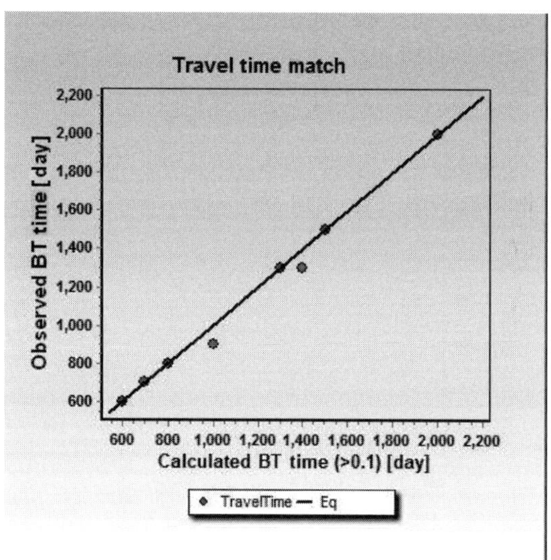

Figure 8.9 Arrival time match based upon the final inversion result.

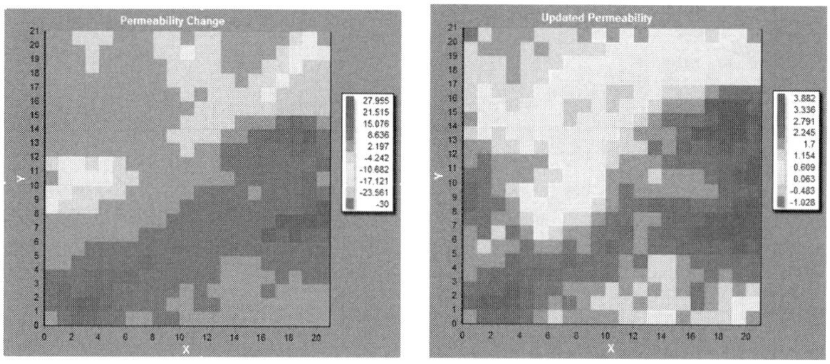

(a) Permeability changes during the inversion.

(b) Final permeability model.

Figure 8.10 Permeabilities resulting from the inversion.

match is followed by an amplitude match, using the semi-analytic fractional flow sensitivities given by Equation (6.131)

$$\frac{\delta f_j}{\delta k_i} = -\sum_{n=1}^{N_j} \frac{1}{t} \frac{df_w}{dS} S_o{}' \left(\frac{\tau}{t}\right) \int_{v_i} \frac{1}{k_o |\mathbf{q}_o|} dr.$$

The updated permeabilities resulting from the inversion are plotted in Figure 8.10(b). A comparison of the final updated permeability distribution with the reference permeability distribution indicates that the inversion is able to reproduce much of the large-scale features of the reference permeability field.

The reader is urged to solve the software example problems given throughout the book to get an appreciation of the power and applicability of the methods.

References

Abbaszadeh, D. M., and Brigham, W. E. 1984. Analysis of well-to-well tracer flow to determine reservoir layering. *Journal of Petroleum Technology*, **36**, 1753–1762.

Abramowitz, M., and Stegun, I. A. 1972. *Handbook of Mathematical Functions*. New York: Dover Publications Inc.

Ajo-Franklin, J., Daley, T., Butler-Veytia, B., Peterson, J., Wu, Y., Kelly, B., and Hubbard, S. 2011. SEG-2011-3727. *Multi-level continuous active source seismic monitoring (ML-CASSM): Mapping shallow hydrofracture evolution at a TCE contaminated site*. Society of Exploration Geophysics Annual Meeting. San Antonio, Texas.

Aki, K., and Richards, P. G. 1980a. *Quantitative Seismology*. San Francisco: W. H. Freeman and Sons.

Aki, K., and Richards, P. G. 1980b. *Quantitative Seismology*. San Francisco: W. H. Freeman and Company.

Anile, A. M., Hunter, J. K., Pantano, P., and Russo, G. 1993. *Ray Methods for Nonlinear Waves in Fluids and Plasmas*. Essex: Longman Scientific and Technical.

Annable, M., Rao, P., Hatfield, K., Graham, W., Wood, A., and Enfield, C. 1998. Partitioning tracers for measuring residual NAPL: Field-scale test results. *Journal of Environmental Engineering*, **124**, 498–503.

Aster, R. C., Borchers, B., and Thurber, C. 2013. *Parameter Estimation and Inverse Problems*. San Diego: Academic Press.

Auriault, J. L. 1980. Dynamic behavior of a porous medium saturated by a Newtonian fluid. *International Journal of Engineering Science*, **18**, 775–785.

Auriault, J. L., Moyne, C., and Amaral Souto, H. P. 2010. On the asymmetry of the dispersion tensor in porous media. *Transport in Porous Media*, **85**, 771–783.

Aziz, K., and Settari, A. 1979. *Petroleum Reservoir Simulation*. Amsterdam: Elsevier.

Backus, G. E., and Mulcahy, M. 1976. Moment tensors and other phenomenological descriptions of seismic sources I- Continuous displacements. *Geophysical Journal of the Royal Astronomical Society*, **46**, 341–371.

Barenblatt, G. I. 1979. *Similarity, Self-Similarity, and Intermediate Asymptotics*. New York: Consultants Bureau.

Bear, J. 1961. On the tensor form of dispersion in porous media. *Journal of Geophysical Research*, **66**, 1185–1197.

Bear, J. 1972. *Dynamics of Fluids in Porous Media*. New York: Dover Publications.

Bear, J., Corapcioglu, M. Y., and Balakrishna, J. 1984. Modeling of centrifugal filtration in unsaturated deformable porous media. *Advances in Water Resources*, **7**, 150–167.

Behrens, R., Condon, P., Haworth, W., Bergeron, M., Wang, Z., and Ecker, C. 2002. 4D seismic monitoring of water influx at bay marchand: The practical use of 4D in an imperfect world. *SPE Reservoir Evaluation and Engineering*, **5**, 410–420.

Bell, J. B., Trangenstein, J. A., and Shubin, G. R. 1986. Conservation laws of mixed type describing three-phase flow in porous media. *SIAM Journal on Applied Mathematics*, **46**, 1000–1017.

Belytschko, T., Liu, W. K., Moran, B., and Elkhodary, K. I. 2014. *Nonlinear Finite Elements for Continua and Structures*. West Sussex: Wiley and Sons.

Benamou, J. D. 1996. Big ray tracing: Multivalued travel time field computation using viscosity solutions of the eikonal equation. *Journal of Computational Physics*, **128**, 463–474.

Bender, C. M., and Orszag, S. A. 1978. *Advanced Mathematical Methods for Scientists and Engineers*. New York: McGraw-Hill Book Company.

Berryman, J. G. 1986. Effective medium approximation for elastic constants of porous solids with microscopic heterogeneity. *Journal of Applied Physics*, **59**, 1136–1140.

Berryman, J. G. 1995. Mixture theory for rock properties. In: Ahrens, T. J. (ed), *Rock Physics and Phase Relations*. Washington, D. C.: American Geophysical Union.

Berryman, J. G., and Milton, G. W. 1991. Exact results for generalized Gassmann's equations in composite porous media with two constituents. *Geophysics*, **56**, 1950–1960.

Berryman, J. G., Thigpen, L., and Chin, R. C. Y. 1988. Bulk elastic wave propagation in partially saturated porous solids. *Journal of the Acoustical Society of America*, **84**, 360–373.

Bertrand, A., Folstad, P. G., Lyngnes, B., Buizard, S., Hoeber, H., Pham, N., PS., De, and Grandiierrepont. 2014. Ekofisk life-of-field seismic: Operations and 4D processing. *The Leading Edge*, **33**, 142–148.

Biot, M. A. 1941. General theory of three-dimensional consolidation. *Journal of Applied Physics*, **12**, 155–164.

Biot, M. A. 1956a. Theory of propagation of elastic waves in a fluid-saturated porous solid. I. Low-frequency range. *Journal of the Acoustical Society of America*, **28**, 168–178.

Biot, M. A. 1956b. Theory of propagation of elastic waves in a fluid-saturated porous solid. II. Higher-frequency range. *Journal of the Acoustical Society of America*, **28**, 179–1191.

Biot, M. A. 1962a. Generalized theory of acoustic propagation in porous dissipative media. *Journal of the Acoustical Society of America*, **34**, 1254–1264.

Biot, M. A. 1962b. Mechanics of deformation and acoustic propagation in porous media. *Journal of Applied Physics*, **33**, 1482–1498.

Bishop, A. W., and Blight, G. E. 1963. Some aspects of effective stress in saturated and partially saturated soils. *Geotechnique*, **13**, 177–197.

Bracewell, R. N. 2000. *The Fourier Transform and Its Applications*. Boston: McGraw-Hill Book Company.

Bratvedt, F., Gimse, T., and Tegnander, C. 1996. Streamline computations for porous media flow including gravity. *Transport in Porous Media*, **25**, 63–78.

Brauchler, R., Hu, R., Vogt, T., Al-Halbouni, D., Heinrichs, T., Ptak, T., and Sauter, M. 2010. Cross-well slug interference tests: An effective characterization method for resolving aquifer heterogeneity. *Journal of Hydrology*, **384**, 33–45.

Bruining, H., Darwish, M., and Rijnks, A. 2012. Computation of the longitudinal and transverse dispersion coefficients in an adsorbing porous medium using homogenization. *Transport in Porous Media*, **91**, 833–859.

Buckley, S. E., and Leverett, M. C. 1942. Mechanism of fluid displacement in sands. *Transactions of the American Society of Mechanical Engineers*, **146**, 107–116.

Burridge, R., and Keller, J. B. 1981. Poroelasticity equations derived from microstructure. *Journal of the Acoustical Society of America*, **70**, 1140–1146.

Butler, J. J., Garnett, E., and Healey, J. M. 2003. Analysis of slug test in formations of high hydraulic conductivity. *Groundwater*, **41**, 620–630.

Calvert, R. 2005. *Insights and methods for 4D reservoir monitoring and characterization.* Distinguished Instructor Short Course 8: EAGE/SEG.

Carbonell, H. B. G., and Whitaker, S. 1983. Dispersion in pulsed systems - II Theoretical developments for passive dispersion in porous media. *Chemical Engineering Science*, **38**, 1795–1802.

Carcione, J. M., Morency, C., and Santos, J. E. 2010. Computational poroelasticity – A review. *Geophysics*, **75**, A229–A243.

Chapman, C. H. 2004. *Fundamentals of Seismic Wave Propagation.* Cambridge: Cambridge University Press.

Charlaix, E., Kushnick, A. P., and Stokes, J. P. 1988. Experimental study of dynamic permeability in porous media. *Physical Review Letters*, **61**, 1595–1598.

Chen, Z. X. 1988. Some invariant solutions to two-phase fluid displacement problems including capillary effects. *SPE Reservoir Engineering*, **30**, 691–700.

Cheng, H., Datta-Gupta, A., and He, Z. 2005. A comparison of travel-time and amplitude inversion for scale production data integration into geologic models: Sensitivity, non-linearity, and practical implications. *SPE Journal*, **10**, 75–90.

Chorin, A. J., and Marsden, J. E. 1993. *A Mathematical Introduction to Fluid Mechanics.* Berlin: Springer-Verlag.

Cleary, R. W., and Ungs, M. J. 1978. Groundwater pollution and hydrology, mathematical models, and computer programs. *Water Resources Program Report*, **78**, 1–34.

Cole, J. D. 1951. On a quasilinear parabolic equation occurring in aerodynamics. *Quarterly of Applied Mathematics*, **9**, 225–236.

Cole, J. D., and Kevorkian, J. 1963. Uniformly valid asymptotic approximations for certain nonlinear differential equations. Pages 113–120 of: *Nonlinear Differential Equations and Nonlinear Mechanics.* New York: Academic Press.

Courant, R., and Friedrichs, K. O. 1948. *Supersonic Flow and Shock Waves.* New York: Interscience Publishers.

Courant, R., and Hilbert, D. 1962a. *Methods of Mathematical Physics.* New York: John Wiley and Sons.

Courant, R., and Hilbert, D. 1962b. *Methods of Mathematical Physics.* New York: Interscience.

Coussy, O. 2010. *Mechanics and Physics of Porous Solids.* Chichester: John Wiley and Sons.

Crandall, M. G., and Lions, P. 1983. Viscosity solutions of Hamilton-Jacobi equations. *Transactions of the American Mathematical Society*, **277**, 1–43.

Crandall, M. G., and Majda, A. 1980. Monotone difference approximations for scalar conservation laws. *Mathematics of Computation*, **34**, 1–21.

Crandall, M. G., Evans, L. C., and Lions, P. 1984. Some properties of viscosity solutions of Hamilton-Jacobi equations. *Transactions of the American Mathematical Society*, **282**, 487–502.

Crane, M. J., and Blunt, M. J. 1999. Streamline-based simulation of solute transport. *Water Resources Research*, **35**, 3061–3078.

Crank, J. 1975. *The Mathematics of Diffusion.* London: Oxford University Press.

Dafermos, C. M. 1972. Polygonal approximation of solutions of the initial value problem for a conservation law. *Journal of Mathematical Analysis and Applications*, **38**, 33–41.

Dafermos, C. M. 2000. *Hyperbolic Conservation Laws in Continuum Physics*. New York: Springer.

Dagan, G. 1982. Stochastic modeling of groundwater flow by unconditional and conditional probabilities, 2, The solute transport. *Water Resources Research*, **18**, 835–848.

Daley, T. M., Ajo-Franklin, J. B., and Doughty, C. 2011. Constraining the reservoir model of an injected CO_2 plume with crosswell CASSM at the Frio-II brine pilot. *International Journal of Greenhouse Gas Control*, **5**, 1022–1030.

Datta-Gupta, A., and King, M. J. 1995. A semianalytic approach to tracer flow modeling in heterogeneous permeable media. *Advances in Water Resources*, **18**, 9–24.

Datta-Gupta, A., and King, M. J. 2007. *Streamline Simulation: Theory and Practice*. Richardson, Texas: Society of Petroleum Engineers.

Datta-Gupta, A., Yoon, S., Vasco, D. W., and Pope, G. A. 2002. Inverse modeling of partitioning interwell tracer tests: A streamline approach. *Water Resources Research*, **38**(6), 1–15.

Davis, H. F. 1967. *Introduction to Vector Analysis*. Boston, Massachuetts: Allyn and Bacon.

de Boer, R. 2000. *Theory of Porous Media*. Berlin: Springer.

de Josselin de Jong, G. 1958. Longitudinal and transverse diffusion in granular deposits. *Transactions of the American Geophysical Union*, **39**, 67–74.

de Marsily, G. 1986. *Quantitative Hydrogeology*. San Diego: Academic Press.

Debnath, L. 2005. *Nonlinear Partial Differential Equations for Scientists and Engineers*. Boston: Birkhauser.

Deschamps, T., and Chen, L. D. 2001. Fast extraction of minimal paths in 3D images and applications to virtual endoscopy. *Medical Image Analysis*, **5**, 281–299.

Dingle, R. B. 1973. *Asymptotic Expansions: Their Derivation and Interpretation*. London: Academic Press.

Dorny, C. N. 1983. *A Vector Space Approach to Models and Optimization*. Malabar, Florida: Robert E. Krieger Publishing Company.

Drumheller, D. S. 1978. Theoretical treatment of a porous solid using mixture theory. *International Journal of Solids and Structures*, **14**, 441–456.

Drumheller, D. S. 1998. *Introduction to Wave Propagation in Nonlinear Fluids and Solids*. Cambridge: Cambridge University Press.

Dutta, N. C., and Ode, H. 1979a. Attenuation and dispersion of compressional waves in fluid-filled porous rocks with partial gas saturation (White model) – Part I: Biot theory. *Geophysics*, **44**, 1777–1788.

Dutta, N. C., and Ode, H. 1979b. Attenuation and dispersion of compressional waves in fluid-filled porous rocks with partial gas saturation (White model) – Part II: Results. *Geophysics*, **44**, 789–805.

Dvorkin, J., Mavko, G., and Nur, A. 1995. Squirt flow in fully saturated rocks. *Geophysics*, **60**, 97–107.

Erdélyi, A. 1956. *Asymptotic Expansions*. New York: Dover Publications Inc.

Euler, L. 1754. De seriebus divergentibus. *Novi Commentarii Acad. Sci. Petropolitanae*, **5**, 205–237.

Falls, A. H., and Schulte, W. M. 1992. Features of three component, three phase displacement in porous media. *SPE Reservoir Engineering*, **7**, 426–432.

Faucette, W. J. 1996. A geometric interpretation of the solution of the general quartic polynomial:. *The American Mathematical Monthly*, **103**, 51–57.

Ferretti, A. 2014. *Satellite InSAR Data: Reservoir Monitoring from Space*. Netherlands: European Association of Geoscientists and Engineers.

Fick, A. 1855. Ueber Diffusion. *Annalen der Physik und Chemie von J. C. Pogendorff*, **94**, 59–86.

Fletcher, R. 2000. *Practical Methods of Optimization*. Cornwall: John Wiley and Sons.

Fokas, A. S., and Yortsos, Y. C. 1982. On the exactly solvable equation $S_t = \left[(\beta S + \gamma)^{-2} S_x\right]_x + \alpha\,(\beta S + \gamma)^{-2}\, S_x$ occurring in two-phase flow in porous media. *SIAM J. Appl. Math*, **42**, 318–332.

Formal, S., and Sethian, J. A. 2002. Fast-phase space computation of multiple arrivals. *Proceedings of the National Academy of Sciences*, **99**, 7329–7334.

Fourier, J. B. 1822. *Theorie Analytique de la Chaleur*. New York: Dover Publications, English translation by A. Freeman (1955).

Frenkel, J. 1944. On the theory of seismic and seismoelectric phenomena in a moist soil. *Journal of Physics*, **8**(September), 879–887.

Fujita, Y., Datta-Gupta, A., and King, M. J. 2015. A comprehensive reservoir simulator for unconventional reservoirs based on the Fast Marching Method and diffusive time of flight. SPE 173269 In: *Proceedings of Society of Petroleum Engineers Reservoir Simulation Symposium*. Society of Petroleum Engineers, Houston, Texas. *SPE Journal* 2016. http://dx.doi.org/10.2118/173269-PA.

Fung, Y. C. 1969. *A First Course in Continuum Mechanics*. New Jersy: Prentice-Hall.

Gardner, C. S., and Morikawa, G. K. 1960. Similarity in the asymptotic behavior of collision-free hydromagnetic waves and water waves. *Courant Institute of Mathematical Sciences*, **9082**, 1–30.

Garg, S. K. 1971. Wave propagation effects in a fluid-saturated porous solid. *Journal of Geophysical Research*, **76**, 7947–7962.

Garg, S. K., and Nayfeh, A. H. 1986. Compressional wave propagation in liquid and/or gas saturated elastic porous media. *Journal of Applied Physics*, **60**, 3045–3055.

Gassmann, F. 1951a. Elastic waves through a packing of spheres. *Geophysics*, **16**, 673–685.

Gassmann, F. 1951b. Ueber die elastizitat poroser medien. *Viertel. Naturforsch. Ges. Zurich*, **96**, 1–23.

Gelhar, L. W., and Axness, C. L. 1983. Three-dimensional stochastic analysis of macrodispersion in aquifers. *Water Resources Research*, **19**, 161–180.

Gelhar, L. W., Gutjahr, A. L., and Naff, R. L. 1979. Stochastic analysis of macrodispersion in a stratified aquifer. *Water Resources Research*, **15**, 1387–1397.

Gautier, Y., Nœtinger, B., and Roggero, F. 2001. History matching using a streamline-based approach and exact sensitivity coefficients. SPE 716626 In: *Proceedings of the Society of Petroleum Engineers Annual Technical Conference and Exhibition*, New Orleans, Louisiana.

Gibson-Poole, C. M., and Raikes, S. 2010. Enhanced understanding of CO_2 storage at Krechba from 3D seismic. Pages 10–13 of: *Proceedings of the 9th Annual Conference on Carbon Capture and Sequestration*. Pittsburgh, May, 2010: Pennsylvania.

Gill, P. E., Murray, W., and Wright, M. H. 1982. *Practical Optimization*. New York: Academic Press.

Godunov, S. K. 1959. Finite difference method for numerical computation of discontinuous solutions of the equations of fluid dynamics. *Matematicheskii Sbornik*, **47**, 271–306.

Goldstein, H. 1950. *Classical Mechanics*. Reading, MA: Addison-Wesley.

Golub, G. H., and Van Loan, C. F. 1989. *Matrix Computations*. Baltimore: The John Hopkins University Press.

Gradshteyn, I. S., and Ryzhik, I. M. 1965. *Tables of Integrals, Series, and Products*. New York: Academic Press.

Green, A. E., and Naghdi, P. M. 1965. A dynamical theory of interacting continua. *International Journal of Engineering Science*, **3**, 231–341.

Guzman, R. E., and Fayers, F. J. 1997. Solutions to the three-phase Buckley-Leverett problem. *SPE Journal*, **2**, 301–311.

Hanyga, A., and Seredynska, M. 1999. Some effects of the memory kernel singularity on wave propagation and inversion in poroelastic media Forward problems – I. *Geophysical Journal International*, **137**, 319–335.

Hart, R. D. and St. John, C. M. 1986. Formulation of a fully coupled thermal-mechanical-fluid flow model for nonlinear geologic systems. *International Journal of Rock Mechanics and Mining Sciences and Geomechanics Abstracts*, **23** (3), 213–224.

Harten, A. 1983. High resolution schemes for hyperbolic conservation laws. *Journal of Computational Physics*, **49**, 357–393.

Harten, A., and Osher, S. 1987. Uniformly high-order accurate nonoscillatory schemes I. *SIAM Journal on Numerical Analysis*, **24**, 279–309.

He, Z., Yoon, S., Datta-Gupta, A. 2002. Streamline-based production data integration under changing field conditions. *SPE Journal*, **7**(4), 423–436.

Helfferich, F. G. 1981. Theory of multicomponent, multiphase displacement in porous media. *SPE Journal*, **21**, 51–62.

Hohl, D., Jimenez, E., and Datta-Gupta, A. 2006. Field experiences with history matching an offshore turbidite reservoir using inverse modeling. SPE 101983 In: *Proceedings of the SPE Annual Technical Conference and Exhibition*. San Antonio, Texas.

Holden, H., and Risebro, N. H. 2002. *Front Tracking for Hyperbolic Conservation Laws*. New York: Springer-Verlag.

Holden, H., Holden, L., and Hoegh-Krohn, R. 1988. A numerical method for first order nonlinear scalar conservation laws in one-dimension. *Computers and Mathematics with Applications*, **15**, 595–602.

Hopf, E. 1950. The partial differential equation $u_t + uu_x = \mu u_{xx}$. *Communications on Pure and Applied Mathematics*, **3**, 201–230.

Huang, X., Meister, L., and Workman, R. 1998. Improving production history matching using time-lapse seismic data. *The Leading Edge*, **17**, 1430–1433.

Hubbert, M. K. 1940. The theory of ground-water motion. *The Journal of Geology*, **48**(8), 785–944.

Jaeger, J. C., Cook, N. G. W., and Zimmerman, R. W. 2007. *Fundamentals of Rock Mechanics*. Oxford: Blackwell Publishing.

Javandel, I., Doughty, C., and Tsang, C. F. 1984. *Groundwater Transport: Handbook of Models*. Washington, D. C.: American Geophysical Union Monograph.

Jeffrey, A., and Kawahara, T. 1982. *Asymptotic Methods in Nonlinear Wave Theory*. Boston: Pitman Advanced Publishing Program.

Jin, M., Delshad, M., Varadrajan, D., McKinney, D., Pope, G. A., Sepehrnoori, K., Tilburg, C. E., and Jackson, R. E. 1995. Estimation and remediation performance assessment of subsurface nonaqueous phase liquids. *Water Resources Research*, **31**, 1201–1211.

Johnson, D. L. 2001. Theory of frequency dependent acoustics in patchy-saturated porous media. *Journal of the Acoustical Society of America*, **110**, 682–694.

Johnson, D. L., Koplik, J., and Dashen, R. 1987. Theory of dynamic permeability and tortuosity in fluid-saturated porous media. *Journal of Fluid Mechanics*, **176**, 379–402.

Juanes, R., and Patzek, T. W. 2004. Analytical solution to the Riemann problem of three-phase flow in porous media. *Transport in Porous Media*, **55**, 47–70.

Kam, D., and Datta-Gupta, A. 2016. Streamline-based Transport Tomography With Distributed Water Arrival Times, *SPE Reservoir Evaluation and Engineering*, http://dx.doi.org/10.2118/169105-PA.

Karasaki, K., Long, J., and Witherspoon, P. 1988. Analytical models of slug tests. *Water Resources Research*, **24**, 115–126.

Karpman, V. I. 1975. *Non-Linear Waves in Dispersive Media*. Oxford: Pergamon Press.

Keller, J. B. 1954. Geometrical acoustics I. The theory of weak shock waves. *Journal of Applied Physics*, **25**, 938–947.

Keller, J. B. 1977. Effective behavior of heterogeneous media. Pages 213–224 of: Landman, U. (ed), *Statistical Mechanics and Statistical Methods in Theory and Applications*. New York. **23**: Plenum.

Kevorkian, J., and Cole, J. D. 1996. *Multiple Scale and Singular Perturbation Methods*. New York: Springer-Verlag.

Kim, J. U., Datta-Gupta, A., Jimenez, E., and Hohl, D. 2010. A dual scale approach to production data integration into high resolution geologic models. *Journal of Petroleum Science and Engineering*, **71**, 147–159.

King, M. J., and Datta-Gupta, A. 1998. Streamline simulation: A current perspective. *In Situ*, **22**, 91–140.

Kline, M., and Kay, I. W. 1965. *Electromagnetics Theory and Geometrical Optics*. New York: Interscience Publishers.

Koch, D. L., and Brady, J. F. 1987. The symmetry properties of the effective diffusivity tensor in anisotropic porous media. *Physics of Fluids*, **30**, 642–650.

Kosten, C. W., and Zwikker, C. 1941. Extended theory of the absorption of sound by compressible wall-coverings. *Physica*, **8**, 968–978.

Kravtsov, Y. A., and Orlov, Y. I. 1990. *Geometrical Optics of Inhomogeneous Media*. Berlin: Springer-Verlag.

Lake, L. W. 1989. *Enhanced Oil Recovery*. Englewook: Prentice Hall.

Lanczos, C. 1962. *The Variational Principles of Mechanics*. New York: Dover Publications.

Landro, M. 2001. Discrimination between pressure and fluid saturation changes from time-lapse seismic data. *Geophysics*, **66**, 836–844.

Lawson, C. L., and Hanson, R. J. 1974. *Solving Least Squares Problems*. Englewood Cliffs, New Jersey: Prentice-Hall.

Le Veque, R. J. 1990. *Numerical Methods for Conservation Laws*. Basel: Birkhauser-Verlag.

Lee, W. J. 1982. *Well Testing*. Richardson, Texas: Society of Petroleum Engineers.

Levy, T. 1979. Propagation of waves in a fluid-saturated porous elastic solid. *International Journal of Engineering Science*, **17**, 1005–1014.

Lie, K. a., and Juanes, R. 2005. A front-tracking method for the simulation of three-phase flow in porous media. *Computational Geosciences*, **9**, 29–59.

Lighthill, M. J. 1958. *The Fourier Transform and Generalized Functions*. Cambridge: Cambridge Tracts on Mechanics and Applied Mathematics.

Lindstedt, A. 1883. Beitrag zur Integration der Differentialgleichungen der Storungsteorie. *Mémoires de l'Académie impériale des sciences de St. Pétersbourg*, **31**(4).

Liu, Y., and Si, B. C. 2008. Analytic modeling of one-dimensional diffusion in layered systems with position-dependent diffusion coefficients. *Advances in Water Resources*, **31**, 251–268.

Lo, W. C., Sposito, G., and Majer, E. 2002. Immiscible two-phase fluid flows in deformable porous media. *Advances in Water Resources*, **25**, 1105–1117.

Logan, J. D. 2008. *An Introduction to Nonlinear Partial Differential Equations*. Hoboken: John Wiley and Sons.

Ludwig, J. B. 1966. Uniform asymptotic expansions at a caustic. *Communications in Pure and Applied Mathematics*, **19**, 215–250.

Luenberger, D. G. 1973. *Introduction to Linear and Nonlinear Programming*. Reading Massachusetts: Addison-Wesley.

Luke, J. C. 1966. A perturbation method for nonlinear dispersive wave problems. *Proceedings of the Royal Society of London A: Mathematical, Physical and Engineering Sciences*. **292**(1430), 403–412.

Luneburg, R. K. 1966. *Mathematical Theory of Optics*. Berkeley: University of California Press.

Ma, X., Al-Harbi, M., Datta-Gupta, A., and Efendiev, Y. 2008. An efficient two-stage sampling method for uncertainty quantification in history matching geological models. *SPE Journal*, **3**(1), 77–87.

MacBeth, C., and Al-Maskeri, Y. 2006. Extraction of permeability from time-lapse seismic data. *Geophysical Prospecting*, **54**, 333–349.

MacNeal, R. H. 1953. An asymmetric finite difference network. *Quarterly of Applied Mathematics*, **2**, 295–310.

Marsden, J. E., and Tromba, A. J. 1976. *Vector Calculus*. San Francisco: W. H. Freeman and Company.

Maslov, V. P., and Omel'yanov, G. A. 2001. *Geometric Asymptotics for Nonlinear PDE. I*. Providence: American Mathematical Society.

Masson, Y. J., and Pride, S. R. 2007. Poroelastic finite difference modeling of seismic attenuation and dispersion due to mesoscopic-scale heterogeneity. *Journal of Geophysical Research*, **112**, 1–17.

Masson, Y. J., and Pride, S. R. 2010. Finite-difference modeling of Biot's poroelastic equations across all frequencies. *Geophysics*, **75**, N33–N41.

Masson, Y. J., and Pride, S. R. 2011. Seismic attenuation due to patchy saturation. *Journal of Geophysical Research*, **116**, 1–17.

Masson, Y. J., Pride, S. R., and Kihei, K. T. 2006. Finite difference modeling of Biot's poroelastic equations at seismic frequencies. *Journal of Geophysical Research*, **111**, 1–13.

Mavko, G., and Nur, A. 1975. Melt squirt in the asthenosphere. *Journal of Geophysical Research*, **80**, 1444–1448.

Mavko, G., and Nur, A. 1979. Wave attenuation in partially saturated rocks. *Geophysics*, **44**, 161–178.

Mavko, G., Mukerji, T., and Dvorkin, J. 1998. *The Rock Physics Handbook: Tools for Seismic Analysis in Porous Media*. Cambridge: Cambridge University Press.

McWhorter, D. B. 1971. *Infiltration affected by flow of air*. Hydrol. Paper 49. Colorado State University, Fort Collins.

McWhorter, D. B., and Sunada, D. K. 1990. Exact integral solution for two-phase flow. *Water Resources Research*, **26**(3), 399–413.

Menke, W. 1989. *Geophysical Data Analysis: Discrete Inverse Theory*. San Diego: Academic Press.

Miller, W. 1977. *Symmetry and Separation of Variables*. Reading: Addison-Wesley.

Miura, R. M., and Kruskal, M. D. 1974. Application of a nonlinear WKB method to the Korteweg-DeVries equation. *SIAM Journal on Applied Mathematics*, **26**(2), 376–395.

Morency, C., Yang, L., and Tromp, J. 2011. Acoustic, elastic and poroelastic simulations of CO_2 sequestration crosswell monitoring based on spectral-element and adjount methods. *Geophysical Journal International*, **185**, 955–966.

Morland, L. W. 1972. A simple constitutive theory for a fluid-saturated porous solid. *Journal of Geophysical Research*, **77**, 890–900.

Müller, T. M., Gurevich, B., and Lebedev, M. 2010. Seismic wave attenuation and dispersion resulting from wave-induced flow in porous rocks – A review. *Geophysics*, **75**, 147–75.

Narasimhan, T. N., and Witherspoon, P. A. 1976. An integrated finite difference method for analyzing fluid flow in porous media. *Water Resources Research*, **12**, 57–64.

Needham, T. 2000. *Visual Complex Analysis*. Oxford: Oxford University Press.

Nickalls, R. W. D. 2009. The quartic equation: Invariants and Euler's solution revealed. *The Mathematical Gazette*, **93**, 66–75.

Nikolaevskii, V. N. 1959. Konvektivnaya diffusiya v poristykh sredakh. *Prikladnaya Matematika i Mekhanika*, **23**, 1042–1050.

Noble, B., and Daniel, J. W. 1977. *Applied Linear Algebra*. Englewood Cliffs: Prentice-Hall.

Nocedal, J., and Wright, S. J. 2006. *Numerical Optimization*. New York: Springer.

Nolet, G. 1987. Seismic wave propagation and seismic tomography. Pages 1–23 of: Nolet, G. (ed), *Seismic Tomography*. Dordrecht: D. Reidel.

Norris, A. N. 1993. Low frequency dispersion and attenuation in partially saturated rocks. *Journal of the Acoustical Society of America*, **94**, 359–370.

O'Connell, R. J., and Budiansky, B. 1977. Viscoelastic properties of fluid-saturated cracked solids. *Journal of Geophysical Research*, **82**, 5719–5735.

Oleninik, O. A. 1957. On the uniqueness of the generalized solution of the Cauchy problem for a non-linear system of equations occurring in mechanics. *Uspekhi Matematicheskikh Nauk*, **78**, 169–176.

Osher, S., and Fedkiw, R. 2003. *Level Set Methods and Dynamic Implicit Surfaces*. New York: Springer.

Paige, C. C., and Saunders, M. A. 1982. LSQR: An algorithm for sparse linear equations and sparse linear systems. *ACM Transactions on Mathematical Software*, **8**, 195–209.

Panday, S., and Corapcioglu, M. Y. 1989. Reservoir transport equations by compositional approach. *Transport in Porous Media*, **4**, 369–393.

Panton, R. L. 2005. *Incompressible Flow*. Hoboken, New Jersey: Wiley.

Parker, R. L. 1994. *Geophysical Inverse Theory*. Princeton: Princeton University Press.

Peaceman, D. W. 1977. *Fundamentals of Numerical Reservoir Simulation*. Amsterdam: Elsevier Scientific Publishing.

Plyushchchenkov, B. D., and Turchaninov, V. 2000. Acoustic logging modeling by refined Biot's equations. *International Journal of Modern Physics C*, **12**, 305–396.

Poincaré, H. 1886. *Acta Math*, **8**, 295–344.

Pollock, D. W. 1988. Semianalytical computation of path lines for finite-difference models. *Groundwater*, **26**, 743–750.

Pope, G. A. 1980. The application of fractional flow theory to enhanced oil recovery. *SPE Journal*, **20**, 191–205.

Press, W. H., Teukolsky, S. A., Vetterling, W. T., and Flannery, B. P. 1992. *Numerical Recipes*. Cambridge: Cambridge University Press.

Pride, S. R. 2005. Relationships between seismic and hydrological properties. Pages 253–291 of: Rubin, Y., and Hubbard, S. S. (eds), *Hydrogeophysics*. New York: Springer.

Pride, S. R., and Berryman, J. G. 1998. Connecting theory to experiment in poroelasticity. *Journal of the Mechanics and Physics of Solids*, **46**, 719–747.

Pride, S. R., and Berryman, J. G. 2003a. Linear dynamics of double-porosity and dual-permeability materials. I. Governing equations and acoustic attenuation. *Phyical Review E*, **68**, 036603–1–10.

Pride, S. R., and Berryman, J. G. 2003b. Linear dynamics of double-porosity and dual-permeability materials. II. Fluid transport equations. *Phyical Review E*, **68**, 036604–1–10.

Pride, S. R., Berryman, J. G., and Harris, J. M. 2004. Seismic attenuation due to wave-induced flow. *Journal of Geophysical Research*, **109**, 1–19.

Pride, S. R., Gangi, A. F., and Morgan, F. D. 1992. Deriving the equations of motion for isotropic motion. *Journal of the Acoustical Society of America*, **92**, 3278–3290.

Pride, S. R., Morgan, F. D., and Gangi, A. F. 1993. Drag forces of porous-medium acoustics. *Physical Review B*, **47**, 4964–4978.

Pruess, K., and Narasimhan, T. N. 1982. On fluid reserves and the production of superheated steam from fractured, vapor-dominated geothermal reservoirs. *Journal of Geophysical Research*, **87**, 9329–9339.

Pruess, K., Oldenburg, C., and Moridis, G. 1999. *TOUGH2 User's Guide, Version 2.0*. Tech. rept. 43134. Lawrence Berkeley National Laboratory. LBNL Report.

Rey, A., Bhark, E., Gao, K., Datta-Gupta, A., and Gibson, R. 2012. Streamline-based integration of time-lapse seismic and production data into petroleum reservoir models. *Geophysics*, **77**, M73–M87.

Rice, J. R., and Cleary, M. P. 1976. Some basic stress diffusion solutions for fluid-saturated elastic porous media with compressible constituents. *Reviews of Geophysics and Space Physics*, **14**, 227–241.

Roach, G. F. 1970. *Green's Functions: Introductory Theory with Applications*. New York: Van Nostrand Reinhold Company.

Rucci, A., Vasco, D. W., and Novali, F. 2010. Fluid pressure arrival time tomography: Estimation and assessment in the presence of inequality constraints, with an application to production at the Krechba field, Algeria. *Geophysics*, **75**, O39–O55.

Rucci, A., Vasco, D. W., and Novali, F. 2013. Monitoring the geologic storage of carbon dioxide using multicomponent SAR interferometry. *Geophysical Journal International*, **193**, 197–208.

Saad, Y. 2003. *Iterative Methods for Sparse Linear Systems*. Philadelphia: Society for Industrial and Applied Mathematics.

Sachdev, P. L. 2000. *Self-Similarity and Beyond: Exact Solutions of Nonlinear Problems*. Boca Raton: Chapman and Hall.

Saffman, P. G. 1959. A theory of dispersion in porous media. *Journal of Fluid Mechanics*, **6**, 321–349.

Santos, J. E., Corbero, J. M., and Douglas, J. 1990. Static and dynamic behavior of a porous solid saturated by a two-phase fluid. *Journal of the Acoustical Society of America*, **87**, 1428–1438.

Scheidegger, A. E. 1954. Statistical hydrodynamics in porous media. *Journal of Applied Physics*, **25**, 994–1001.

Scheidegger, A. E. 1957. On the theory of flow of miscible phases in porous media. *Compte rendu d'Assemblée Générale Extraordinaire. Toronto, International Association of Scientific Hydrology*, **2**, 236–242.

Scheidegger, A. E. 1961. General theory of dispersion in porous media. *Journal of Geophysical Research*, **66**, 3273–3278.

Schey, H. M. 1973. *Div, Grad, Curl, and All That*. New York: W. W. Norton and Company.

Schmid, K. S., Geiger, S., and Sorbie, K. S. 2011. Semianalytical solutions for cocurrent and countercurrent imbibition, and dispersion of solutes in immiscible two-phase flow. *Water Resources Research*, **47**, 1–16.

Schwartz, F. W. 1977. Macroscopic dispersion in porous media: The controlling factors. *Water Resources Research*, **13**, 743–752.

Sethian, J. A. 1996. Theory, algorithms, and applications of level set methods for propagating interfaces. *Acta Numerica*, **5**, 309–395.

Sethian, J. A. 1999. *Level Set and Fast Marching Methods*. Cambridge: Cambridge University Press.

Shearer, M., and Trangenstein, J. A. 1989. Loss of real characteristics for models of three-phase flow in a porous medium. *Transport in Porous Media*, **4**, 499–525.

Silvester, J. R. 2000. Determinants of block matrices. *The Mathematical Gazette*, **84**, 460–467.

Slattery, J. C. 1968. Multiphase viscoelastic fluids through porous media. *American Institute of Chemical Engineering Journal*, **14**, 50–56.

Slattery, J. C. 1981. *Momentum, Energy, and Mass Transfer in Continua*. New York: Krieger.

Smith, R. 1981. The early stages of contaminant dispersion in shear flows. *Journal of Fluid Mechanics*, **111**, 107–122.

Sneddon, I. N. 2006. *Elements of Partial Differential Equations*. New York: Dover Publications.

Sommerfeld, A. 1964. *Optics*. New York: Academic Press.

Stakgold, I. 1979. *Green's Functions and Boundary Value Problems*. New York: John Wiley and Sons.

Sun, N. Z. 1994. *Inverse Problems in Groundwater Modeling*. Norwell, Massaschuetts: Kluwer Academic Press.

Sun, N. Z., and Yeh, W. W. G. 1990. Coupled inverse problems in groundwater modeling, 1, Sensitivity analysis and parameter identification. *Water Resources Research*, **26**, 2507–2525.

Sweby, P. K. 1984. High resolution schemes using flux limiters for hyperbolic conservation laws. *SIAM Journal on Numerical Analysis*, **21**, 995–1011.

Tanaka, S., Datta-Gupta, A., and King, M. J. 2013. A novel approach for incorporation of capillarity and gravity into streamline simulation using orthogonal projection. SPE 163640. In: *Proceedigs of Society of Petroleum Engineering Reservoir Simulation Symposium*. The Woodlands, Texas.

Taniuti, T., and Nishihara, K. 1983. *Nonlinear Waves*. London: Pitman Publishing.

Tarantola, A. 1987. *Inverse Problem Theory: Methods for Data Fitting and Model Parameter Estimation*. Amsterdam: Elseview Science Publishers.

Taylor, G. 1953. Dispersion of soluble matter in solvent flowing slowly through a tube. *Proc. Royal. Society, London, Series A*, **219**, 186–203.

Terzaghi, K. 1943. *Theoretical Soil Mechanics*. New York: John Wiley.

Todd, D. K. 1959. *Ground Water Hydrology*. New York: John Wiley.

Trangenstein, J. A. 1988. Numerical analysis of reservoir fluid flow. In: *Lecture Notes in Engineering*, 34, Multiphase Flow in Porous Media, C.A. Brebbia and S.A. Orszag (eds.), Berlin–Heilderberg–New York: Springer-Verlag.

Truesdell, C. 1962. Mechanical basis of diffusion. *Journal of Chemical Physics*, **37**, 2336–2344.

Truesdell, C. 1985. *The elements of Continuum Mechanics*. New York: Springer-Verlag.

Tuncay, K., and Corapcioglu, M. Y. 1997. Wave propagation in poroelastic media saturated by two fluids. *Journal of Applied Mechanics*, **64**, 313–320.

Tura, A., and Lumley, D. E. 1998. Subsurface fluid-flow properties from time-lapse elastic wave reflection data. Pages 125–138 of: *Proceedings of the 43rd Annual Meeting, SPIE*. International Society of Optical Engineering.

Turcotte, D. L., and Schubert, G. 1982. *Geodynamics*. New York: John Wiley and Sons.

Vallis, G. K. 2006. *Atmospheric and Oceanic Fluid Dynamics*. Cambridge: Cambridge University Press.

van Gestel, J. P., Kimmedal, J. H., Barkved, O. I., Mundal, I., Bakke, R., and Best, K., D. 2008. Continuous seismic surveillance of Valhal field. *The Leading Edge*, **27**, 1616–1621.

Vasco, D. W. 2004a. An asymptotic solution for two-phase flow in the presence of capillary forces. *Water Resources Research*, **40**, 1–13.

Vasco, D. W. 2004b. Estimation of flow properties using surface deformation and head data: A trajectory-based approach. *Water Resources Research*, **40**, 1–14.

Vasco, D. W. 2011. On the propagation of a coupled saturation and pressure front. *Water Resources Research*, **47**, 1–21.

Vasco, D. W. 2013. On the propagation of a disturbance in a smoothly varying heterogeneous porous medium saturated with three fluid phases. *Geophysics*, **78**, L1–L26.

Vasco, D. W., and Datta-Gupta, A. 1999. Asymptotic solutions for solute transport: A formalism for tracer tomography. *Water Resources Research*, **35**, 1–16.

Vasco, D. W., and Datta-Gupta, A. 2001a. Asymptotics, saturation fronts, and high resolution reservoir characterization. *Transport in Porous Media*, **42**, 315–350.

Vasco, D. W., and Datta-Gupta, A. 2001b. Asymptotics, streamlines, and reservoir modeling: A pathway to production tomography. *The Leading Edge*, 142–148.

Vasco, D. W., and Finsterle, S. 2004. Numerical trajectory calculations for the efficient inversion of transient flow and tracer observations. *Water Resources Research*, **40**, 1–17.

Vasco, D. W., Daley, T. M., and Bakulin, A. 2014. Utilizing the onset of time-lapse changes: a robust basis for reservoir monitoring and characterization. *Geophysical Journal International*, **197**, 542–556.

Vasco, D. W., Ferretti, A., and Novali, F. 2008a. Estimating permeability from quasi-static deformation: Temporal variations and arrival time inversion. *Geophysics*, **73**, O37–O52.

Vasco, D. W., Karasaki, K., and Kishida, K. 2001. A coupled inversion of pressure and surface displacement. *Water Resources Research*, **37**, 3071–3089.

Vasco, D. W., Karasaki, K., and Nakagome, O. 2002. Monitoring production using surface deformation: the Hijiori test site and the Okuaizu geothermal field, Japan. *Geothermics*, **31**, 303–342.

Vasco, D. W., Keers, H., and Karasaki, K. 2000. Estimation of reservoir properties using transient pressure data: An asymptotic approach. *Water Resources Research*, **36**, 3447–3465.

Vasco, D. W., Pride, S. R., and Commer, M. 2016. Trajectory-based modeling of fluid transport in a medium with smoothly-varying heterogeneity. *Water Resources Research*, (in press).

Vasco, D. W., Yoon, S., and Datta-Gupta, A. 1999. Integrating dynamic data into highresolution reservoir models using streamline-based analytic sensitivity coefficients. *SPE Journal*, **4**, 389–399.

Vasco, D. W., Bakulin, A., Baek, H., and Johnson, L. R. 2015. Reservoir characterization based upon the onset of time-lapse amplitude changes. *Geophysics*, **80**, 1–14.

Vasco, D. W., Keers, H., Khazanehdari, J., and Cooke, A. 2008b. Seismic imaging or reservoir flow properties: Resolving water influx and reservoir permeability. *Geophysics*, **73**, O1–O13.

Vasco, D. W., Datta-Gupta, A., Behrens, R., Condon, P., and Rickett, J. 2004. Seismic imaging of reservoir flow properties: Time-lapse amplitude changes. *Geophysics*, **69**, 1425–1442.

Vasco, D. W., Rucci, A., Ferretti, A., Novali, F., Bissell, R., Ringrose, P., Mathieson, A., and Wright, I. 2010. Satellite-based measurements of surface deformation reveal flow associated with the geological storage of carbon dioxide. *Geophysical Research Letters*, **37**, 1–5.

Virieux, J., Flores-Luna, C., and Gibert, D. 1994. Asymptotic theory for diffusive electromagnetic imaging. *Geophysical Journal International*, **119**, 857–868.

Voyiadjis, G. Z., and Song, C. R. 2006. *The Coupled Theory of Mixtures in Geomechanics with Applications*. New York: Springer.

Wang, H. F. 2000. *Theory of Linear Poroelasticity*. Princeton: Princeton University Press.

Warren, J. E., and Skiba, F. F. 1964. Macroscopic dispersion. *Transactions of the American Institute of Mining, Metallurgy, and Petroleum Engineering*, **231**, 215–230.

Watanabe, S., Han, J., Datta-Gupta, A., and King, M. J. 2014. Streamline-based time lapse seismic data integration incorporating pressure and saturation effects. *Journal of Petroleum Technology*, **66**(4), 122–126.

Whitaker, S. 1969. Advances in the theory of fluid motion in porous media. *Industrial and Engineering Chemistry*, **61**, 14–28.

White, J. E. 1975. Computed seismic speeds and attenuation in rocks with partial gas saturation. *Geophysics*, **40**, 224–232.

Whitham, G. B. 1974. *Linear and Nonlinear Waves*. New York: John Wiley and Sons.

Wilmanski, K. 2006. A few remarks on Biot's model and linear acoustics of poroelastic saturated materials. *Soil Dynamics and Earthquake Engineering*, **26**, 509–536.

Yeh, W. W. G. 1990. Review of parameter identification procedures in groundwater hydrology: The inverse problem. *Water Resources Research*, **22**, 95–108.

Yoon, S., Malallah, A. H., Datta-Gupta, A., Vasco, D. W., and Behrens, R. A. 2001. A multiscale approach to production data integration using streamline models. *SPE Journal*, **6**, 182–192.

Yortsos, Y. C., and Fokas, A. S. 1983. An analytical solution for linear waterflood including the effects of capillary pressure. *SPE Journal*, **23**, 115–124.

Zhang, R., Vasco, D., Daley, T. M., and Harbert, W. 2015. Characterization of a fracture zone using seismic attributed at the In Salah CO_2 storage project. *Interpretation*, **3**, SM37–SM46.

Zhang, Y., Bansal, N., Fujita, Y., Datta-Gupta, A., King, M. J., and Sankaran, S. 2014. From streamlines to Fast Marching: Rapid simulation and performance assessment of shale gas reservoirs using diffusive time of flight as a spatial coordinate. SPE 168997 In: *Proceedings of Society of Petroleum Engineers Unconventional Resources Conference*, The Woodlands, Texas. *SPE Journal*, 2016. http://dx.doi.org/10.2118/168997-PA.

Zhou, C., Cai, W., Luo, Y., Schuster, G., and Hassanzadeh, S. 1995. Acoustic wave equation travel time and waveform inversion of crosshole seismic data. *Geophysics*, **60**, 765–774.

Zhou, M., and Sheng, P. 1989. First-principles calculations of dynamic permeability in porous media. *Physical Review B*, **39**, 12027–12039.

Zimmerman, R. W. 2000. Coupling in poroelasticity and thermoelasticity. *International Journal of Rock Mechanics and Mining Science*, **37**, 79–87.

Index